Lecture Notes in Computer Science　　　10267

Commenced Publication in 1973
Founding and Former Series Editors:
Gerhard Goos, Juris Hartmanis, and Jan van Leeuwen

More information about this series at http://www.springer.com/series/7412

Jesús Ariel Carrasco-Ochoa
José Francisco Martínez-Trinidad
José Arturo Olvera-López (Eds.)

Pattern Recognition

9th Mexican Conference, MCPR 2017
Huatulco, Mexico, June 21–24, 2017
Proceedings

 Springer

Editors
Jesús Ariel Carrasco-Ochoa
National Institute of Astrophysics, Optics,
  and Electronics
Puebla
Mexico

José Arturo Olvera-López
Autonomous University of Puebla
Puebla
Mexico

José Francisco Martínez-Trinidad
National Institute of Astrophysics, Optics,
  and Electronics
Puebla
Mexico

ISSN 0302-9743   ISSN 1611-3349 (electronic)
Lecture Notes in Computer Science
ISBN 978-3-319-59225-1   ISBN 978-3-319-59226-8 (eBook)
DOI 10.1007/978-3-319-59226-8

Library of Congress Control Number: 2017941496

LNCS Sublibrary: SL6 – Image Processing, Computer Vision, Pattern Recognition, and Graphics

Printed on acid-free paper

This Springer imprint is published by Springer Nature
The registered company is Springer International Publishing AG
The registered company address is: Gewerbestrasse 11, 6330 Cham, Switzerland

Preface

The Mexican Conference on Pattern Recognition 2017 (MCPR 2017) was the ninth event in the series organized by the Computer Science Department of the National Institute for Astrophysics Optics and Electronics (INAOE) of Mexico. This year the conference was jointly organized with the University of Puebla, under the auspices of the Mexican Association for Computer Vision, Neurocomputing and Robotics (MACVNR), which is a member society of the International Association for Pattern Recognition (IAPR). MCPR 2017 was held in Huatulco, Mexico, during June 21–24, 2017.

This conference aims to provide a forum for the exchange of scientific results, practice, and new knowledge, as well as promoting collaboration among research groups in pattern recognition and related areas in Mexico and around the world.

As in previous years, MCPR 2017 attracted not only Mexican researchers but also worldwide participation. We received contributions from 16 countries. In total, 55 manuscripts were submitted, out of which 29 were accepted for publication in these proceedings and for presentation at the conference. Each of these submissions was strictly peer-reviewed by at least two members of the Program Committee, all of them experts in their respective fields of pattern recognition, which resulted in these excellent conference proceedings.

Beside the presentation of the selected contributions, we were very honored to have as invited speakers the following internationally recognized researchers:

- Prof. Ajith Abraham, Machine Intelligence Research Labs, USA
- Prof. Sudeep Sarkar, Department of Computer Science, University of South Florida, USA
- Prof. Eduardo Francisco Morales Manzanares, Department of Computer Science, National Institute for Astrophysics Optics and Electronics, Mexico

These distinguished researchers gave keynote addresses on various pattern recognition topics and also presented enlightening tutorials during the conference. To all of them, we express our appreciation for these presentations.

We would like to thank all the people who devoted so much time and effort to the successful running of the conference. In particular, we extend our gratitude to all the authors who contributed to the conference. We are also very grateful for the efforts and the quality of the reviews of all Program Committee members and additional reviewers. Their work allowed us to maintain the high quality of the conference and provided a conference program of high standard.

We are sure that MCPR 2017 provided a fruitful forum for the Mexican pattern recognition researchers and the broader international pattern recognition community.

June 2017

Jesús Ariel Carrasco-Ochoa
José Francisco Martínez-Trinidad
José Arturo Olvera-López

Organization

MCPR 2017 was sponsored by the Computer Science Department of the National Institute of Astrophysics, Optics and Electronics (INAOE).

General Conference Co-chairs

Jesús Ariel Carrasco-Ochoa	National Institute of Astrophysics, Optics, and Electronics (INAOE), Mexico
José Francisco Martínez-Trinidad	National Institute of Astrophysics, Optics, and Electronics (INAOE), Mexico
José Arturo Olvera-López	Autonomous University of Puebla (BUAP), Mexico

Local Arrangements Committee

Cerón Benítez Gorgonio	National Institute of Astrophysics, Optics, and Electronics (INAOE), Mexico
Cervantes Cuahuey Brenda Alicia	National Institute of Astrophysics, Optics, and Electronics (INAOE), Mexico

Program Committee

A. Asano	Kansai University, Japan
I. Batyrshin	Mexican Petroleum Institute, Mexico
J.M. Benedi	Universidad Politécnica de Valencia, Spain
Chia-Yen Chen	National University of Kaohsiung, Taiwan
H.J. Escalante-Balderas	INAOE, Mexico
J. Facon	Pontifícia Universidade Católica do Paraná, Brazil
M. García-Borroto	CUJAE, Cuba
A. Gelbukh	CIC-IPN, Mexico
L. Goldfarb	University of New Brunswick, Canada
H. Gomes	Universidade Federal de Campina Grande, Brazil
P. Gómez-Gil	INAOE, Mexico
J.A. González-Bernal	INAOE, Mexico
L. Heutte	Université de Rouen, France
L. Igual	University of Barcelona, Spain
X. Jiang	University of Münster, Germany
M. Kampel	Vienna University of Technology, Austria
R. Klette	University of Auckland, New Zealand
V. Kober	CICESE, Mexico
D. Laurendeau	Université Laval, Canada
M.S. Lazo-Cortés	Universidad de las Ciencias Informaticas, Cuba

Contents

X   Contents

Image Processing and Analysis

Robotics and Remote Sensing

Natural Language Processing and Recognition

Applications of Pattern Recognition

Pattern Recognition and Artificial Intelligence Techniques

An Algorithm for Computing Goldman Fuzzy Reducts

J. Ariel Carrasco-Ochoa, Manuel S. Lazo-Cortés[(✉)],
and José Fco. Martínez-Trinidad

Instituto Nacional de Astrofísica, Óptica y Electrónica,
Sta. Ma. Tonantzintla, Puebla, Mexico
{ariel,mlazo,fmartine}@inaoep.mx

Abstract. Feature selection and attribute reduction have been tackled in the Rough Set Theory through fuzzy reducts. Recently, Goldman fuzzy reducts which are fuzzy subsets of attributes were introduced. In this paper, we introduce an algorithm for computing all Goldman fuzzy reducts of a decision system, this algorithm is the first one reported for this purpose. The experiments over standard and synthetic data sets show that the proposed algorithm is useful for datasets with up to twenty attributes.

1 Introduction

Goldman fuzzy reducts were introduced in [3] as a new kind of reducts. Goldman fuzzy reducts are inspired by an idea developed by R. S. Goldman in the framework of Testor Theory [2]. Additionally in [3] the application of Goldman fuzzy reducts for solving supervised classification problems was also discussed. These reducts are fuzzy in the sense that each Goldman fuzzy reduct is a fuzzy subset of the set of attributes, this means that every attribute belonging to a Goldman fuzzy reduct has associated a membership degree to this attribute subset. This membership degree represents the ability to discern that each attribute in the Goldman fuzzy reduct has.

The fundamental result of the present work is to introduce a first algorithm for computing all Goldman fuzzy reducts.

This document is organized as follows. Section 2 provides basic concepts related to Goldman fuzzy reducts. Section 3 introduces the proposed algorithm for computing Goldman fuzzy reducts. Besides, some experiments applying the proposed algorithm over real and synthetic databases are shown. Finally, our conclusions are summarized in Sect. 4.

2 Goldman Fuzzy Reducts

In many data analysis applications, information and knowledge are stored and represented as a decision table which provides a convenient way to describe a finite set of objects within a universe through a finite set of attributes [5,6].

© Springer International Publishing AG 2017
J.A. Carrasco-Ochoa et al. (Eds.): MCPR 2017, LNCS 10267, pp. 3–12, 2017.
DOI: 10.1007/978-3-319-59226-8_1

In Rough Set Theory, a decision table is a matrix representation of a decision system, in this matrix rows represent objects while columns specify attributes. Formally, a decision system is defined as a 4-tuple $DS = (U, A_t^* = A_t \cup \{d\}, \{V_a \mid a \in A_t^*\}, \{I_a \mid a \in A_t^*\})$, where U is a finite non-empty set of objects, A_t^* is a finite non-empty set of attributes, d denotes the decision attribute, V_a is a non-empty set of values of $a \in A_t^*$, and $I_a : U \to V_a$ is a function that maps an object of U to one value in V_a.

In practice, it is common that decision tables contain descriptions of a finite sample U of objects from a larger (possibly infinite) universe $\mathcal{U}$, where values of descriptive attributes are always known for all objects from $\mathcal{U}$, but the decision attribute is a hidden function, except for those objects from the sample U. The main problem of learning theory is to generalize the decision function to the whole universe $\mathcal{U}$.

Let us define for each attribute a in A_t a real valued dissimilarity function for comparing pairs of values $\varphi_a : V_a \times V_a \to [0, 1]$ in such a way that 0 is interpreted as the minimal difference and 1 is interpreted as the maximum possible difference.

Applying these dissimilarity functions to all possible pairs of objects belonging to different classes in DS, a [0,1]-pairwise dissimilarity matrix can be built. We will denote such dissimilarity matrix as DM. We assume that DS is consistent, that is, there is not a pair of indiscernible objects belonging to different classes, this means that DM does not have a row containing only zeros.

We will refer to the value corresponding to row ρ_i in the column associated to attribute a_j in DM as $\mu_{\rho_i}(a_j)$.

Definition 1. *(Goldman fuzzy reduct) Let $A_t = \{a_1, a_2, ..., a_n\}$ and let $T = \{a_{r_1} | \mu_T(a_{r_1}), ..., a_{r_s} | \mu_T(a_{r_s})\}$ be a fuzzy subset of A_t such that $\forall p \in \{1, 2, ..., s\}$ $\mu_T(a_{r_p}) \neq 0$. T is a Goldman fuzzy reduct with respect to DS if:*

(i) $\forall \rho_i \in DM$ (being ρ_i the i-th row in DM) $\exists a_{r_p} | \mu_T(a_{r_p}) \in T$ such that $\mu_T(a_{r_p}) \leq \mu_{\rho_i}(a_{r_p})$.
(ii) $\forall p \in \{1, 2, ..., s\}, T \setminus \{a_{r_p} | \mu_T(a_{r_p})\}$ does not fulfill condition (i).
(iii) $\forall T'$ such that $T \subset T'$ and $supp(T) = supp(T')$ (it means that $\forall p \in \{1, 2, ..., s\}$ $\mu_T(a_{r_p}) \leq \mu_{T'}(a_{r_p})$ and for at least one index the inequality is strict) T' does not fulfill condition (i).

We will denote the set of all Goldman fuzzy reducts of a decision system by $\Psi^*(DS)$. $\Psi(DS)$ will denote the set of all fuzzy subsets of A_t satisfying condition (i) in Definition 1.

According to the above definition, a Goldman fuzzy reduct is a fuzzy subset of attributes such that this sub-set of attributes and their corresponding membership degrees, are able to discern all pairs of objects belonging to different classes (condition (i)). Condition (ii) means that if a fuzzy singleton $\{a_{r_p} | \mu_{r_p}\}$ is eliminated from a Goldman fuzzy reduct, the resulting subset is not anymore a Goldman fuzzy reduct. Condition (iii) means that if the membership degree of any attribute in T is increased, then the resulting subset is not anymore a Goldman fuzzy reduct. This definition is supported by the *subset-based definition of reduct* [7].

Suppose A is a finite set and $\mathfrak{p}(A)$ is the power set of A. Let $\mathbb{P}$ be a unary predicate on $\mathfrak{p}(A)$. $\mathbb{P}(S)$ stands for the statement that subset S fulfills property $\mathbb{P}$. The values of $\mathbb{P}$ are computed by an evaluation $\mathfrak{e}$ with reference to certain available data, for example, a decision system. For a subset $S \in \mathfrak{p}(A)$, $\mathbb{P}(S)$ is true if S fulfills property $\mathbb{P}$, otherwise, it is false. In this way, a conceptual definition of reduct is given based on an evaluation $\mathfrak{e}$ as follows.

Definition 2. *(Subset-based definition [7]). Given an evaluation $\mathfrak{e}$ of $\mathbb{P}$, a subset R of A is a reduct if R fulfills the following conditions:*

(a) existence: $\mathbb{P}_{\mathfrak{e}}(A)$.
(b) sufficiency: $\mathbb{P}_{\mathfrak{e}}(R)$.
(c) minimization: $\forall B \subset R \ (\neg \mathbb{P}_{\mathfrak{e}}(B))$.

These three conditions reflect the fundamental characteristics of a reduct. Condition of existence (a) ensures that a reduct of S exists. Condition of sufficiency (b) expresses that a reduct R of A is sufficient for preserving the property $\mathbb{P}$ of A. Condition of minimization (c) expresses that a reduct is a minimal subset of A fulfilling property $\mathbb{P}$ in the sense that none of the proper subsets of R fulfills property $\mathbb{P}$.

For our convenience, we consider $\mathfrak{P}(A)$ as the set of all fuzzy subsets of A, instead of the classical power set $\mathfrak{p}(A)$. Besides, we consider the partial order $\preceq$ defined in [3] instead of the classic inclusion:

Let $t_1, t_2 \in \mathfrak{P}(A)$, then we say that $t_1 \preceq t_2$ iff
$(t_1 \cap t_2) \cup ((supp(t_1) \setminus supp(t_2)) \cap t_1) \cup ((supp(t_2) \setminus supp(t_1)) \cap t_2) = t_2.$[1]

In [3], it was proved that $\preceq$ is a partial order over $\mathfrak{P}(A)$, as well as, that $T \in \Psi(DS)$ is a Goldman fuzzy reduct with respect to DS if T is minimal for the relation $\preceq$ defined over $\Psi(DS)$.

Example 1. Let $A = \{a_1, a_2, a_3, a_4\}$, and let $\mathfrak{P}(A)$ the set of all subsets of A, including fuzzy subsets. Let t_1, t_2 and $t_3 \in \mathfrak{P}(A)$, $t_1 = \{a_1|0.5, a_2|0.4, a_3|1\}$, $t_2 = \{a_1|0.5, a_2|0.4, a_3|1, a_4|0.6\}$, $t_3 = \{a_1|0.5, a_2|0.4, a_3|0.8, a_4|0.6\}$. According to the definition of $\preceq$ we have, for example, that $t_1 \preceq t_2$ and $t_2 \preceq t_3$ since $(t_1 \cap t_2) \cup ((supp(t_1) \setminus supp(t_2)) \cap t_1) \cup ((supp(t_2) \setminus supp(t_1)) \cap t_2) = \{a_1|0.5, a_2|0.4, a_3|1\} \cup (\emptyset \cap \{a_1|0.5, a_2|0.4, a_3|1\}) \cup (a_4 \cap \{a_1|0.5, a_2|0.4, a_3|1, a_4|0.6\}) = \{a_1|0.5, a_2|0.4, a_3|1, a_4|0.6\} = t_2$ and $(t_2 \cap t_3) \cup ((supp(t_2) \setminus supp(t_3)) \cap t_2) \cup ((supp(t_3) \setminus supp(t_2)) \cap t_3) = \{a_1|0.5, a_2|0.4, a_3|0.8, a_4|0.6\} \cup (\emptyset \cap \{a_1|0.5, a_2|0.4, a_3|1, a_4|0.6\}) \cup (\emptyset \cap \{a_1|0.5, a_2|0.4, a_3|0.8, a_4|0.6\}) = \{a_1|0.5, a_2|0.4, a_3|0.8, a_4|0.6\} = t_3$

Let A be, as before, the set of condition attributes in DS, and $S = \{a_{r_1}|\mu_{r_1}, a_{r_2}|\mu_{r_2}, ..., a_{r_s}|\mu_{r_s}\}$ a fuzzy subset of A, $(0 < \mu_{r_p} \leq 1, p = \{1, 2, ...s\})$. Let $A^o = \{a_1|\mu_1^o, a_2|\mu_2^o, ..., a_n|\mu_n^o\}$, being $\mu_j^o = min\{\mu_{\rho_i}(a_j) \neq 0\}$ for all rows in DM, $1 \leq j \leq n$.

---

[1] Since $supp(t_1)$ and $supp(t_2)$ are not fuzzy, $\setminus$ denotes the classic set difference operation in Set Theory; on the other hand, $\cup$ and $\cap$ are, respectively, the union and intersection operations as classically defined in Fuzzy Set Theory.

Let us consider the following predicate $\mathbb{P}$:

$\mathbb{P}(S) \equiv \forall \rho_i \in DM \; \exists x_{r_p} | \mu_{r_p} \in S$ such that $\mu_{r_p} \leq \mu_{\rho_i}(a_j)$.

Notice that A^o fulfills the property $\mathbb{P}$ by construction, unless there exists a zero row in DM, but this is not possible since we have assumed that DS is consistent. Then, we have $\mathbb{P}_e(A^o)$.

On the other hand, let $T = \{a_{r_1} | \mu_T(a_{r_1}), ..., a_{r_s} | \mu_T(a_{r_s})\}$ be a Goldman fuzzy reduct, then from condition (i) in Definition 1 it follows that T also has the property $\mathbb{P}$, i.e. $\mathbb{P}_e(T)$. Finally, taking into account that minimal elements according to $\preceq$ in $\Psi(DS)$ are Goldman fuzzy reducts, it follows that $\forall B \neq T; \; B \preceq T \implies [\neg \mathbb{P}_e(B)]$.

Then we have that if T is a Goldman fuzzy reduct, it satisfies Definition 2.

Example 2. Consider the following matrix:

$$BM = \begin{matrix} & a_1 & a_2 & a_3 \\ & \begin{pmatrix} 0.2 & 0.5 & 0.7 \\ 0.5 & 0.4 & 0.8 \\ 0.3 & 0.2 & 0.9 \end{pmatrix} \end{matrix}$$

For this matrix the set of Goldman fuzzy reducts is $\Psi^*(BM) = \{\{a_1|0.2\}, \{a_2|0.2\}, \{a_3|0.7\}, \{a_1|0.3, a_2|0.5\}, \{a_2|0.4, a_3|0.9\}, \{a_2|0.5, a_3|0.8\}, \{a_1|0.5, a_2|0.5, a_3|0.9\}\}$

It is not difficult to prove that Definition 1 is equivalent to the next definition, which is easier to be verified.

Definition 3. $T = \{a_{r_1}|\mu_{r_1}, a_{r_2}|\mu_{r_2}, ..., a_{r_s}|\mu_{r_s}\}$ *is a Goldman fuzzy reduct with respect to DS iff T satisfies condition (i) of Definition 1 and $\forall a_{r_i} \in supp(T) \; \exists \rho_j$ in $DM : [\mu_T(a_{r_i}) = \mu_{\rho_j}(a_{r_i}) \wedge \forall (p \neq i)\mu_T(a_{r_p}) > \mu_{\rho_j}(a_{r_p})]$.*

Definition 3 indicates that each attribute, together with its membership degree, of a Goldman fuzzy reduct, is indispensable for covering some rows in DM. This necessary condition can be verified for each attribute at the same time of condition (i) of Definition 1; with a low additional cost. Moreover, using Definition 3, verifying conditions (ii) and (iii) of Definition 1 can be done with a very low computational cost.

3   GFR Algorithm

The proposed algorithm, called GFR, is inspired by the MSLC algorithm [4] for computing reducts, which is based on the binary discernibility matrix.

GFR is able to compute all Goldman fuzzy reducts of a decision table, but it can also be used to find just one reduct, or a certain number of them.

First, the search space is conveniently ordered such that the pruning strategy goes through all the possible fuzzy subsets of attributes to decide whether or not they are Goldman fuzzy reducts but discarding from the analysis some subsets.

It is not difficult to prove that if ρ_i and ρ_j are rows of DM, and $\forall p \in \{1, 2, ..., n\}$ $\mu_{\rho_i}(a_p) \leq \mu_{\rho_j}(a_p)$ and for at least one attribute the inequality is strict, then ρ_j is not needed for computing Goldman fuzzy reducts, in this case we say that ρ_j is a superfluous row. Thus, we can filter DM eliminating all superfluous rows. The new matrix is called basic matrix and will be denoted as BM.

Notice that, since only superfluous rows are eliminated, the Goldman fuzzy reducts computed from BM are exactly the same as those computed from DM. Moreover, a direct consequence of the last fact is that Definition 3 also applies for BM.

The GFR algorithm traverses the searching space in ascending order according to the Boolean representation of the attribute subsets, and the membership degrees in each column are also considered in ascending order. Let m_i and M_i be the minimum and the maximum values for the attribute a_i in BM respectively, we have that:

(1) The first fuzzy subset of attributes in the order will be $\{a_n|m_n\}$ corresponding to the Boolean tuple 00...01 with the minimum possible membership degree for the attribute a_n.
(2) The last fuzzy subset of attributes in the order will be $\{a_1|M_1, a_2|M_2, ..., a_n|M_n\}$, corresponding to the Boolean tuple 11...1 with the maximum possible membership degrees for each attribute.
(3) Given a fuzzy subset of attributes $T = \{a_{r_1}|\mu_T(a_{r_1}), ..., a_{r_s}|\mu_T(a_{r_s})\}$; $r_1 < ... < r_s$, the next fuzzy subset in the order is calculated as follows:
  (a) Find k such that $k = max\{j \mid \mu_T(a_{r_j}) \neq M_{r_j}\}$, k is the greatest index among the attributes in T for which the membership degree is not the maximum.
  (b) If $k < s$ then for all attributes from $a_{r_{k+1}}$ to a_{r_s} in the current combination, the corresponding minimum membership degree m_{r_j} ($j = k+1, ..., s$) is assigned. In any case, for a_{r_k} the next value (in ascending order) of the membership degrees in the corresponding column of BM is assigned.
  (c) If $\forall j \in \{1, ..., s\}$ $\mu_T(a_{r_j}) = M_{r_j}$ then the next combination of attributes is generated and to each attribute the corresponding minimum membership degree is assigned.

Example 3. The traversal order of GFR over the searching space for the matrix BM in Example 2 comprises the 63 combinations shown in Table 1.

The pruning strategy of the GFR algorithm is based on the following facts:

(1) If T is a Goldman fuzzy reduct and $T \subseteq T'$ with $supp(T) = supp(T')$, then T' is not a Goldman fuzzy reduct.
(2) Let $T = \{a_{r_1}|\mu_T(a_{r_1}), ..., a_{r_s}|\mu_T(a_{r_s})\}$ be a fuzzy subset of attributes. We say that a_{r_j} covers a row ρ_s of BM iff $\mu_T(a_{r_j}) \leq \mu_{\rho_s}(a_{r_j})$ and $\forall p < j$ $\mu_T(a_{r_p}) > \mu_{\rho_s}(a_{r_p})$. We say that a_{r_j} generates redundancy if it does not cover any row of BM or if the number of rows that cannot be covered without using attributes subsequent to a_{r_j} is less than the number of attributes

Table 1. Traversal order for the BM of Example 1

n	Vector	Membership degrees	n	Vector	Membership degrees
1	001	$a_3\|0.7$	33	110	$a_1\|0.3, a_2\|0.5$
2	001	$a_3\|0.9$	34	110	$a_1\|0.5, a_2\|0.2$
4	010	$a_2\|0.2$	35	110	$a_1\|0.5, a_2\|0.4$
5	010	$a_2\|0.4$	36	110	$a_1\|0.5, a_2\|0.5$
6	010	$a_2\|0.5$	37	111	$a_1\|0.2, a_2\|0.2, a_3\|0.7$
7	011	$a_2\|0.2, a_3\|0.7$	38	111	$a_1\|0.2, a_2\|0.2, a_3\|0.8$
8	011	$a_2\|0.2, a_3\|0.8$	39	111	$a_1\|0.2, a_2\|0.2, a_3\|0.9$
9	011	$a_2\|0.2, a_3\|0.9$	40	111	$a_1\|0.2, a_2\|0.4, a_3\|0.7$
10	011	$a_2\|0.4, a_3\|0.7$	41	111	$a_1\|0.2, a_2\|0.4, a_3\|0.8$
11	011	$a_2\|0.4, a_3\|0.8$	42	111	$a_1\|0.2, a_2\|0.4, a_3\|0.9$
12	011	$a_2\|0.4, a_3\|0.9$	43	111	$a_1\|0.2, a_2\|0.5, a_3\|0.7$
13	011	$a_2\|0.5, a_3\|0.7$	44	111	$a_1\|0.2, a_2\|0.5, a_3\|0.8$
14	011	$a_2\|0.5, a_3\|0.8$	45	111	$a_1\|0.2, a_2\|0.5, a_3\|0.9$
15	011	$a_2\|0.5, a_3\|0.9$	46	111	$a_1\|0.3, a_2\|0.2, a_3\|0.7$
16	100	$a_1\|0.2$	47	111	$a_1\|0.3, a_2\|0.2, a_3\|0.8$
17	100	$a_1\|0.3$	48	111	$a_1\|0.3, a_2\|0.2, a_3\|0.9$
18	100	$a_1\|0.5$	49	111	$a_1\|0.3, a_2\|0.4, a_3\|0.7$
19	101	$a_1\|0.2, a_3\|0.7$	50	111	$a_1\|0.3, a_2\|0.4, a_3\|0.8$
20	101	$a_1\|0.2, a_3\|0.8$	51	111	$a_1\|0.3, a_2\|0.4, a_3\|0.9$
21	101	$a_1\|0.2, a_3\|0.9$	52	111	$a_1\|0.3, a_2\|0.5, a_3\|0.7$
22	101	$a_1\|0.3, a_3\|0.7$	53	111	$a_1\|0.3, a_2\|0.5, a_3\|0.8$
23	101	$a_1\|0.3, a_3\|0.8$	54	111	$a_1\|0.3, a_2\|0.5, a_3\|0.9$
24	101	$a_1\|0.3, a_3\|0.9$	55	111	$a_1\|0.5, a_2\|0.2, a_3\|0.7$
25	101	$a_1\|0.5, a_3\|0.7$	56	111	$a_1\|0.5, a_2\|0.2, a_3\|0.8$
26	101	$a_1\|0.5, a_3\|0.8$	57	111	$a_1\|0.5, a_2\|0.2, a_3\|0.9$
27	101	$a_1\|0.5, a_3\|0.9$	58	111	$a_1\|0.5, a_2\|0.4, a_3\|0.7$
28	110	$a_1\|0.2, a_2\|0.2$	59	111	$a_1\|0.5, a_2\|0.4, a_3\|0.8$
29	110	$a_1\|0.2, a_2\|0.4$	60	111	$a_1\|0.5, a_2\|0.4, a_3\|0.9$
30	110	$a_1\|0.2, a_2\|0.5$	61	111	$a_1\|0.5, a_2\|0.5, a_3\|0.7$
31	110	$a_1\|0.3, a_2\|0.2$	62	111	$a_1\|0.5, a_2\|0.5, a_3\|0.8$
32	110	$a_1\|0.3, a_2\|0.4$	63	111	$a_1\|0.5, a_2\|0.5, a_3\|0.9$

after it. This means that one of the attributes subsequent to a_{r_j} does not cover any row.

If T is not a Goldman fuzzy reduct, but it satisfies condition (i) in Definition 1, then all the combinations of attributes that follow in the order and maintain the same membership degree in the attribute that generates redundancy with the smallest index are not Goldman fuzzy reducts.

(3) If T does not fulfill condition (i) in Definition 1 and $T \subseteq T'$ with $supp(T) = supp(T')$, then T' also does not fulfill condition (i) in Definition 1.

(4) If $T = \{a_{r_p} | \mu_T(a_{r_p})\}$ fulfills condition (i) in Definition 1, then T is a Goldman fuzzy reduct, and consequently a_{r_p} does not appear in any other Goldman fuzzy reduct with this membership degree $\mu_T(a_{r_p})$.

(5) If $T = \{a_{r_p} | \mu_T(a_{r_p})\}$ does not fulfill condition (i) in Definition 1, then no singleton $\{a_{r_p} | \mu_{r_p}\}$ is a Goldman fuzzy reduct.

The GFR algorithm is as follows:

1. FOR $i = 1, ..., n$

$m_i = min_{j=1,...,m}\{\mu_{\rho_j}(a_i)\}$; $M_i = max_{j=1,...,m}\{\mu_{\rho_j}(a_i)\}$; $T = \{a_n | m_n\}$

2. Verify whether T is a Goldman fuzzy reduct using Definition 3

   IF YES

      save T

      IF T is a singleton $T = \{a_k | \mu_T(a_k)\}$

         $m_k = min_{j=1,...,m}\{\mu_{\rho_j}(a_k)$ such that $\mu_{\rho_j}(a_k) > \mu_T(a_k)\}$

         take the next combination of attributes assigning to each attribute a_p

         its minimum membership degree m_p

      ELSE

         $k = max\{j$ such that $\mu_T(a_j) \neq m_j\}$

         $r = max\{j < k$ such that $\mu_T(a_j) \neq M_j\}$

         IF k or r does not exist

            take the next combination of attributes assigning to each attribute a_p

            its minimum membership degree m_p

         ELSE

            assign to x_k its minimum membership degree m_k

            assign to x_r the next membership degree

            assign to x_j for $j = r + 1, ..., k - 1$ its minimum membership degree m_j

   IF NOT

      IF T fulfills condition (i) of Definition 1

         Let k be the smallest index among the attributes that generate redundancy

         IF x_k does not cover any row

            $r = max\{j < k$ such that $\mu_T(a_j) \neq M_j\}$

         ELSE

            $r = max\{j \leq k$ such that $\mu_T(a_j) \neq M_j\}$

         IF r does not exist

            take the next combination of attributes assigning to each attribute a_p

            its minimum membership degree m_p

         ELSE

            assign to x_r the next membership degree

            assign to x_j for $j \geq r + 1$ its minimum membership degree m_j

      ELSE (T does not fulfill condition (i) of Definition 1)

         IF T is a singleton $T = \{a_k | \eta_k\}$

            take the next combination of attributes assigning to each attribute a_p

            its minimum membership degree m_p

         ELSE

            $k = max\{j$ such that $\mu_T(a_j) \neq m_j\}$

$$r = max\{j < k \text{ such that } \mu_T(a_j) \neq M_j\}$$
IF k or r does not exist
    take the next combination of attributes assigning to each attribute a_p
    its minimum membership degree m_p
ELSE
    assign to x_k its minimum membership degree m_k
    assign to x_r the next membership degree
    assign to x_j for $j = r + 1, ..., k - 1$ its minimum membership degree m_j
3. IF $T \neq \{a_1|M_1, a_2|M_2, ..., a_n|M_n\}$ GO TO 2
4. END

3.1   Experimental Results

To illustrate the behavior of the proposed algorithm, ten datasets were selected from the UCI Machine Learning Repository [1]. For each dataset (decision system), we computed its dissimilarity matrix, and then its respective basic matrix, and our proposed algorithm GFR was applied over this basic matrix. All experiments were carried out on an Intel(R) Core(TM) Duo CPU T5800 @ 2.00 GHz 64-bit system with 4 GB of RAM running on the Windows 10 System. Table 2 contains information about the selected datasets and it also shows the amount of Goldman fuzzy reducts as well as the runtime taken by our algorithm for each dataset.

Table 2. Amount of Goldman fuzzy reducts and the runtime taken by GFR

Dataset	Descriptive attributes	Classes	Objects	BM rows	Goldman fuzzy reducts	Runtime (s)
Balance Scale	4	3	625	4	1	<1
Contraceptive Method Choice	9	3	1473	9	1	<1
Iris	4	3	150	18	20	<1
Yeast	8	10	1484	73	91	<1
Ecoli	7	8	336	130	405	<1
Heart Disease Cleveland	13	2	303	146	405	<1
Labor	16	2	57	151	1151	61
Liver Disorder	6	2	345	219	800	1
Glass	9	6	214	294	3411	189
Pima Indians Diabetes	8	2	768	1212	21657	51490

As we can see in Table 2, the amount of rows in the basic matrix (see BM rows column) highly influences the time needed by our algorithm for computing all the Goldman fuzzy reducts. For this reason, in a second experiment, we evaluate the scalability of GFR; for this experiment we randomly generate a basic matrix with 20 rows and 5 columns (attributes), each cell in the matrix has a value between 0.0 and 1.0. We increased the number of columns of this matrix by adding a column at a time up to 20 columns.

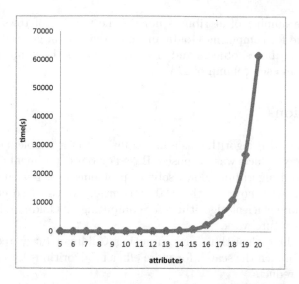

Fig. 1. Runtime of GFR vs number of attributes using synthetic basic matrices with 20 rows.

Table 3 shows the results of this experiment, for each generated matrix, Table 3 includes the number of Goldman fuzzy reducts calculated and the time required (in seconds). Figure 1 shows a graph of runtime (in seconds) vs. number of attributes. Clearly, we can see the exponential dependence of the runtime in terms of the number of attributes.

Since for GFR the size of the search space is $\prod_{i=1}^{n}(d_i + 1)$, being d_i the number of different non zero values in column i in BM, from Fig. 1 we can see that for an amount of attributes greater than 20 the time needed by the algorithm grows drastically. In addition, from Table 2 we observe that the runtime depends

Table 3. Scalability of the GFR algorithm regarding the number of attributes using synthetic basic matrices with 20 rows.

Attributes	Goldman fuzzy reducts	Runtime (s)	Attributes	Goldman fuzzy reducts	Runtime (s)
5	12	<1	13	5204	65
6	81	<1	14	8005	203
7	130	<1	15	14865	657
8	349	<1	16	35553	2161
9	584	1	17	52479	4901
10	1139	2	18	69988	10778
11	1904	8	19	95595	21976
12	3668	22	20	189621	61564

not only on the number of attributes nor on the number of rows in the basic matrix. Runtime for computing Goldman fuzzy reducts, in general, depends on the number of attributes, objects and classes, as well as the number of different non zero values in each column of BM.

4   Conclusions

In this paper, the GFR algorithm as a first solution to the problem of computing all Goldman fuzzy reducts was proposed. Based on our experiments, we conclude that the proposed algorithm allows solving problems in a reasonable time for decision systems with no more than 20 attributes. We can also conclude that the search for more efficient algorithms for computing all Goldman fuzzy reducts is mandatory as future work.

GFR is the first algorithm for computing Goldman fuzzy reducts, which opens as a new branch the search for more efficient algorithms for computing all Goldman fuzzy reducts.

References

1. Bache, K., Lichman, M.: UCI Machine Learning Repository. http://archive.ics.uci.edu/ml. Irvine, CA, University of California, School of Information and Computer Science (2013)
2. Goldman, R.S.: Problems of fuzzy test theory. Avtomat. Telemech. **10**, 146–153 (1980). (in Russian)
3. Lazo-Cortés, M.S., Martínez-Trinidad, J.F., Carrasco-Ochoa, J.A.: A glance to the goldman's testors from the point of view of rough set theory. In: Martínez-Trinidad, J.F., Carrasco-Ochoa, J.A., Ayala-Ramírez, V., Olvera-López, J.A., Jiang, X. (eds.) MCPR 2016. LNCS, vol. 9703, pp. 189–197. Springer, Cham (2016). doi:10.1007/978-3-319-39393-3_19
4. Lazo-Cortés, M.S., Martínez-Trinidad, J.F., Carrasco-Ochoa, J.A., Sanchez-Diaz, G.: A new algorithm for computing reducts based on the binary discernibility matrix. Intell. Data Anal. **20**(2), 317–337 (2016)
5. Pawlak, Z.: Rough sets. Int. J. Comput. Inform. Sci. **11**(5), 341–356 (1982)
6. Pawlak, Z.: Rough Sets: Theoretical Aspects of Reasoning About Data. Kluwer Academic Publishers, Dordrecht (1991)
7. Yao, Y.Y.: The two sides of the theory of rough sets. Knowl. Based Syst. **80**, 67–77 (2015)

A Parallel Genetic Algorithm for Pattern Recognition in Mixed Databases

Angel Kuri-Morales[✉] and Javier Sagastuy-Breña

Instituto Tecnológico Autónomo de México, Río Hondo no. 1, 01000 Mexico D.F., Mexico
{akuri,jsagastu}@itam.mx

Abstract. Structured data bases may include both numerical and non-numerical attributes (categorical or CA). Databases which include CAs are called "mixed" databases (MD). Metric clustering algorithms are ineffectual when presented with MDs because, in such algorithms, the similarity between the objects is determined by measuring the differences between them, in accordance with some predefined metric. Nevertheless, the information contained in the CAs of MDs is fundamental to understand and identify the patterns therein. A practical alternative is to encode the instances of the CAs numerically. To do this we must consider the fact that there is a limited subset of codes which will preserve the patterns in the MD. To identify such pattern-preserving codes (PPC) we appeal to neural networks (NN) and genetic algorithms (GA). It is possible to identify a set of PPCs by trying out a bounded number of codes (the individuals of a GA's population) and demanding the GA to identify the best individual. Such individual is the best practical PPC for the MD. The computational complexity of this task is considerable. To decrease processing time we appeal to multi-core architectures and the implementation of multiple threads in an algorithm called ParCENG. In this paper we discuss the method and establish experimental bounds on its parameters. This will allow us to tackle larger databases in much shorter execution times.

Keywords: Categorical databases · Neural networks · Genetic algorithms · Parallel computation

1 Introduction

Cluster Analysis is the name given to a diverse collection of techniques that can be used to classify objects in a structured database. The classification will depend upon the particular method used because it is possible to measure similarity and dissimilarity (distance between the objects in the DB) in a number of ways. Once having selected the distance measure we must choose the clustering algorithm. There are many methods available. Five classical ones are (a) Average Linkage Clustering, (b) Complete Linkage Clustering, (c) Single Linkage Clustering, (d) Within Groups Clustering, (e) Ward's Method [1]. Alternative methods, based on computational intelligence, are (f) K-Means, (g) Fuzzy C-Means, (h) Self-Organizing Maps, (i) Fuzzy Learning Vector Quantization [2]. All of these methods have been designed to tackle the analysis of strictly numerical databases, i.e. those in which all the attributes are directly expressible as numbers.

© Springer International Publishing AG 2017
J.A. Carrasco-Ochoa et al. (Eds.): MCPR 2017, LNCS 10267, pp. 13–21, 2017.
DOI: 10.1007/978-3-319-59226-8_2

If any of the attributes is non-numerical (i.e. categorical) none of the methods in the list is applicable. Clustering of categorical attributes (i.e., attributes whose domain is not numeric) is a difficult, yet important task: many fields, from statistics to psychology deal with categorical data. In spite of its importance, the task of categorical clustering has received relatively scant attention. Much of the published algorithms to cluster categorical data rely on the usage of a distance metric that captures the separation between two vectors of categorical attributes, such as the Jaccard coefficient [3]. An interesting alternative is explored in [4] where COOLCAT, a method which uses the notion of entropy to group records, is presented. It is based on information loss minimization. Another reason for the limited exploration of categorical clustering techniques is its inherent difficulty.

In [5] a different approach is taken by (a) Preserving the patterns embedded in the database and (b) Pinpointing the codes which preserve such patterns. These two steps result in the correct identification of a set of PPCs. The first issue implies the use of an algorithm which is certifiably capable of pattern identification. Multilayer perceptron networks (MLP) have been mathematically proven to do so [6]. On the other hand, we need a method which guarantees the correct identification of a set of codes. These stem from a very large ensemble and should minimize the approximation error implicit in the practical implementation of point (a) above. Genetic algorithms (GA) have been proven to always attain the global optimum of an arbitrary function [7] and, furthermore, a specific GA called the Eclectic GA (or EGA) has been shown to solve the optimization problem [8] efficiently. The resulting algorithm is called CENG (Categorical Encoding with Neural Networks and Genetic Algorithms) and its parallelized version ParCENG.

The rest of the paper is organized as follows. In Sect. 2 we briefly describe (a) Pseudo-binary encoding as an alternative to our approach, (b) The optimization problem CENG solves, (c) The basic tenets of multi-layer perceptron (MLP) networks and the EGA. In Sect. 3 we present the global methodology resulting in the parallel version of CENG, for which see [5]. In Sect. 4 we present some experimental results and, finally, in Sect. 5 we present our conclusions.

2 Encoding Mixed Databases

As stated in the introduction, the basic idea is to apply clustering algorithms designed for strictly numerical databases (ND) to MDs by encoding the instances of categorical variables with a number. This is by no means a new concept. MDs, however, offer a particular challenge when clustering is attempted because it is, in principle, impossible to impose a metric on CAs. There is no way in which numerical codes may be assigned to the CAs in general.

2.1 Pseudo-Binary Encoding

A common alternative is to replace every CA variable by a set of binary variables, each corresponding to the instances of the category. The CAs in the MD are replaced by numerical ones where every categorical variable is replaced by a set of t binary numerical codes. An MD (with c categories and n numerical variables) will be replaced by an ND with $n - c + ct$ variables. This approach suffers from the following limitations:

(a) The number of attributes of ND will be larger that of MD. In many cases this leads to unwieldy databases which are more difficult to store and handle.

(b) The type of coding system selected implies a subjective choice since all pseudo-binary variables may be assigned any two values (typically "0" denotes "absence"; "1" denotes "presence"). This choice is subjective. Any two different values are possible. Nevertheless, the mathematical properties of ND will vary with the different choices, thus leading to clusters which depend on the way in which "presence" or "absence" is encoded.

(c) Finally, with this sort of scheme the pseudo-binary variables do no longer reflect the essence of the idea conveyed by category c. A variable corresponding to the t-th instance of the category reflects the way a tuple is "affected" by belonging to the t-th categorical value, which is correct. But now the original issue "How does the behavior of the individuals change according to the category?" is replaced by "How does the behavior of the individuals change when the category's value is t?" The two questions are not interchangeable.

2.2 Pattern Preserving Codes

An alternative goal is to assign codes (which we call Pattern Preserving Codes or PPCs) to each and all the instances of every class (category) which will preserve the patterns present for a given MD.

Consider a set of n-dimensional tuples (say U) whose cardinality is m. Assume there are n unknown functions of n-1 variables each, which we denote with

$$f_k(v_1, \dots, v_{k-1}, v_{k+1}, \dots, v_n); k = 1, \dots, n$$

Let us also assume that there is a method which allows us to approximate f_k (from the tuples) with F_k. Denote the resulting n functions of $n - 1$ independent variables with F_i, thus

$$F_k = f(v_1, \dots, v_{k-1}, v_{k+1}, \dots, v_n); k = 1, \dots, n \qquad (1)$$

The difference between f_k and F_k will be denoted with ε_k such that, for attribute k and the m tuples in the database

$$\varepsilon_k = max[abs(f_{ki} - F_{ki})]; i = 1, \dots, m \qquad (2)$$

Our contention is that the PPCs are the ones which minimize ε_k for all k. This is so because only those codes which retain the relationships between variable k and the remaining $n - 1$ variables AND do this for ALL variables in the ensemble will preserve the whole set of relations (i.e. patterns) present in the data base, as in (3).

$$\Xi = min[max(\varepsilon_k; k = 1, \dots, n)] \qquad (3)$$

Notice that this is a multi-objective optimization problem because complying with condition k in (2) for any given value of k may induce the non-compliance for a different possible k. Using the min-max expression of (3) equates to selecting one point in the Pareto's front [10].

To achieve the purported goal we must have a tool which is capable of identifying the F_k's in (1) and the codes which attain the minimization of (3). This is possible using NNs and GAs.

Cybenko [6] proved the Universal Approximation Theorem which states that MLP networks with a single hidden layer are sufficient to compute a uniform ε approximation to a given training set represented by the set of inputs $x_1,...,x_{mO}$ and a desired (target) output $f(x_1,...,x_{mO})$ where m_O denotes the number of independent variables and $F(x_1,...,x_{mO})$ is the output of the NN, thus

$$\left| F(x_1, \ldots, x_{mO}) - f(x_1, \ldots, x_{mO}) \right| < \varepsilon \qquad (4)$$

Rudolph [7] proved that the canonical GA (CGA) maintaining the best solution found over time after selection converges to the global optimum. This result ensures that CGA is an optimization tool which will always reach the best global value.

However, its time of convergence is not bounded. In [8] a statistical survey was conducted wherein 5 evolutionary algorithms were tested vs. a very large ($>10^6$) number of problems. The relative performance of these algorithms is shown in Fig. 1. The five algorithms selected for this study were: CHC (Cross generational elitist selection, Heterogeneous recombination, and Cataclysmic mutation GA), RMH (Random Mutation Hill Climber), SGA (Statistical GA), TGA (eliTist CGA) and EGA (Eclectic GA).

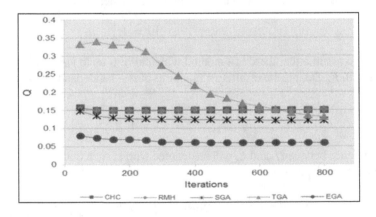

Fig. 1. Relative performance of different GAs.

Notice that EGA reached values very close (within 5%) to the global known optimum and did so in less than 400 generations. Therefore, from Rudolph's theorem and the previous results we have statistical proof that the best practical set of codes for the categorical instances will be found by EGA in a short number of generations.

3   General Methodology

The above mentioned tools are computationally intensive and when applied in CENG, yield an effective algorithm, albeit with relatively large execution times even on small databases.

3.1   The CENG Algorithm

The general algorithm for CENG is as follows:

- Specify the mixed database MD.
- MD is analyzed to determine c and t (see above). The size of the genome (the number of bits needed to encode the solution with a precision of 2^{-22}) is $L = 22\ ct$. EGA randomly generates a set of PS strings of size L. This is population P_0. Every one of the PS strings is an individual of EGA.

For $i=1$ to G
    For $j=1$ to PS
        From individual j, ct numerical codes are extracted and ND_{ij} is generated replacing the categorical instances by their numerical counterparts. Numerical variables are left undisturbed. MLP_{ij}'s architecture (corresponding to individual j) is determined as described above.
        For $k=1$ to $n-1$
            MLP_{ij} is fed with a data matrix in which the k-th attribute of ND is taken as a variable dependent on the remaining $n-1$. MLP_{ij} is trained and the maximum error e_{ijk} (i.e. the one resulting by feeding the already trained MLP_{ij} with all tuples vs. the dependent variable) is calculated.
        endFor
        Fitness(j) =[max (e_{ijk})]
    endFor
    if $Fitness(j)<0.01$ EGA ends. Otherwise the PS individuals of P_i are selected, crossed over and mutated. This is the new P_i.
endFor

EGA yields the codes for every one of the ct instances of the categorical variables and ND: a version of MD in which all categorical instances are replaced by the proper numerical codes.

3.2   ParCENG

As opposed to other optimization tools, however, GAs are prone to easy parallelization. Every individual's evaluation is independent of every other one's, so that a multi-thread implementation of every individual's fitness evaluation is natural. So much so, that we may *a priori* determine that execution time will decrease basically as the inverse of the number of threads (cores).

Nevertheless, there is still the need to fine-tune the parameters of ParCENG. Two main issues affecting execution time were explored: (a) The parameters of the EGA and (b) Those of the NNs.

From the algorithm presented in Sect. 3.1, the process of training the $n - 1$ MLPs in the innermost loop of CENG is independent for every individual in the generation. Thus, it is natural to spawn PS parallel processes which will execute this portion of the algorithm.

3.3 ParCENG's Configuration Parameters

The maximum number of generations of the EGA has a direct and linear impact on the overall execution time of the algorithm. It becomes fundamental to let the algorithm run for enough generations to reach a global minimum but not for any more than that.

During the evolution of each generation in the EGA, every individual must train and evaluate $n - 1$ MLPs on the codes defined by the genome of the individual. The training process of an MLP using the backpropagation algorithm [9] depends on the number of epochs on the training data, which in turn also has a linear impact on the execution time required to effectively train it. Therefore, the MLPs should be trained using no more epochs than needed to correctly approximate the patterns present in ND. See Sect. 4.

4   Experiments

4.1 Parallel CENG

The training process of the $n - 1$ MLPs for each individual in a generation was programmed to be executed in parallel. This approach takes advantage of multi-core architectures by using the total number of cores available on the machine the code is executing on. In Fig. 2 we show the execution times for a database consisting of 3 numerical variables, 10 categorical variables with 4 instances each, and 500 tuples.

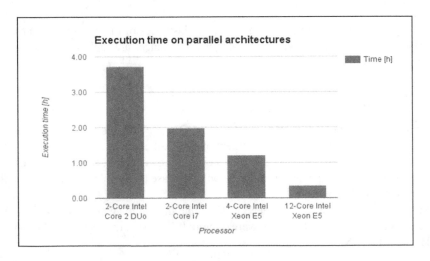

Fig. 2. ParCENG's execution times

Parallel CENG was run for the same database on 4 different architectures, as shown in Table 1.

Table 1. Execution times for ParCENG.

Processor	Clock Rate	Physical cores	Virtual cores
Intel Core 2 Duo	2.66 GHz	2	NA
Intel Core i3	1.7 GHz	2	4
Intel Xeon E5	2.6 GHz	4	8
Intel Xeon E5	2.7 GHz	12	24

For this experiment, the maximum number of generations for the EGA was set to 100 and the number of epochs the MLPs were trained for was set to 1000.

4.2 Determining the Number of Generations for the EGA

We ran ParCENG for 1,000 generations and recorded the fitness obtained after each generation for 3 databases of different sizes and with a different number of variables. Each of the databases included 3 numerical variables. However, one also contained 3 categorical variables with 4 instances each and consisted of 700 tuples, another contained 5 categorical variables with 4 instances each and 500 tuples. A last one contained 9 categorical variables with 4 instances each and 300 tuples. These databases were chosen since they reflected similar execution times in the sequential case.

In Fig. 3 we can see that database size in terms of variables and tuples directly affects the number of generations in which the EGA converges. We may see that $G = 800$ generations are enough. G is sensitive to the size of the input and complexity of the patterns present in it. It should be statistically determined depending on the data CENG works on.

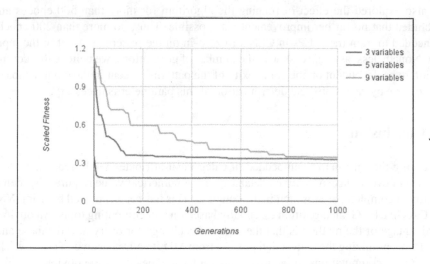

Fig. 3. Fitness varying maximum number of generations for the EGA

4.3 Determining the Number of Epochs to Train the MLPs

To determine the minimum number of epochs needed to train the MLPs in CENG, we ran the algorithm varying only the number of epochs. CENG was set to run for 100 generations and was fed a database consisting of 3 numerical variables, and 10 categorical variables with 4 instances each. The algorithm was then configured to run varying the number of epochs from 20 to 500 in steps of 20. Each configuration was run on 10 databases with tuples varying from 100 to 1,000 in steps of 100. The results are shown in Fig. 4.

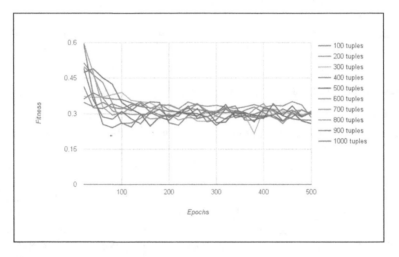

Fig. 4. Fitness varying number of epochs for MLP training

After 200 epochs, there appears to be no further improvement in terms of final fitness. We also explored the effect of running the algorithm for more than 500 epochs and concluded that no further improvement was possible. Thus, no more than 200 epochs are needed to train the MLPs in CENG to zero-in on the patterns present in the input data. Nonetheless, as is the case with the number of generations, we believe this configuration to be dependent on the complexity of the patterns present in the input database. Hence, case by case statistical determination of this parameter is necessary.

5   Conclusions

Preserving the information embedded in categorical attributes while replacing every instance of every class of a mixed database with a numerical value requires the identification of complex patterns tacit in the database. This has been achieved by using NNs and GAs in CENG, an algorithm essentially based on soft computing tools. An obvious disadvantage of the method is that the best codes change for every new database and, therefore, optimizing the conversion process from MDs to NDs becomes a priority. The algorithm is computationally intensive and we have appealed to the obvious possible parallel evaluation of the individuals in a version called ParCENG.

By using n physical cores execution time decreased almost linearly with n. However, to exploit the parallel alternative to its utmost we ought to fine-tune the parameters of, both, the EGA and the NNs. From the experiments conducted we were able to set upper bounds of the number of epochs to train the NNs as well as the number of generations of the EGA. These bounds are particular to ParCENG and, when selected as per our results, lead to execution times (for complex databases which originally necessitated of hours of CPU time) of a few minutes. This result is hardly surprising. However, the quantitative determination of the relation between the execution times and the parameters involved in ParCENG allows us to determine its behavior *a priori*. Thus, ParCENG specifications are freed from the purely heuristic estimation of its parameters. This will allow us to tackle much larger databases setting ParCENG's parameters accordingly.

One has to keep in mind that ParCENG offers a very interesting alternative to condition mixed databases into purely numerical ones. Thus, the resulting data, while retaining the patterns originally contained in the categories, are now susceptible to be applied to numerically cluster the data.

References

1. Norusis, M.: SPSS 16.0 statistical procedures companion. Prentice Hall Press, Upper Saddle River (2008)
2. Goebel, M., Gruenwald, L.: A survey of data mining and knowledge discovery software tools. ACM SIGKDD Explor. Newslett. **1**(1), 20–33 (1999)
3. Sokal, R.R.: The principles of numerical taxonomy: twenty-five years later. Comput.-Assist. Bacterial Syst. **15**, 1 (1985)
4. Barbará, D., Yi, L., Julia C.: COOLCAT: an entropy-based algorithm for categorical clustering. In: Proceedings of the Eleventh International Conference on Information and Knowledge Management, pp. 582–589. ACM (2002)
5. Kuri-Morales, A.F.: Categorical encoding with neural networks and genetic algorithms. In: Zhuang, X., Guarnaccia, C. (eds.) WSEAS Proceedings of the 6th International Conference on Applied Informatics and. Computing Theory, pp. 167–175, 01 July 2015. ISBN 9781618043139, ISSN: 1790-5109
6. Cybenko, G.: Approximation by superpositions of a sigmoidal function. Math. Control, Signals Syst. **2**(4), 303–314 (1989)
7. Rudolph, G.: Convergence analysis of canonical genetic algorithms. IEEE Trans. Neural Networks **5**(1), 96–101 (1994)
8. Castro, F., Gelbukh, A., González, M. (eds.): MICAI 2013. LNCS (LNAI), vol. 8266. Springer, Heidelberg (2013). doi:10.1007/978-3-642-45111-9
9. Widrow, B., Lehr, M.A.: 30 years of adaptive neural networks: perceptron, madaline, and backpropagation. Proc. IEEE **78**(9), 1415–1442 (1990)
10. Deb, K., Agrawal, S., Pratap, A., Meyarivan, T.: A fast elitist non-dominated sorting genetic algorithm for multi-objective optimization: NSGA-II. In: Schoenauer, M., Deb, K., Rudolph, G., Yao, X., Lutton, E., Merelo, J.J., Schwefel, H.-P. (eds.) PPSN 2000. LNCS, vol. 1917, pp. 849–858. Springer, Heidelberg (2000). doi:10.1007/3-540-45356-3_83

Extending Extremal Polygonal Arrays for the Merrifield-Simmons Index

Guillermo De Ita Luna[1], J. Raymundo Marcial-Romero[2(✉)], J.A. Hernández[2],
Rosa Maria Valdovinos[2], and Marcelo Romero[2]

[1] Facultad de Ciencias de la Computación, BUAP, Puebla, Mexico
deita@cs.buap.mx
[2] Facultad de Ingeniería, UAEM, Toluca, Mexico
{jrmarcialr,xoseahernandez,rvaldovinos,mromeroh}@uaemex.mx

Abstract. Polygonal array graphs have been widely investigated, and they represent a relevant area of interest in mathematical chemistry because they have been used to study intrinsic properties of molecular graphs. For example, to determine the Merrifield-Simmons index of a polygonal array A_n that is the number of independent sets of that graph, denoted as $i(A_n)$.

In this paper we consider the problem of extending an initial polygonal array A_n adding a new polygon p to form A_{n+1}, for minimizing or maximizing the Merrifield-Simmons index $i(A_{n+1}) = i(A_n \cup p)$. Our method does not require to compute $i(A_n)$ or $i(A_n \cup p)$, explicitly.

Keywords: Counting the number of independent sets · Enumerative algorithms · Efficient counting · Merrifield-Simmons index

1 Introduction

Counting problems are not only mathematically interesting, but also they arise in many applications. Regarding hard counting problems, the computation of the number of independent sets of a graph G, denoted as $i(G)$, has been a key in determining the frontier between efficient and intractable counting problems.

Polygonal array graphs have been widely investigated, and they represent a relevant area of interest in mathematical chemistry because they have been used to study intrinsic properties of molecular graphs. In addition, it is also of great importance to recognize substructures of those compounds and learn messages from the graphic model by clear elucidation of their structures and properties [6].

There are several works analyzing extremal values for the number of independent sets (known in mathematical chemistry as the Merrifield-Simmons index) on chain graphs [6,7]. Merrifield and Simmons showed the correlation between $i(G)$ and boiling points on polygonal chain graphs that represent chemical molecules. This index is a typical example of an invariant graph used in mathematical chemistry for quantifying relevant details of molecular structures.

© Springer International Publishing AG 2017
J.A. Carrasco-Ochoa et al. (Eds.): MCPR 2017, LNCS 10267, pp. 22–31, 2017.
DOI: 10.1007/978-3-319-59226-8_3

In 1993, Gutman discussed the extremal hexagonal chains according to three topological invariants: Hosoya index, largest eigenvalue and Merrified-Simmons index. His work greatly stimulated the study of extremal polygonal chains.

On his seminal paper [3], Gutman showed extremal linear chains for Merrifield index in the particular case of hexagonal chains. He conjectured that the chain with the smallest Merrifield-Simmons index is unique and it corresponds to the zig-zag polyphenegraph.

L.Z. Zhang et al. [4,5] showed Gutman's conjecture. They showed that the minimum value for the Merrifield-Simmons index is achieved by the zig-zag polyphenegraph. Later on, Cao et al. [2] showed extremal poligonal chains for k-matchings (Hosoya index), considering the topology of polygonal arrays that provide maximum as well as minimum values for the Hosoya index. Their demonstrations are based on the use of the Z-polynomial (Z-counting polynomial).

The previous mentioned works are concerned to hexagonal chains, or to the Hosoya index. We consider in this paper, results for the Merrifield-Simmon index for any kind of polygonal arrays. We consider the problem of extending a polygonal array A by adding a new polygon p, preserving the structure of polygonal arrays and such that $i(A \cup p)$ is maximum. Instead of computing the Z-polynomial for the Merrifield-Simmon index as Cao et al. [2] suggested, our proofs are based on properties that are derived from the product between Fibonacci's numbers with complementary indexes values. In fact, our method does not require compute $i(G)$ explicity, and it can be adapted to compute other intrinsic properties on molecular graphs.

2   Polygonal Chains

Let $G = (V, E)$ be a molecular graph. Denote by $n(G, k)$ the number of ways in which k mutually independent vertices can be selected in G. By definition, $n(G, 0) = 1$ and $n(G, 1) = |V(G)|$, for all graph G. Furthermore, $i(G) = \sum_{k \geq 0} n(G, k)$ is the *Merrifield-Simmons index* of G, that is, exactly the number of independent sets of G.

A polygonal chain is a graph $P_{k,t}$ obtained by identifying a finite number of t congruent regular polygons such that each basic polygon, except the first and the last one, is adjacent to exactly two basic polygons. When each polygon in $P_{k,t}$ has the same number of k nodes, then $P_{k,t}$ becomes a linear array of t k-gons. A special class of polygonal chains is the class of hexagonal chains, which are chains formed by n 6-gons. Hexagonal systems play an important role in mathematical chemistry as they are natural representations of catacondensed benzenoid hydrocarbons.

The propensity of carbon atoms to form compounds, made of hexagonal arrays fused along the edges gives a relevant importance to the study of chemical properties of benzenoid hydrocarbons. Those graphs have been widely investigated and represent a relevant area of interest in mathematical chemistry, since it is used for quantifying relevant details of the molecular structure of the benzenoid hydrocarbons [1,6].

Let $H_n = h_1 h_2 \cdots h_n$ be a polygonal chain with n basic polygons, where each h_i and h_{i+1} have exactly one common edge $e_i, i = 1, 2, \ldots, n - 1$. A polygonal chain with at least two polygons has two end-polygons, h_1 and h_n, while $h_2, \ldots, h_{n-1}$ are the internal polygons of the chain. In a polygonal chain, each vertex has degree either 2 or 3. The vertices of degree 3 are exactly the end points of the common edges between two consecutive polygons.

The distance $d_G(x; y)$ from a vertex x to another vertex y is the minimum number of edges in an $x - y$ path of G. Similarly, we define the distance between two edges e_1, e_2 on the graph G as the minimum number of edges in an $e_1 - e_2$ path of G.

Let H_n be a polygonal chain with n basic polygons joined by one common edge between two consecutive basic polygons. If for each pair of consecutive joining edges e_i and e_{i+1} of the polygonal chain it holds that $d_{H_n}(e_i, e_{i+1}) = 2$ then H_n is known as a linear polygonal chain (L_n), and if $d_{H_n}(e_i, e_{i+1}) = 1$ for each pair of consecutive common edges, then H_n is known as a zig-zag polygonal chain (Z_n), see Fig. 1 for an example.

In recent years, several works have been done for determining the extremal graphs corresponding to the Hosoya and Merrifield-Simmons indexes [2,6,7]. For many graph classes that have been studied so far, the graph that minimizes the Merrifield-Simmons index is also the one that maximizes the Hosoya index, and vice versa, although its relation is still not totally understood.

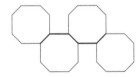

Fig. 1. Example of a zig-zag octagonal chain

3   Fibonacci Properties to Compute Indices of Polygonal Chains

It is known that for any simple path P_n of size n (P_n has n vertices), P_n fulfills $i(P_n) = F_{n+2}$, where F_n is the nth-Fibonacci number with initial values $F_0 = 0, F_1 = 1$.

Let us consider an isolated vertex as a linear path of size 1, since in this way $i(P_1) = F_3 = 2$. Thus, the size of any simple path is the number of vertices in it. Let $k > 0$ be a constant integer and let P_i and P_j be two disjointed simple paths, such that $i + j = k$. It is known that $i(P_i \cdot P_j) = i(P_i) \cdot i(P_j) = F_{i+2} \cdot F_{j+2}$. We want to determine for which pair (i, j), $i, j \geq 2$, the value of $i(P_i) \cdot i(P_j)$ is maximum when $j + i = k$, for any $k > 0$ constant.

We apply Binet's formula to state the computation of Fibonnaci numbers, i.e. $F_t = \frac{a^t - b^t}{a - b}$, where $a = \left(\frac{1+\sqrt{5}}{2}\right)$ and $b = \left(\frac{1-\sqrt{5}}{2}\right)$ are roots of the polynomial $r^2 - r - 1 = 0$. Furthermore, $a + b = 1$ and $ab = -1$.

Let us define the sequence $\beta_{t,s}$, with $t \geq 1$ and $t \geq s \geq 1$ as follows

$$\beta_{t,s} = F_s F_{t-s} \tag{1}$$

We firstly show that the sequence $\beta_{t,s}$ becomes symmetric when $s > \lfloor \frac{t}{2} \rfloor$,

Lemma 1. *For all j, such that $1 \leq j \leq \lfloor \frac{t}{2} \rfloor - 2$, the sequence $\beta_{t,s}$ satisfies the following*

1. $\beta_{t,\lfloor \frac{t}{2} \rfloor - j} = \beta_{t,\lfloor \frac{t}{2} \rfloor + j}$ *if t is even,*
2. $\beta_{t,\lfloor \frac{t}{2} \rfloor - j} = \beta_{t,\lfloor \frac{t}{2} \rfloor + j + 1}$ *if t is odd,*

Proof. From Eq. (1) we have

$$(a-b)^2 \beta_{t,\lfloor \frac{t}{2} \rfloor - j} = \left(a^{\lfloor \frac{t}{2} \rfloor - j} - b^{\lfloor \frac{t}{2} \rfloor - j} \right) \left(a^{t-\lfloor \frac{t}{2} \rfloor + j} - b^{t-\lfloor \frac{t}{2} \rfloor + j} \right)$$

$$= a^t - a^{\lfloor \frac{t}{2} \rfloor - j} b^{t-\lfloor \frac{t}{2} \rfloor + j} - a^{t-\lfloor \frac{t}{2} \rfloor + j} b^{\lfloor \frac{t}{2} \rfloor - j} + b^t$$

$$(a-b)^2 \beta_{t,\lfloor \frac{t}{2} \rfloor + j} = a^t - a^{\lfloor \frac{t}{2} \rfloor + j} b^{t-\lfloor \frac{t}{2} \rfloor - j} - a^{t-\lfloor \frac{t}{2} \rfloor - j} b^{\lfloor \frac{t}{2} \rfloor + j} + b^t$$

Thus,

$$(a-b)^2 \beta_{t,\lfloor \frac{t}{2} \rfloor - j} = a^t - a^{\lfloor \frac{t}{2} \rfloor - j} b^{t-\lfloor \frac{t}{2} \rfloor + j} - a^{t-\lfloor \frac{t}{2} \rfloor + j} b^{\lfloor \frac{t}{2} \rfloor - j} + b^t$$

$$- a^{\lfloor \frac{t}{2} \rfloor + j} b^{t-\lfloor \frac{t}{2} \rfloor - j} - a^{t-\lfloor \frac{t}{2} \rfloor - j} b^{\lfloor \frac{t}{2} \rfloor + j}$$

$$a^{\lfloor \frac{t}{2} \rfloor + j} b^{t-\lfloor \frac{t}{2} \rfloor - j} + a^{t-\lfloor \frac{t}{2} \rfloor - j} b^{\lfloor \frac{t}{2} \rfloor + j}$$

$$= (a-b)^2 \beta_{t,\lfloor \frac{t}{2} \rfloor + j} + a^{\lfloor \frac{t}{2} \rfloor + j} b^{t-\lfloor \frac{t}{2} \rfloor - j} + a^{t-\lfloor \frac{t}{2} \rfloor - j} b^{\lfloor \frac{t}{2} \rfloor + j}$$

$$- a^{\lfloor \frac{t}{2} \rfloor - j} b^{t-\lfloor \frac{t}{2} \rfloor + j} - a^{t-\lfloor \frac{t}{2} \rfloor + j} b^{\lfloor \frac{t}{2} \rfloor - j}$$

$$= (a-b)^2 \beta_{t,\lfloor \frac{t}{2} \rfloor + j} + a^{\lfloor \frac{t}{2} \rfloor - j} b^{t-\lfloor \frac{t}{2} \rfloor - j} \left[a^{2j} - b^{2j} \right]$$

$$+ a^{t-\lfloor \frac{t}{2} \rfloor - j} b^{\lfloor \frac{t}{2} \rfloor - j} \left[b^{2j} - a^{2j} \right]$$

$$= (a-b)^2 \beta_{t,\lfloor \frac{t}{2} \rfloor + j} + \left(a^{2j} - b^{2j} \right) \left(a^{\lfloor \frac{t}{2} \rfloor - j} b^{t-\lfloor \frac{t}{2} \rfloor - j} - a^{t-\lfloor \frac{t}{2} \rfloor - j} b^{\lfloor \frac{t}{2} \rfloor - j} \right)$$

$$= (a-b)^2 \beta_{t,\lfloor \frac{t}{2} \rfloor + j} + \frac{\left(a^{2j} - b^{2j} \right) a^{2\lfloor \frac{t}{2} \rfloor} b^{2\lfloor \frac{t}{2} \rfloor}}{a^{\lfloor \frac{t}{2} \rfloor + j} b^{\lfloor \frac{t}{2} \rfloor + j}} \left(b^{t-2\lfloor \frac{t}{2} \rfloor} - a^{t-2\lfloor \frac{t}{2} \rfloor} \right)$$

but t is even, that is $t = 2r$ for some $r \in \mathbb{Z}$ and $t - 2\lfloor \frac{t}{2} \rfloor = 0$. Therefore, $b^{t-2\lfloor \frac{t}{2} \rfloor} - a^{t-2\lfloor \frac{t}{2} \rfloor} = b^0 - a^0 = 0$. □

Thus, we might assume that $2 \leq s \leq \lfloor \frac{t}{2} \rfloor$ then $s - 1 < t - s$. Now, we show that if s is even in $\beta_{t,s}$ then the sequence is increasing.

Lemma 2. *The sequence $\beta_{t,2}$, $\beta_{t,4}$, $\beta_{t,6},...$ that is $\{\beta_{t,2p}\}_{t,p}$ satisfies $\beta_{t,2p} < \beta_{t,2(p+1)}$ for every $p \in \{1, 2, ..., \lfloor \frac{t}{4} \rfloor\}$ and all t.*

Proof. We have,

$$\beta_{t,2p} = \frac{\left(a^{2p} - b^{2p}\right)\left(a^{t-2p} - b^{t-2p}\right)}{(a-b)^2}$$

$$= \frac{a^t - a^{2p}b^{t-2p} - a^{t-2p}b^{2p} + b^t}{(a-b)^2}$$

$$\beta_{t,2(p+1)} = \frac{\left(a^{2p+2} - b^{2p+2}\right)\left(a^{t-2p-2} - b^{t-2p-2}\right)}{(a-b)^2}$$

$$\beta_{t,2p+2} = \frac{a^t - a^{2p+2}b^{t-2p-2} - a^{t-2p-2}b^{2p+2} + b^t}{(a-b)^2}$$

$$(a-b)^2\beta_{t,2p+2} = a^t - a^{2p+2}b^{t-2p-2} - a^{t-2p-2}b^{2p+2} + b^t$$
$$- a^{2p}b^{t-2p} - a^{t-2p}b^{2p}$$
$$+ a^{2p}b^{t-2p} + a^{t-2p}b^{2p}$$
$$= (a-b)^2\beta_{t,2p} + a^{2p}b^{t-2p} + a^{t-2p}b^{2p}$$
$$- a^{2p+2}b^{t-2p-2} - a^{t-2p-2}b^{2p+2}$$
$$= (a-b)^2\beta_{t,2p} + \left(1 - a^2b^{-2}\right)a^{2p}b^{t-2p}$$
$$+ a^{t-2p}b^{2p}(1 - a^{-2}b^2)$$
$$= (a-b)^2\beta_{t,2p} + \left(b^2 - a^2\right)a^{2p+2}b^{t-2p}$$
$$+ \left(a^2 - b^2\right)a^{t-2p}b^{2p+2}$$

$$\beta_{t,2(p+1)} = \beta_{t,2p} + \frac{k}{(a-b)}\left[a^{t-2p}b^{2p+2} - a^{2p+2}b^{t-2p}\right]$$
$$= \beta_{t,2p} + ka^{2p+2}b^{2p+2}\left[a^{t-4p-2} - b^{t-4p-2}\right]$$
$$= \beta_{t,2p} + k\frac{a^{t-4p-2} - b^{t-4p-2}}{a-b}$$
$$= \beta_{t,2p} + kF_{t-2(2p+1)}$$

and as $k, F_{t-2(2p+1)} > 0$, the proof is complete.   □

On the other hand, if s is odd in $\beta_{t,s}$ then the sequence is decreasing, as the following lemma shows.

Lemma 3. *The sequence* $\{\beta_{t,2p+1}\}_{t,p}$ *satisfies* $\beta_{t,2p+1} > \beta_{t,2p+3}$ *for every* $p \in \{0, 2, ..., \lfloor\frac{t}{4}\rfloor - 1\}$ *and all* t.

Proof.

$$\beta_{t,2p+1} = \frac{\left(a^{2p+1} - b^{2p+1}\right)\left(a^{t-2p-1} - b^{t-2p-1}\right)}{(a-b)^2}$$

$$= \frac{a^t - a^{2p+1}b^{t-2p-1} - a^{t-2p-1}b^{2p+1} + b^t}{(a-b)^2}$$

and

$$\beta_{t,2p+3} = \frac{\left(a^{2p+3} - b^{2p+3}\right)\left(a^{t-2p-3)} - b^{t-2p-3}\right)}{(a-b)^2}$$

$$(a-b)^2\beta_{t,2p+3} = a^t - a^{2p+3}b^{t-2p-3} - a^{t-2p-3}b^{2p+3} + b^t$$
$$- a^{2p+1}b^{t-2p-1} - a^{t-2p-1}b^{2p+1}$$
$$+ a^{2p+1}b^{t-2p-1} + a^{t-2p-1}b^{2p+1}$$

$$= (a-b)^2\beta_{t,2p+1} + a^{2p+1}b^{t-2p-1} + a^{t-2p-1}b^{2p+1}$$
$$- a^{2p+3}b^{t-2p-3} - a^{t-2p-3}b^{2p+3}$$
$$= (a-b)^2\beta_{t,2p+1} + a^{2p+1}b^{t-2p-1} + a^{t-2p-1}b^{2p+1}$$
$$- a^{2p+1}b^{t-2p-1}a^2b^{-2} - a^{t-2p-1}b^{2p+1}a^{-2}b^2$$
$$= (a-b)^2\beta_{t,2p+1} + a^{2p+1}b^{t-2p-1}(1 - a^2b^{-2})$$
$$+ a^{t-2p-1}b^{2p+1}(1 - a^{-2}b^2)$$

Using the fact that $ab = -1$ then $a^2b^2 = 1$, and

$$\beta_{t,2p+3} = \beta_{t,2p+1} + \frac{(a^2 - b^2)a^{t-2p-1}b^{2p+3} - (a^2 - b^2)a^{2p+3}b^{t-2p-1}}{(a-b)^2}$$
$$= \beta_{t,2p+1} + k\frac{a^{t-2p-1}b^{2p+3} - a^{2p+3}b^{t-2p-1}}{a-b}$$
$$= \beta_{t,2p+1} + k(ab)^{2p+3}\frac{a^{t-4(p+1)} - b^{t-a(p+1)}}{a-b}$$
$$= \beta_{t,2p+1} - kF_{k-t-4(p+1)}$$

and as $k, F_{k-t-4(p+1)} > 0$, the proof is complete.   □

The main Theorem of the section is the following:

Theorem 1. *For any integers* $t, s, k \geq 1$,

1. $\min_s \{F_s F_{t-s}\} = F_2 F_{t-2} = F_{t-2}$
2. $\max_s \{F_s F_{t-s}\} = F_1 F_{t-1} = F_{t-1}$

Proof. It is enough to proof that $\beta_{t,\lfloor\frac{t}{2}\rfloor} < \beta_{t,\lfloor\frac{t}{2}\rfloor-1}$ which can be easily done by an algebraic manipulation and the rest of the proof follows from above lemmas.   □

4   Extremal Topologies on Polygonal Graphs

In this section, we show how the previous results about the product of Fibonacci numbers can be used for determining the structure of polygonal arrays G such that when adding a new polygon, maximize or minimize $i(G)$.

Let h be a polygon with n sides. Let P_i and P_j be two different linear paths of sizes i and j, respectively, such that $i + j = k$ becomes a constant. Let G be the resulting graph formed by joining P_i and P_j to the end-nodes of any edge $e = \{x, y\} \in E(h)$, as it is illustrated in Fig. 2. Notice that e can be any edge of the polygon since in fact, the initial polygon is a cycle and all of its edges are indistinguishable.

We show that $i(G)$ is maximum under the restriction $|P_i| + |P_j| = k$ when $i = 2$ (P_i has exactly two vertex and only one edge) and $j = k - 2$ (a path of $k - 2$ vertices).

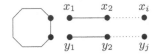

Fig. 2. A base graph

Lemma 4. *Let h be a polygon of n sides and $e = \{x, y\} \in E(h)$. Let $P_i = \{x, x_1, \ldots, x_{i-1}\}$ and $P_j = \{y, y_1, \ldots, y_{j-1}\}$ be two disjointed paths such that $V(P_i) \cap V(h) = \{x\}$, $V(P_j) \cap V(h) = \{y\}$ and $i + j = k$. If $G = h \cup P_i \cup P_j$ then the maximum for $i(G)$ is achieved when $i = 2$ and $j = k - 2$.*

Proof. Applying the division edge rule on the edge $e = \{x, y\}$, we obtain that $i(G) = i(G - e) - i(G - (N[x] \cup N[y]))$. Notice that $(G - e)$ is a linear path of size $n + (i - 1) + (j - 1) = n + k - 2$; therefore, $i(G - e) = F_{n+k}$. Furthermore, $i(G - e)$ is invariant with respect to the selected position of the edge $e \in E(h)$.

On the other hand, $(G - (N[x] \cup N[y]))$ is formed by three disjointed paths: P_{i-2}, P_{j-2} and the path that results from eliminating e and its two adjacent edges from h. Let us denote this last path as P_{n-4}. Then, $i(G - (N[x] \cup N[y]))) = F_{n-2} \cdot F_i \cdot F_j$. In fact, the result of this product does not depend on the initial position of e in h because the three resulting paths will have same sizes independently of the position of e in h.

Then, maximizing $i(G)$ is equivalent to minimizing $F_i \cdot F_j$ since these are the unique sizes on i and j that can vary under the restriction $i + j = k$. According to Theorem 1, $F_i \cdot F_j$ has a minimum value when $F_i = F_2$ and $F_j = F_{k-2}$ which means that the resulting path P_{i-2} after removing $N[x]$ from G should be empty and the resulting path P_{j-2} after removing $N[y]$ from G should have $k - 4$ vertices. $\qquad \square$

Lemma 5. *Let h be a polygon of n sides and $e = \{x, y\} \in E(h)$. Let $P_i = \{x, x_1, \ldots, x_{i-1}\}$ and $P_j = \{y, y_1, \ldots, y_{j-1}\}$ be two disjointed paths such that $V(P_i) \cap V(h) = \{x\}$, $V(P_j) \cap V(h) = \{y\}$ and $i + j = k$. If $G = h \cup P_i \cup P_j$ then the minimum for $i(G)$ is achieved when $i = 1$ and $j = k - 1$.*

Proof. Similar to Lemma 4. $\qquad \square$

Furthermore, the maximum and minimum values for $i(G)$ are achieved independently of the number of edges in the polygon h. Consequently, our results are fulfilled for any polygon joined with two disjointed paths in the end-nodes of one of its edges. We show in Fig. 3 the extremal topologies for $i(G)$ for the class of graphs: $G = h \cup P_i \cup P_j$.

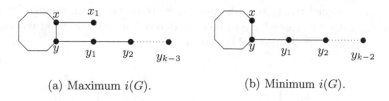

(a) Maximum $i(G)$.   (b) Minimum $i(G)$.

Fig. 3. Extremal topologies for $i(G)$, $G = h \cup P_i \cup P_j$.

Let $A_n : h_1, \ldots, h_n, n \geq 1$ be a polygonal array and p a new polygon of k edges. We denote as e_i the common edge between the polygon h_i and h_{i+1}. We want to extend A_n to A_{n+1} joining p to A_n in such way that $i(A_{n+1})$ is either minimum or maximum into the set of possible edges $e \in E(h_n)$ to be selected for joining it with p.

Let us, enumerate the edges of h_n as $b_0, b_1, \ldots, b_{k-1}$, where b_0 is the common edge between h_n and h_{n-1} and the numeration is according to the clockwise direction. We also consider that h_n has more than 5 edges. We must select $e \in E(h_n)$ such that $i(A_n \cup p)$ is maximum into the set of possible selections of edges of h_n. For example, e can not be anyone from b_0, b_1, b_{k-1} because if we join p to any one of those edges, then $A_n \cup p$ losses the structure of polygonal arrays.

Theorem 2. *Let A_n be a polygonal array whose last polygon h_n has k edges and p be a polygon. Let $b_0, b_1, \ldots, b_{k-1}$ be the edges of h_n such that b_0 is the common edge between h_n and h_{n-1} and the numeration is clockwise. If $A_{n+1} = A_n \cup p$ is a polygonal array then $\max\{i(A_{n+1})\}$ is achieved when p is joined to A_n at b_3, that is a distance 2 from the common edge between h_{n-1} and h_n.*

Proof. Applying the division edge rule on $e = \{x, y\} \in E(h_n)$, we obtain

$$i(A_{n+1}) = i(A_{n+1} - e) - i(A_{n+1} - (N[x] \cup N[y])) \tag{2}$$

The term $i(A_{n+1} - e)$ is invariant from e because its values does not depend on the selected position of e to join p to A_n.

$A_{n+1} - (N[x] \cup N[y])$ is a graph formed by two connected components: a linear path P_{k-5} with $k - 5$ edges and then $i(P_{k-5}) = F_{k-2}$ that is an invariant value independent to the position of $e \in E(h_n)$. And the other connected component, denoted by G, that is a polygonal array where one of the edges in the last polygon is joined to two disjointed paths P_i and P_j. Then, $i(A_{n+1} - (N[x] \cup N[y])) = i(G) * F_{k-2}$, and that value is minimum if in G is preserved a maximum number of edges from A_n. It means that $(N[x] \cup N[y])$ contains a minimum number of edges from p and h_n. As $\delta(x) = \delta(y) = 3$ then the minimum number of edges to be contained in $(N[x] \cup N[y])$ is 7, meaning that e has to be a distance 2 from the common edge e_{n-1} between h_{n-1} and h_n, and also $|P_i| = 0$ and $|P_j| = k - 6$, in order to maximize $i(A_{n+1})$.   □

For the minimum Merrifield-Simmon index, it is not a sufficient condition to state that p is joined to A_n to a distance one from the common edge between h_{n-1} and h_n since two cases have to be considered, Fig. 4 shows the two possible configurations in a octahedral polygonal array.

(a)   (b)

Fig. 4. In the octahedral polygonal array on the left the shortest path for visiting e_1, e_2, e_3 is a path, while it is a tree in the octahedral polygonal array on the right

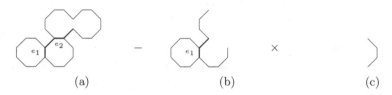

(a)   (b)   (c)

Fig. 5. Application of the edge division rule to the graph Fig. 4a to compute the Merrifield-Simmon index

Applying the edge division rule to e_3, the decomposition to compute the Merrifield-Simmon index on the graphs Fig. 4a and b are shown in Figs. 5 and 6, respectively. The left graphs of Figs. 5a and 6a are invariant, similar to the right graphs (5c and 6c) of the same figures. By Lemma 5, the graph of Fig. 6b has minimum Merrifield-Simmon index than the graph of Fig. 5b hence the polygonal array has minimum Merrifield-Simmon index when the shortest path of e_1, e_2 and e_3 is a path.

Theorem 3. *Let A_n be a polygonal array whose last polygon h_n has k edges and p be a polygon. Let e_i be the edge joining h_i to h_{i+1}. If $A_{n+1} = A_n \cup p$ is a polygonal array then $\min\{i(A_{n+1})\}$ is achieved when p is joined to A_n at distance 1 from e_n and e_{n-1}, e_n, e_p forms a path.*

Proof. Similar to the proof of Theorem 2.   □

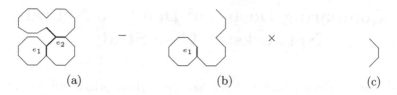

(a)   (b)   (c)

Fig. 6. Application of the edge division rule to the graph Fig. 3b to compute the Merrifield-Simmon index

5   Conclusions

We have shown how properties of Fibonacci numbers can be used for the computation of Merrifield-Simmon index. We have shown that, as expected, when adding a new polygon at distance 2, the maximum number of independent sets is obtained among all the possible combinations. Dually, the minimal number of independent sets is obtained when the distance is 1 and a path is built from the joining edges. Our method does not require to compute explicitly, the number of independent sets of the involved graphs.

References

1. Došlić, T., Måløy, F.: Chain hexagonal cacti: matchings and independent sets. Discrete Math. **310**, 1676–1690 (2010)
2. Yuefen, C., Fuji, Z.: Extremal polygonal chains on k-matchings. MATCH Commun. Math. Comput. Chem. **60**, 217–235 (2008)
3. Gutman, I.: Extremal hexagonal chains. J. Math. Chem. **12**, 197–210 (1993)
4. Zhang, L.Z.: The proof of Gutman's conjectures concerning extremal hexagonal chains. J. Syst. Sci. Math. Sci. **18**(4), 460–465 (1998)
5. Zhang, L.Z., Zhang, F.: Extremal hexagonal chains concerning k-matchings and k-independent sets. J. Math. Chem. **27**, 319–329 (2000)
6. Wagner, S., Gutman, I.: Maxima and minima of the Hosoya index and the Merrifield-Simmons index. Acta Applicandae Mathematicae **112**(3), 323–346 (2010)
7. Deng, H.: Catacondensed benzenoids and phenylenes with the extremal thirdorder Randic Index1. Comm. Math. Comp. Chem. **64**, 471–496 (2010)

Comparing Deep and Dendrite Neural Networks: A Case Study

Gerardo Hernández[1], Erik Zamora[2], and Humberto Sossa[1(✉)]

[1] Instituto Politécnico Nacional - CIC,
Av. Juan de Dios Batiz S/N, Gustavo A. Madero, 07738 Mexico City, Mexico
ghernandez_a13@sagitario.cic.ipn.mx, hsossa@cic.ipn.mx
[2] Instituto Politécnico Nacional - UPIITA,
Av. Instituto Politécnico Nacional 2580, Barrio la Laguna Ticoman,
Gustavo A. Madero, 07340 Mexico City, Mexico
ezamorag@ipn.mx

Abstract. In this paper, a comparative study between two different neural network models is performed for a very simple type of classificaction problem in 2D. The first model is a deep neural network and the second is a dendrite morphological neuron. The metrics to be compared are: training time, classification accuracies and number of learning parameters. We also compare the decision boundaries generated by both models. The experiments show that the dendrite morphological neurons surpass the deep neural networks by a wide margin in terms of higher accuracies and a lesser number of parameters. From this, we raise the hypothesis that deep learning networks can be improved adding morphological neurons.

1 Introduction

In the area of Artificial Intelligence there is a great diversity of algorithms for pattern classification, and one of the most important is the Multi-Layer Perceptron (MLP) which through a training process adjusts the hyperplanes of each neuron in each layer to separate the classes of some dataset [22,25,26]. The training is often based on gradient descent and back-propagation [22]. This model since its appearance in 1961 [25] has been widely used in the area of pattern recognition. However, there are other classification algorithms such as Dendrite Morphological Neuron (DMN) which use a training algorithm completely different from back-propagation [22], in the sense that they do not try to approximate a hyperplane through an iterative training process, analyzing each sample of the training set. Instead, this type of neuron analyzes the elements as a complete set and based on lattice operations generate hyperboxes. They are able to classify the different classes of the training set with a higher rate.

The success of Deep Neural Networks (DNN) is well known for recognizing objects in images [12] and speech in audio [9]. The mathematical operations employed in these neurons remain the same as those of a MLP [22]: sums, multiplications and some well-known non-linear functions. Furthermore, convolutions

© Springer International Publishing AG 2017
J.A. Carrasco-Ochoa et al. (Eds.): MCPR 2017, LNCS 10267, pp. 32–41, 2017.
DOI: 10.1007/978-3-319-59226-8_4

are used for reducing the number of learning parameters [13]. So, the novelty of the last 10 years has focused on more computing power, more layers, more data and the dropout [14,23]. These last two are to avoid overfitting in deep models that previously prevented the MLPs from giving better results than the Support Vector Machines (SVM) [3]. It is important to note that these developments are not related to the mathematical structure. This leads us to ask if there are other mathematical operations that can improve the recognition performance. In this paper, we started a research project in that direction. In particular, we compared DNNs with DMNs for a specific type of problem: multi-class spirals with several loops in 2D. Even when this classification problem is artificial, it is useful for studying the essential properties of the two models. A very first analysis was published in [27]; here we extend the analysis for deeper models and more classes. As classification tools, both models are subjects for comparison in terms of percentage of classification and training times, which depend directly on the number of parameters that constitute the model.

The rest of the paper is organized as follows. Section 2 provides a brief description of works that have proposed a different mathematical structure from the mainstream of neural networks. Sections 3 and 4 present the architecture of DNNs and DMNs, respectively. Section 5 discusses the experimental results. Then, in Sect. 6 we give our conclusions and future work.

2 Previous Work

Currently, there are few studies aimed at improving the mathematical structure of deep neural networks. However, before the term "deep learning" was born, we could find several papers with interesting proposals. Pessoa and Maragos [18] combined linear with rank filters. This architecture has shown that it can recognize digits in images, generating similar or better results compared to classical MLPs in shorter training times. Ivakhnenko [11] proposes a multilayer of polynomials to approximate the decision boundary for clasification problems. This was the first deep learning model published in literature. Dubin and Rumelhart [4] introduce product units into neural networks. These units add complexity to the model in order to use less layers. Other mathematical structures have been proposed such as: higher-order neural networks (NNs) [5], sigma-pi NNs [8], second-order NNs [17], functionally expanded NNs [10], wavelet NNs [29] and Bayesian NNs [16]. Glorot investigated more effective ways of training very deep neural networks using ReLUs as activation functions, achieving results comparable to the state-of-the-art [6]. Bengio [2] argues that in order to learn complex functions through training by gradient descent, it is necessary to use deep architectures. In [1] Bengio also analyzes and considers alternatives to training by standard gradient descent, due to the trade-off between efficient learning and latching on information. In this paper, we evaluate the performance of the DNNs with that of the DMNs to show some limitations of the DNNs and how morphological operations could improve deep learning.

3   Deep Neural Networks

"A deep learning architecture is a multilayered stack of simple modules with multiple non-linear layers" [14] (usually between 5 and 20 layers), and each layer contains a n_i number of modules, where i is the layer number, each module is a neuron with some activation function such as sigmoid or tanh. So an MLP and its generalization a DNN are defined by a set of neurons divided into layers: an input, one or more intermediate and an output layer. Thus, the DNN architectures that are constructed to classify the datasets are neural networks which have an i number of intermediate layers and a n_i number of neurons per layer, and the numbers of neurons per layer n_{i-1} and n_i are not necessarily the same. In our experiments we used the Rectified Linear Unit (ReLU) due to better results in DNN according to [6,14,15], so that a neuron is defined by:

$$f(x) = \max\left(0, w^T x\right), \tag{1}$$

where x is the input vector of N dimensions and w is the weights vector that multiplies the input vector. In the output layer, the activation function is changed by a softmax, which is commonly used to predict the probabilities associated with a multinoulli distribution [7], which is defined by

$$softmax(x)_i = \frac{\exp(x_i)}{\sum_{j=1}^n \exp(x_j)}, \tag{2}$$

The general DNN architecture is shown in Fig. 1. It is also common practice to vary the number of neurons contained in each layer of the DNN. The training method used for the DNN is Nesterov gradient descent with a mini-batch size of 64 and a moment of 0.9, which helps us to a more stable and fast convergence.

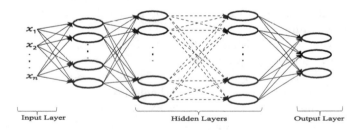

Fig. 1. Architecture of a DNN, where the number of hidden layers is another hyper-parameter.

4   Dendrite Morphological Neurons

A DMN segments the input space into hyperboxes of N dimensions. The output y of a neuron is a scalar given by

$$y = \underset{k}{argmax}(d_{n,k}), \tag{3}$$

where n is the dendrite number, k is the class number, and $d_{n,k}$ is the scalar output of a dendrite given by

$$d_{n,k} = \min_i \left(\min \left(x - w_{min}^n, w_{max}^n - x \right) \right), \tag{4}$$

where x is the input vector, w_{min} and w_{max} are dendrite weight vectors. The min operations together check if x is inside the hyperbox limited by w_{min} and w_{max} as the extreme points (see Fig. 2). If $d_{n.k} > 0$, x is inside the hyperbox, If $d_{n,k} = 0$, x is somewhere in the hyperbox boundary; otherwise, it is outside. A good property of DMN is that they can create complex non-linear decision boundaries that separate classes with only one neuron [20, 21]. The reader can consult [28] for more information.

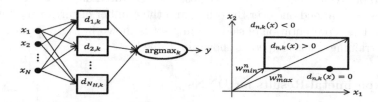

Fig. 2. Dendrite morphological neuron and an example of a hyperbox in 2D generated by its dendrite weights. The hyperbox divides the input space for classification purposes.

The training goal is to determine the number of hyperboxes and their weights needed to classify an input pattern. The regularized divide and conquer training method [28] consists of only two steps. The algorithm begins by opening an initial hyperbox H_0 that encloses all the samples with a margin distance M respect to each side of H_0 to have a better noise tolerance. Next the divide and conquer strategy is executed in a recursive way. The algorithm chooses a training sample x to generate a sub-hyperbox H_{sub} around it. Next it extracts the samples $(X_{H_{sub}}, T_{H_{sub}})$ from (X, T) that are enclosed in H_{sub}, where X is a training samples set represented as a matrix $X \epsilon \Re^{NxQ_{train}}$, Q_{train} is the number of training samples and the target class for each sample is contained in vector $T \epsilon \Re^{1xQ_{train}}$. The recursion divides H_0 until the error rate $E_\%$ in the hyperbox H is less or equal to the hyper-parameter E_0. The error rate is defined as $E_\% = \frac{|X_{mode}|}{|X|}$, where X_{mode} is the set of the most repeated training class [19]. At the end of the recursion process, the deepest hyperbox is assigned to the ruling class, which is set to the statistical mode of T. The recursive closing procedure is executed by appending all generated sub-hyperboxes with their corresponding classes. The hyperboxes with a common hyperface are joined. A complete description of this training method can be found in [24, 28].

5 Experiments

The experiments were designed with the aim of comparing the performance of the two neural networks, taking as a starting point the same training set. The aspects evaluated are the classification accuracy in the validation set, the training time, the number of parameters necessary for the network to correctly classify the training set, and the decision boundaries.

5.1 Spiral Datasets

The training set is a set of synthetic data, designed to test the ability of the two types of neural networks in the unraveling of the hyperplanes, that is, the synthetic data is generated with a high rate of entanglement, and a low degree of overlap between classes. For this purpose the generated data spiral consists of 1 to 5 classes wrapped one over the other, and the number of turns vary between 1 and 10. The representation of said training set is shown in Fig. 3 in such a way that the training set is shaped as shown in Table 1.

5.2 Experimetal Results for DNNs

In order to classify the patterns presented in the Sect. 5.1 the DNN architecture varies in depth the number of neurons per layer, as well as the number of hidden layers, leaving the hyper-parameters fixed to the following values, learning

Table 1. Datasets for spirals with different number of classes $N_C = \{2, 3, 4, 5, 10\}$ and increasing number of loops $N_L = \{1, 2, ..., 10\}$. The number of training patterns is Q_{train} and the number of validation patterns is Q_{val}.

Name	N_C	N_L	Q_{train}	Q_{val}	Name	N_C	N_L	Q_{train}	Q_{val}
1.1	2	1	10000	2500	2.1	3	1	75000	18750
1.2	2	2	10000	2500	2.2	3	2	75000	18750
1.3	2	3	10000	2500	2.3	3	3	75000	18750
1.4	2	4	10000	2500	2.4	3	4	90000	22500
1.5	2	5	10000	2500	2.5	3	5	90000	22500
1.6	2	6	40000	10000					
1.7	2	7	50000	12500	Name	N_C	N_L	Q_{train}	Q_{val}
1.8	2	8	60000	15000	3.1	4	1	120000	30000
1.9	2	9	60000	15000	3.2	4	2	120000	30000
1.10	2	10	60000	15000	3.3	4	3	120000	30000
					3.4	4	4	120000	30000

Name	N_C	N_L	Q_{train}	Q_{val}	Name	N_C	N_L	Q_{train}	Q_{val}
4.1	5	1	150000	37500	5.1	10	1	100000	25000
4.2	5	2	150000	37500	5.2	10	2	100000	25000
4.3	5	3	150000	37500	5.3	10	3	100000	25000
4.4	5	4	150000	37500	5.4	10	4	100000	25000
4.5	5	5	150000	37500	5.5	10	5	100000	25000

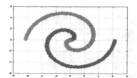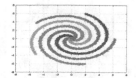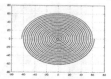

Fig. 3. Spiral of two interlaced spin classes (left), spiral of five classes of one spin per class (center), spiral of two classes with 10 turns each class.

Table 2. Experimental results for DNNs.

Dataset	N_p	T_a	V_a	T_t
1.1	102	0.9947	0.9944	55.68
1.2	402	0.9956	0.9953	47.88
1.3	1332	0.9986	0.998	202.63
1.4	15552	0.998	0.988	335.67
1.5	55952	0.9564	0.9475	590.43
1.6	76152	0.9038	0.8857	3722.79
1.7	211952	0.8302	0.81154	1026.20
1.8	234602	0.776	0.7721	1185.75
1.9	257252	0.7544	0.7306	1508.66
1.10	279902	0.7278	0.7083	2476.12

Dataset	N_p	T_a	V_a	T_t
2.1	252	0.9886	0.9882	35.46
2.2	402	0.9844	0.9915	64.35
2.3	11332	0.9933	0.9745	329.06
2.4	35752	0.9971	0.9917	361.26
2.5	55952	0.8696	0.8416	689.51

Dataset	N_p	T_a	V_a	T_t
3.1	252	0.9839	0.9836	45.46
3.2	502	0.9892	0.9812	205.03
3.3	11432	0.9933	0.9943	201.45
3.4	45852	0.9956	0.9806	437.41
3.5	55952	0.8696	0.8416	565.82

Dataset	N_p	T_a	V_a	T_t
4.1	252	0.9808	0.9794	311.89
4.2	502	0.9831	0.9595	256.71
4.3	25652	0.9934	0.9916	523.81
4.4	35752	0.993	0.9556	749.26
4.5	66052	0.9052	0.8131	314.50

Dataset	N_p	T_a	V_a	T_t
5.1	45852	0.9477	0.9447	78.57
5.2	45852	0.9635	0.9444	91.54
5.3	96352	0.9558	0.9176	455.31
5.4	126652	0.9364	0.8397	598.68
5.5	106542	0.8427	0.7573	352.98

rate of 0.1, Nesterov momentum of 0.9 and batch size of 64. The value of the hyper-parameters was obtained by performing classification tests by varying the values of the learning rate in a range of $[1, 0.001]$, with increments of 0.01. Table 2 summarizes the resulting architectures applied to each training set; the column "Dataset" specifies the number of the training set used, column N_p specifies the number of parameters in the neural network model, column T_a specifies the percentage of classification on the training set, column V_a shows the classification percentage on the validation set obtained by that neural network model, and column T_t shows the total training and validation time. Figure 5 shows the classification accuracies for each neural network, number of classes and number of loops of each training set; showing better results for DMN over DNN models.

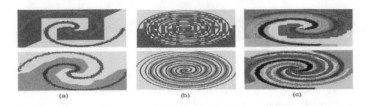

Fig. 4. Decision boundary generated by DMN (first row) and by DNN (second row).

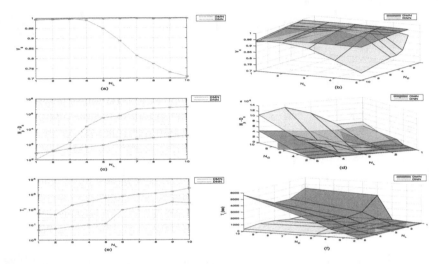

Fig. 5. (a) Classification percentages for the dataset 1 (2 classes, 10 loops), (b) classification ratios for datsets 2–5 (2–10 classes, 1–5 loops). (c) and (d) number of parameters used to classify each dataset. (e) and (f) classification times used to classify each dataset. (N_L, number of loops, N_C, number of classes).

5.3 Experimetal Results for DMNs

In the same way as in Sect. 5.2, in Table 3 the architecture of the DMN is presented; the first column shows the training set number used and the third column G_i shows the index of generalization of the DMN.

5.4 Decision Boundaries

This section compares the decision boundaries generated by the two types of neural network architectures (DNN and DMN) on the same training sets specified in Sect. 5.1. As we observe, the nature of each algorithm is very different, generating approximations to hyperplanes/hyperboxes, which yield similar results. However, for the specific dataset used, we can observe that the generation of hyperboxes of variable size best models the training set with a higher classification rate and less parameters in the DMN model. These results can be observed in Fig. 4. Each pair of images grouped by column, shows the decision

Table 3. Experimental results for DMNs.

Dataset	N_p	G_i	T_a	V_a	T_t
1.1	264	0.0066	0.9968	0.9924	4.56
1.2	356	0.0089	0.9972	0.9936	5.62
1.3	544	0.0136	0.9964	0.994	7.77
1.4	692	0.0173	0.9977	0.9972	9.84
1.5	888	0.0222	0.999	0.9948	11.83
1.6	1764	0.011	0.9974	0.9965	95.43
1.7	2216	0.0111	0.9976	0.9973	148.02
1.8	2684	0.0447	0.9978	0.9955	161.81
1.9	3080	0.0128	0.9976	0.9967	309.37
1.10	3748	0.0156	0.9987	0.9975	268.05

Dataset	N_p	G_i	T_a	V_a	T_t
2.1	2432	0.0203	0.9979	0.9836	111.62
2.2	3344	0.0111	0.9977	0.9908	315.09
2.3	2636	0.0088	0.9978	0.9935	266.36
2.4	3500	0.0097	0.9979	0.9941	388.60
2.5	3900	0.0108	0.9977	0.9946	653.01

Dataset	N_p	G_i	T_a	V_a	T_t
3.1	4244	0.0265	0.9946	0.9766	279.00
3.2	6892	0.0144	0.994	0.9837	1083.17
3.3	5944	0.0124	0.9973	0.99	954.77
3.4	6020	0.0125	0.9955	0.9897	1080.68
3.5	6864	0.0143	0.9977	0.9928	1212.82

Dataset	N_p	G_i	T_a	V_a	T_t
4.1	17632	0.0294	0.9946	0.9715	3953.38
4.2	10872	0.0181	0.9978	0.9853	1968.76
4.3	9288	0.0155	0.9954	0.9864	1714.53
4.4	10172	0.017	0.9978	0.9906	1902.82
4.5	11080	0.0185	0.9964	0.9906	2150.75

Dataset	N_p	G_i	T_a	V_a	T_t
5.1	30812	0.077	0.9921	0.9339	4335.73
5.2	24168	0.0604	0.9942	0.9616	3226.45
5.3	28024	0.0701	0.9941	0.9648	3267.14
5.4	36572	0.0914	0.9981	0.965	4417.48
5.5	45544	0.1139	0.998	0.9642	5690.73

boundary generated by the DMN (top) and the DNN (bottom). As can be seen in column (b), the decision boundaries are best defined by the DMN (column (b), top) than the decision boundaries generated by the DNN (column (b), bottom).

6 Conclusion and Future Work

Linear filters with non-linear activation functions (and back-propagation) are today the battle horses of the neural network community. This leads us to ask the questions: Are there other mathematical structures that produce better results for some problems? What advantages would they have? The motivation of this research is to answer these questions. In this paper, we compare DNNs and DMNs in a very simple 2D classification problem: multi-class spirals with increasing number of loops. We show that the performance of the DMNs surpasses that of the DNNs in terms of higher accuracies and a lesser number of learning parameters. Of course, these results are limited to spiral-like problems, which we specifically designed to test the ability of separation for the two neural architectures. It is clear that the DMN training time is longer than the DNN training time, furthermore, the classification rate is not compromised, that is, the DNNs can be trained in a shorter time, but their validation accuracy is much lower to that obtained by the DMN.

We conclude that this result is due to the nature of both algorithms. The hyperboxes of DMNs make better models for these types of datasets because the divide and conquer training is based on geometrical interpretation of the whole data, and refines the model each recursion step, while training based on gradient descent is a search method in a dark environment only guided by partial dataset information, and local information cost function. From this, we raise the hypothesis that deep learning networks can be improved adding morphological neurons. This is a consideration for future research.

Acknowledgments. E. Zamora and H. Sossa would like to acknowledge the support provided by UPIITA-IPN and CIC-IPN in carrying out this research. This work was economically supported by SIP-IPN (grant numbers 20170836 and 20170693), and CONACYT grant number 65 (Frontiers of Science). G. Hernández acknowledges CONACYT for the scholarship granted towards pursuing his PhD studies.

References

1. Bengio, Y., Simard, P., Frasconi, P.: Learning long-term dependencies with gradient descent is difficult. Trans. Neur. Netw. **5**(2), 157–166 (1994)
2. Bengio, Y.: Learning deep architectures for AI. Found. Trends Mach. Learn. **2**(1), 1–127 (2009)
3. Cortes, C., Vapnik, V.: Support-vector networks. Mach. Learn. **20**(3), 273–297 (1995)
4. Durbin, R., Rumelhart, D.E.: Product units: a computationally powerful and biologically plausible extension to backpropagation networks. Neural Comput. **1**(1), 133–142 (1989)
5. Giles, C.L., Maxwell, T.: Learning, invariance, and generalization in high-order neural networks. Appl. Opt. **26**(23), 4972–4978 (1987)
6. Glorot, X., Bordes, A., Bengio, Y.: Deep sparse rectifier neural networks. In: Gordon, G.J., Dunson, D.B., Dudik, M. (eds.), AISTATS, vol. 15. JMLR Proceedings, pp. 315–323 (2011). JMLR.org
7. Goodfellow, I., Bengio, Y., Courville, A.: Deep Learning. MIT Press (2016). http://www.deeplearningbook.org
8. Gurney, K.N.: Training nets of hardware realizable sigma-pi units. Neural Networks **5**(2), 289–303 (1992)
9. Hinton, G., Deng, L., Dong, Y., Dahl, G.E., Mohamed, A., Jaitly, N., Senior, A., Vanhoucke, V., Nguyen, P., Sainath, T.N., et al.: Deep neural networks for acoustic modeling in speech recognition: the shared views of four research groups. IEEE Signal Process. Mag. **29**(6), 82–97 (2012)
10. Hussain, A.: A new neural network structure for temporal signal processing. In: IEEE International Conference on Acoustics, Speech, and Signal Processing, ICASSP 1997, Munich, Germany, 21–24 April, pp. 3341–3344 (1997)
11. Ivakhnenko, A.G.: Polynomial theory of complex systems. IEEE Trans. Syst. Man Cybern. **SMC–1**(4), 364–378 (1971)
12. Krizhevsky, A., Sutskever, I., Hinton, G.E.: Imagenet classification with deep convolutional neural networks. In: Advances in Neural Information Processing Systems, pp. 1097–1105 (2012)
13. LeCun, Y., Bengio, Y.: The handbook of brain theory and neural networks. In: Convolutional Networks for Images, Speech, and Time Series, pp. 255–258. MIT Press, Cambridge (1998)
14. LeCun, Y., Bengio, Y., Hinton, G.: Deep learning. Nature **521**(7553), 436–444 (2015)

15. LeCun, Y., Bottou, L., Orr, G.B., Müller, K.-R.: Efficient backprop. In: Orr, G.B., Müller, K.-R. (eds.) Neural Networks: Tricks of the Trade. LNCS, vol. 1524, pp. 9–50. Springer, Heidelberg (1998). doi:10.1007/3-540-49430-8_2

16. MacKay, D.J.C.: A practical bayesian framework for backpropagation networks. Neural Comput. **4**(3), 448–472 (1992)

17. Milenkovic, Z., Obradovic, S., Litovski, V.: Annealing based dynamic learning in second-order neural networks. In: IEEE International Conference on Neural Networks, vol. 1, pp. 458–463. IEEE (1996)

18. Pessoa, L.F.C., Maragos, P.: Neural networks with hybrid morphological/rank/linear nodes: a unifying framework with applications to handwritten character recognition. Pattern Recogn. **33**(6), 945–960 (2000)

19. Ritter, G.X., Iancu, L., Urcid, G.: Morphological perceptrons with dendritic structure. In: The 12th IEEE International Conference on Fuzzy Systems, FUZZ-IEEE 2003, St. Louis, Missouri, USA, 25–28 May 2003, pp. 1296–1301 (2003)

20. Ritter, G.X., Urcid, G.: Lattice algebra approach to single-neuron computation. IEEE Trans. Neural Networks **14**(2), 282–295 (2003)

21. Ritter, G.X., Urcid, G.: Learning in lattice neural networks that employ dendritic computing. In: Kaburlasos, V.G., Ritter, G.X. (eds.) Computational Intelligence Based on Lattice Theory. SCI, vol. 67, pp. 25–44. Springer, Heidelberg (2007)

22. Rumelhart, D.E., Hinton, G.E., Williams, R.J.: Parallel distributed processing: Explorations in the microstructure of cognition. In: Learning Internal Representations by Error Propagation, vol. 1, pp. 318–362. MIT Press, Cambridge (1986)

23. Schmidhuber, J.: Deep learning in neural networks: an overview. Neural Networks **61**, 85–117 (2015)

24. Sossa, H., Guevara, E.: Efficient training for dendrite morphological neural networks. Neurocomputing **131**, 132–142 (2014)

25. Van Der Malsburg, C.: Frank Rosenblatt: principles of neurodynamics: perceptrons and the theory of brain mechanisms. In: Palm, G., Aertsen, A. (eds.) Brain Theory. Springer, Heidelberg (1986)

26. Wasserman, P.D., Schwartz, T.J.: Neural networks. II. What are they and why is everybody so interested in them now? IEEE Expert **3**(1), 10–15 (1988)

27. Zamora, E., Sossa,H.: Dendrite morphological neurons trained by stochastic gradient descent. In: IEEE Symposium Series on Computational Intelligence (SSCI), pp. 1–8, December 2016

28. Zamora, E., Sossa, H.: Regularized divide and conquer training for dendrite morphological neurons. In: Mechatronics and Robotics Service: Theory and Applications, Mexican Mechatronics Association, November 2016

29. Zhang, Q., Benveniste, A.: Wavelet networks. IEEE Trans. Neural Networks **3**(6), 889–898 (1992)

A Novel Contrast Pattern Selection Method for Class Imbalance Problems

Octavio Loyola-González[1,2(✉)], José Fco. Martínez-Trinidad[1],
Jesús Ariel Carrasco-Ochoa[1], and Milton García-Borroto[3]

[1] Instituto Nacional de Astrofísica, Óptica y Electrónica, Luis Enrique Erro No. 1,
Sta. María Tonanzintla, 72840 Puebla, Mexico
{octavioloyola,fmartine,ariel}@inaoep.mx
[2] Centro de Bioplantas, Universidad de Ciego de Ávila.,
Carretera a Morón Km 9, 69450 Ciego de ávila, Cuba
octavioloyola@bioplantas.cu
[3] Instituto Superior Politécnico José Antonio Echeverría.,
Calle 114 No. 11901, 19390 Marianao, La Habana, Cuba
mgarciab@ceis.cujae.edu.cu

Abstract. Selecting contrast patterns is an important task for pattern-based classifiers, especially in class imbalance problems. The main reason is that the contrast pattern miners commonly extract several patterns with high support for the majority class and only a few patterns, with low support, for the minority class. This produces a bias of classification results toward the majority class, obtaining a low accuracy for the minority class. In this paper, we introduce a contrast pattern selection method for class imbalance problems. Our proposal selects all the contrast patterns for the minority class and a certain percent of contrast patterns for the majority class. Our experiments performed over several imbalanced databases show that our proposal selects significantly better contrast patterns, obtaining better AUC results, than other approaches reported in the literature.

Keywords: Supervised classification · Pattern selection · Contrast patterns · Imbalanced databases

1 Introduction

Several classifiers have been proposed for supervised classification, among them, an important family is the contrast pattern-based classifiers. A *pattern* is an expression defined in a language that describes a set of objects. For example, a pattern that describes a set of plants can be the following: $[Petal\_Width \in [0.60, 1.60]] \wedge [Roots \leq 10] \wedge [Stem = "Thick"]$. A pattern that appears significantly more in a class than in the remaining classes is named as *contrast pattern*. Finally, a classifier which predicts the class of a query object based on a set of contrast patterns is called: *contrast pattern-based classifier*. It is important to highlight that the pattern-based classifiers, as well as their results, can

© Springer International Publishing AG 2017
J.A. Carrasco-Ochoa et al. (Eds.): MCPR 2017, LNCS 10267, pp. 42–52, 2017.
DOI: 10.1007/978-3-319-59226-8_5

be understood by the experts in the application domain through the patterns associated to each class. Also, contrast pattern-based classifiers have reported significantly better classification results than other popular classification models, like naive bayes, nearest neighbor, bagging, boosting, and SVM [7,11].

In some real-world applications, there are problems where the objects are not equally distributed into the classes, like online banking fraud detection, liver and pancreas disorders, forecasting of ozone levels, prediction of protein sequences, and face recognition. In these applications, there exist significantly fewer objects belonging to a class (commonly labeled as *minority* class) regarding the remaining classes. This problem is known as class imbalance problem [18–20].

Some classifiers, which show good classification results in problems with balanced classes do not necessarily achieve good performance in class imbalance problems. The main reason is that they produce a bias toward the *majority* class (the class with more objects). Accordingly, the accuracy of these classifiers for the minority class could be close to zero [19,20].

On class imbalance problems, some pattern-based classifiers, like CAEP [8], do not achieve good classification results because of contrast patterns from the minority class are fewer and they have low support regarding those contrast patterns from the majority class. Then, some classification strategies, which are based only on the support of the contrast patterns, tend to be biased toward the majority class [2,18,19].

A proposal for supervised classification based on contrast patterns in class imbalance problems is selecting just a subset of good contrast patterns. The idea is to select, for each class, a collection of high-quality patterns. Consequently, at the classification stage, those contrast patterns with low support for the minority class do not become overwhelmed by those contrast patterns with high support for the majority class, which are much more [4,9,19,24,25].

In the literature there are three main approaches for contrast pattern selection: (i) selecting only the best contrast pattern, (ii) selecting the k best contrast patterns, and (iii) selecting all contrast patterns covering the training dataset [4,9,18,24,25]. In this paper, we propose a novel contrast pattern selection by class for class imbalance problems; the idea consists in selecting all the contrast patterns for the minority class and only a certain percent of contrast patterns for the majority class. Our proposal allows obtaining better accuracy results, when the selected contrast patterns are used by a contrast pattern-based classifier, than other contrast pattern selection approaches of the state-of-the-art.

The rest of the paper has the following structure. Section 2 contains a brief description of the main contrast pattern selection approaches reported in the state-of-the-art. Section 3 introduces our proposal for selecting contrast patterns in class imbalance problems. Section 4 provides the experimental setup. Section 5 presents the experimental results as well as a discussion of them. Finally, Sect. 6 provides our conclusions and future work.

2 Related Work

In pattern-based classification, an important task is to select a collection of high-quality patterns for obtaining good classification results [18,27]. Additionally, the fewer patterns, the faster the classification stage and easier to understand the results by experts in the application domain.

Three main approaches have been proposed in the literature for selecting contrast patterns [4,17,18,22,24,25]. These approaches use a quality measure for contrast patterns with the aim of creating a ranking of contrast patterns. A quality measure is a function $q(P, C, \bar{C}) \rightarrow R$, which assigns a higher value to a pattern P when it better discriminates the objects in a class C from the objects in the remaining problem classes $\bar{C}$ [17,18]. Usually, the measure for ranking the contrast patterns depends on the contrast pattern-based classifier to be used (e.g., confidence for association rules, X^2 for decision trees, growth rate for emerging patterns). The three main approaches for selecting contrast patterns are:

Best contrast pattern (Best CP): Select the best contrast pattern, according to the ranking, covering the query object. This approach is used by several rule-based classifiers, like CBA [16], for classifying query objects. A drawback of this approach, in class imbalance problems, is that commonly the best contrast pattern according to the ranking is from the majority class. Consequently, the accuracy of the classifier for the minority class is bad.

Best k contrast patterns (Best k): Select the best k contrast patterns from the ranking which cover the query object. Usually, this approach is used by rule-based classifiers like CPAR [26] and some emerging pattern selection methods, as the one proposed in [17]. Recently in [18] the authors proposed to use a fixed percent of patterns instead of a fixed number of patterns. A disadvantage of this approach in class imbalance problems is that the patterns for the minority class are too few and they have low support, then selecting a few patterns of the minority class could degrade the accuracy of the classifier for the minority class.

Covering the training dataset (Covering): Select the best contrast patterns, according to the ranking, covering all the objects of the training dataset. This approach is used by some rule based-classifier, like ACN [14] and CMAR [15], and some emerging pattern selection methods [11,17,18], showing good accuracy results.

The second and third approaches were studied in [18] for selecting contract patterns in class imbalance problems. The authors tested several quality measures for contrast patterns and they concluded that the best quality measure for ranking contrast patterns in class imbalance problems is *Jaccard* [23]. Based on this conclusion, we will propose a novel contrast pattern selection method, which uses the quality measure *Jaccard* for ranking the contrast patterns.

3 New Contrast Pattern Selection Method

In this section, we introduce a contrast pattern selection method for class imbalance problems.

Usually, in class imbalance problems, contrast pattern mining algorithms extract several patterns with high support for the majority class and only a few patterns, with low support, for the minority class [2,18,19]. This produces that some contrast pattern-based classifiers, like CAEP [8], become biased toward the majority class. Some strategies find a solution by selecting a collection of high-quality patterns, but their main drawback is that some patterns of the minority class, which could help at the classification stage, are discarded [15,16,26]. For solving this problem, we propose to select the patterns by class; selecting all the contrast patterns of the minority class and only a few contrast patterns of the majority class. The main idea is not to discard useful patterns of the minority class and avoiding the selection of many patterns of the majority class, which could overwhelm the patterns of the minority class at the classification stage.

Our pattern selection method can be described by the following steps:

1. Select all the contrast patterns of the minority class to avoid reducing the number of patterns of this class.
2. Rank the contrast patterns of the majority class by using a quality measure for contrast patterns.
3. Select the best k contrast patterns of the majority class. The k value is a percent of the total number of patterns, which is provided by the user.

Commonly, the number of patterns extracted from a database depends on different factors such as the nature of training dataset, the contrast pattern mining algorithm, the a-priori global discretization, among others. Hence, instead of selecting a fixed number of contrast patterns, in our contrast pattern selection method, we propose to select just a percent of patterns. The main reason is that the number of patterns to select could be too high, regarding the amount of mined patterns, which means that the selected patterns could be almost all; or too small, which would lead to reduce more than necessary the number of patterns.

Finally, it is important to highlight that in step 2, we suggest to use the quality measure *Jaccard* [23] because this measure has shown good results for ranking contrast patterns in class imbalance problems [18].

4 Experimental Setup

In order to evaluate the performance of the proposed selection method, we will perform a comparison of our proposal against the three main approaches for selecting contrast patterns reported in the literature. To do this, first, we will extract the patterns by using a contrast pattern miner. After that, we will create a pattern ranking by applying the quality measure *Jaccard* [23] over the collection of patterns previously extracted. Next, we will select the patterns by using the three main pattern selection approaches shown in Sect. 2 and our proposal. Finally, each subset of patterns will be used to build a contrast pattern-based classifier. By doing this, we can detect which selection method attains better classification results. As the contrast pattern miner and the classifier are the same and the only difference are the contrast patterns selected by means of the selection method, then a good or bad performance in the classification results can be attributed to the selection method employed.

Table 1 shows the 95 databases used in our experiments, which were taken from the KEEL dataset repository[1] [1]. For avoiding problems due to data distribution in class imbalance problems, for each database, we performed a distribution optimally balanced stratified five cross-validation, as suggested in [20].

For assessing the performance of our classification results, we used the AUC measure [13] because it is the most used measure for class imbalance problems [18–20]. All our AUC results were averaged over the 5-fold cross validation.

As contrast pattern-based classifier, we selected PBC4cip [19], since, it has reported better AUC results than other state-of-the-art classifiers for class imbalance problems [19].

For mining contrast patterns, we selected the Random Forest miner (RFm) [10] using the Hellinger distance [3] as node splitting measure, as suggested in [19]. The main reason is that RFm has been used jointly with the PBC4cip classifier, obtaining higher accuracies than other state-of-the-art contrast pattern mining algorithms [19].

For selecting contrast patterns, we used the three main approaches reported in the literature (see Sect. 2). For selecting the best k contrast patterns by class, we used the values 10%, 50%, and 80% which have been used in previous studies for class imbalance problems [18]. For selecting contrast patterns using our proposal we used the following k values: 5%, 10%, 15%, 20%, 25%, 30%, 35%, 40%, 45%, 50%, and 80%. By using these values, we can investigate if our proposal is able to attain statistically similar AUC results at using fewer patterns than the number of patterns used in [18].

We also used the Shaffer and Finner post-hoc procedures, and the Friedman test to compare all the classification results, as suggested in [5,6]. Post-hoc results will be shown by using CD (*critical distance*) diagrams [5]. In a CD diagram, the rightmost classifier is the best classifier. The position of the classifier within the segment represents its rank value, and if two or more classifiers share a thick line it means that they have statistically similar behavior.

5   Experimental Results

This section is devoted to analyzing and discussing about the classification results achieved by the contrast pattern selection methods described in Sect. 2, using all the imbalanced databases shown in Table 1.

---

[1] http://www.keel.es/datasets.php.

Table 1. Summary of the imbalanced databases used in our study. Containing the name in the KEEL dataset repository (name), the number of objects (#Objects) and features (#Feat.), and the IR [21].

Name	#Objects	#Feat.	IR	Name	#Objects	#Feat.	IR
glass1	214	9	1.82	ecoli0146vs5	280	6	13.00
ecoli0vs1	220	7	1.86	shuttlec0vsc4	1829	9	13.87
wisconsin	683	9	1.86	yeast1vs7	459	7	14.30
pima	768	8	1.87	glass4	214	9	15.46
iris0	150	4	2.00	ecoli4	336	7	15.80
glass0	214	9	2.06	pageblocks13vs4	472	10	15.86
yeast1	1484	8	2.46	abalone9vs18	731	8	16.40
haberman	306	3	2.78	dermatology6	358	34	16.90
vehicle2	846	18	2.88	zoo3	101	16	19.20
vehicle1	846	18	2.90	glass016vs5	184	9	19.44
vehicle3	846	18	2.99	shuttlec2vsc4	129	9	20.50
glass0123vs456	214	9	3.20	shuttle6vs23	230	9	22.00
vehicle0	846	18	3.25	yeast1458vs7	693	8	22.10
ecoli1	336	7	3.36	glass5	214	9	22.78
newthyroid1	215	5	5.14	yeast2vs8	482	8	23.10
newthyroid2	215	5	5.14	lymphography	148	18	23.67
ecoli2	336	7	5.46	flareF	1066	11	23.79
segment0	2308	19	6.02	cargood	1728	6	24.04
glass6	214	9	6.38	carvgood	1728	6	25.58
yeast3	1484	8	8.10	krvskzeroonevsdraw	2901	6	26.63
ecoli3	336	7	8.60	krvskonevsfifteen	2244	6	27.77
pageblocks0	5472	10	8.79	yeast4	1484	8	28.10
ecoli034vs5	200	7	9.00	winequalityred4	1599	11	29.17
yeast2vs4	514	8	9.08	poker9vs7	244	10	29.50
ecoli067vs35	222	7	9.09	yeast1289vs7	947	8	30.57
ecoli0234vs5	202	7	9.10	abalone3vs11	502	8	32.47
glass015vs2	172	9	9.12	winequalitywhite9vs4	168	11	32.60
yeast0359vs78	506	8	9.12	yeast5	1484	8	32.73
yeast0256vs3789	1004	8	9.14	krvskthreevseleven	2935	6	35.23
yeast02579vs368	1004	8	9.14	winequalityred8vs6	656	11	35.44
ecoli046vs5	203	6	9.15	ecoli0137vs26	281	7	39.14
ecoli01vs235	244	7	9.17	abalone17vs78910	2338	8	39.31
ecoli0267vs35	224	7	9.18	abalone21vs8	581	8	40.50
glass04vs5	92	9	9.22	yeast6	1484	8	41.40
ecoli0346vs5	205	7	9.25	winequalitywhite3vs7	900	11	44.00
ecoli0347vs56	257	7	9.28	winequalityred8vs67	855	11	46.50
yeast05679vs4	528	8	9.35	abalone19vs10111213	1622	8	49.69
vowel0	988	13	9.98	krvskzerovseight	1460	6	53.07
ecoli067vs5	220	6	10.00	winequalitywhite39vs5	1482	11	58.28
glass016vs2	192	9	10.29	poker89vs6	1485	10	58.40
ecoli0147vs2356	336	7	10.59	shuttle2vs5	3316	9	66.67
led7digit02456789vs1	443	7	10.97	winequalityred3vs5	691	11	68.10
ecoli01vs5	240	6	11.00	abalone20vs8910	1916	8	72.69
glass06vs5	108	9	11.00	krvskzerovsfifteen	2193	6	80.22
glass0146vs2	205	9	11.06	poker89vs5	2075	10	82.00
glass2	214	9	11.59	poker8vs6	1477	10	85.88
ecoli0147vs56	332	6	12.28	abalone19	4174	8	129.44
cleveland0vs4	177	13	12.62				

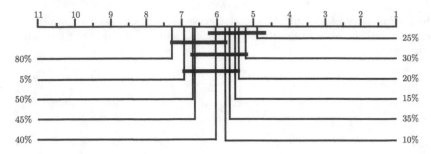

Fig. 1. CD diagram with a statistical comparison (using $\alpha = 0.05$) of the AUC results of our proposal for selecting contrast patterns, using different k values over all the tested databases.

In order to simplify the presentation, a supplementary material website[2] has been created for this paper, which contains several tables from experimental results as well as detailed tables from the statistical test results.

Figure 1 shows a CD diagram with a statistical comparison of the AUC results obtained by our proposal using different k values and considering all the imbalanced databases shown in Table 1. From this figure, we can conclude that our proposal using $k = 25\%$ obtained the best position into the Friedman's ranking. However, the difference of the AUC results of our proposal using $k = 25\%$ against using k as: 10%, 15%, 20%, 30%, 35%, and 40% is not statistically significant. Therefore, we selected $k = 10\%$ since it allows selecting the fewest number of patterns.

Figure 2 shows a CD diagram with a statistical comparison of the AUC results obtained by our proposal, using $k = 10\%$, against the other contrast pattern selection methods reviewed in the Sect. 2, as well as by using all the contrast patterns (All CPs). From this figure, we can see that the AUC results of our proposal against those AUC results archived by Best $k = 50\%$, Best $k = 80\%$, and All CPs are not statistically significant. However, our proposal obtains a better position into the Friedman's ranking than using all the contrast patterns (All CPs). Also, our proposal uses fewer contrast patterns for classification than the other approaches having statistically similar behavior. On the other hand, notice that the contrast pattern selection method Best CP statistically obtained the worst results. This is because, in class imbalance problems, commonly the best contrast pattern according to the ranking comes from the majority class and consequently the accuracy for the minority class is greatly affected.

5.1   Regarding Different Class Imbalance Levels

For studying the effect of the class imbalance level on the contrast pattern selection methods previously analyzed, we divided the databases into equal-frequency groups depending on the IR of each one. For doing this, we used the *Discretize*[3]

---

[2] https://sites.google.com/site/octavioloyola/papers/PSM4MajClass.
[3] Path in WEKA: weka.filters.unsupervised.attribute.Discretize.

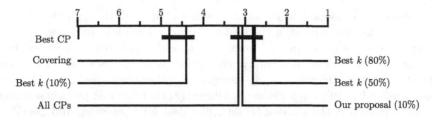

Fig. 2. CD diagram with a statistical comparison (using $\alpha = 0.05$) of the AUC results of our proposal and the other contrast pattern selection methods reported in the literature.

method, taken from the WEKA Data Mining Tool [12], to create six equal-frequency groups depending on the IR of the databases. These groups are shown in Table 1 using horizontal thin lines.

Table 2. Results of the best contrast pattern selection for each bin

Name	Bin interval	#Databases	Best selection method
Bin1	(1.820, 5.300]	16	Best $k = 80\%$
Bin2	(5.300, 9.175]	16	Best $k = 50\%$
Bin3	(9.175, 12.810]	16	Our proposal ($k = 10\%$)
Bin4	(12.810, 23.730]	16	Best $k = 80\%$
Bin5	(23.730, 39.905]	16	Best $k = 50\%$
Bin6	(39.905, 129.440]	15	Best $k = 80\%$

Table 2 shows the best contrast pattern selection method for each bin. From this table, we can conclude that for the less imbalanced databases (Bin1), Bin4 and for the Bin6, the best contrast pattern selection method is Best $k = 80\%$. For Bin3, the best selection method is our proposed method using $k = 10\%$. Finally, for Bin2 and Bin5 the best contrast pattern selection is Best $k = 50\%$. These results help us to select the best contrast pattern selection method depending on the class imbalance level of the database.

6 Conclusions and Future Work

Selecting a collection of high-quality patterns is an important task for pattern-based classification. The main aim is to achieve good classification results using as few patterns as possible in order to obtain a model easier to understand by experts in the application domain. Following this idea, the main contribution of this paper is a new contrast pattern selection method for contrast pattern-based classification in class imbalance problems. Our proposal selects all the patterns of the minority class and based on a ranking computed through the *Jaccard* measure, it selects a percent of the best patterns of the majority class.

From our experiments using several imbalanced databases, we can conclude that our proposal performs significantly better, when it uses the 25% of the contrast patterns of the majority class, regarding other tested percents. Also, our proposal using $k = 10\%$ outperforms significantly other contrast pattern selection methods reported in the state-of-the-art, like Best CP, Covering, and Best $k = 10\%$. Moreover, our proposal using $k = 10\%$ have not statistical differences with other contrast pattern selection methods, like Best $k = 50\%$, Best $k = 80\%$, and All CPs, but these methods need more patterns.

On the other hand, based on our experiments regarding the class imbalance ratio of the databases, we suggest that: if the database has an IR smaller than or equal to 5.3, or its IR ranges in (12.810, 23.730], or its IR ranges in (39.905, 129.440], then Best $k = 80\%$ is the best contrast pattern selection method. If the database has an IR in (9.175, 12.810] then our proposal using $k = 10\%$ is recommended. And finally, if the database has an IR in (5.300, 9.175] or its IR ranges in (23.730, 39.905] then we suggest using the contrast pattern selection method Best $k = 50\%$.

Finally, as future work, we will explore the use of maximal or closed contrast patterns as an alternative for selecting a reduced subset of contrast patterns for classification in class imbalance problems.

Acknowledgment. This work was partly supported by National Council of Science and Technology of Mexico under the scholarship grant 370272.

References

1. Alcalá-Fdez, J., Fernández, A., Luengo, J., Derrac, J., García, S.: KEEL data-mining software tool: data set repository, integration of algorithms and experimental analysis framework. J. Multiple-Valued Logic Soft Comput. **17**(2–3), 255–287 (2011)
2. Alhammady, H.: A novel approach for mining emerging patterns in rare-class datasets. In: Sobh, T. (ed.) Innovations and Advanced Techniques in Computer and Information Sciences and Engineering, pp. 207–211. Springer, Dordrecht (2007)
3. Cieslak, D., Hoens, T., Chawla, N., Kegelmeyer, W.: Hellinger distance decision trees are robust and skew-insensitive. Data Min. Knowl. Disc. **24**(1), 136–158 (2012)
4. Coenen, F., Leng, P.: An evaluation of approaches to classification rule selection. In: Fourth IEEE International Conference on Data Mining, pp. 359–362 (2004)
5. Demšar, J.: Statistical comparisons of classifiers over multiple data sets. J. Mach. Learn. Res. **7**, 1–30 (2006)
6. Derrac, J., García, S., Molina, D., Herrera, F.: A practical tutorial on the use of nonparametric statistical tests as a methodology for comparing evolutionary and swarm intelligence algorithms. Swarm Evol. Comp. **1**(1), 3–18 (2011)
7. Dong, G., Bailey, J.: Contrast Data Mining: Concepts, Algorithms, and Applications. Chapman and Hall/CRC, 1st edn. (2012)

8. Dong, G., Zhang, X., Wong, L., Li, J.: CAEP: classification by aggregating emerging patterns. In: Arikawa, S., Furukawa, K. (eds.) DS 1999. LNCS, vol. 1721, pp. 30–42. Springer, Heidelberg (1999). doi:10.1007/3-540-46846-3_4

9. Fürnkranz, J., Flach, P.: An analysis of stopping and filtering criteria for rule learning. In: Boulicaut, J.-F., Esposito, F., Giannotti, F., Pedreschi, D. (eds.) ECML 2004. LNCS, vol. 3201, pp. 123–133. Springer, Heidelberg (2004). doi:10.1007/978-3-540-30115-8_14

10. García-Borroto, M., Martínez-Trinidad, J.F., Carrasco-Ochoa, J.A.: Finding the best diversity generation procedures for mining contrast patterns. Expert Syst. Appl. **42**(11), 4859–4866 (2015)

11. García-Borroto, M., Martínez-Trinidad, J.F., Carrasco-Ochoa, J.A., Medina-Pérez, M.A., Ruiz-Shulcloper, J.: LCMine: an efficient algorithm for mining discriminative regularities and its application in supervised classification. Pattern Recogn. **43**(9), 3025–3034 (2010)

12. Hall, M., Frank, E., Holmes, G., Pfahringer, B., Reutemann, P., Witten, I.H.: The WEKA data mining software: an update. SIGKDD Expl. **11**(1), 10–18 (2009)

13. Huang, J., Ling, C.X.: Using AUC and accuracy in evaluating learning algorithms. IEEE Trans. Knowl. Data Eng. **17**(3), 299–310 (2005)

14. Kundu, G., Islam, M., Munir, S., Bari, M.: ACN: an associative classifier with negative rules. In: Proceedings of the 11th IEEE International Conference on Computational Science and Engineering, pp. 369–375. IEEE Xplore Press (2008)

15. Li, W., Han, J., Pei, J.: CMAR: accurate and efficient classification based on multiple class-association rules. In: Proceedings of the International Conference on Data Mining, ICDM 2001, pp. 369–376. IEEE (2001)

16. Liu, B., Hsu, W., Ma, Y.: Integrating classification and association rule mining. In: Proceedings of the Fourth International Conference on Knowledge Discovery and Data mining, KDD 1998, pp. 80–86. AAAI (1998)

17. Loyola-González, O., Garcia-Borroto, M., Martínez-Trinidad, J.F., Carrasco-Ochoa, J.A.: An empirical comparison among quality measures for pattern based classifiers. Intell. Data Anal. **18**, S5–S17 (2014)

18. Loyola-González, O., Martínez-Trinidad, J.F., Carrasco-Ochoa, J.A., García-Borroto, M.: Effect of class imbalance on quality measures for contrast patterns: an experimental study. Inform. Sci. **374**, 179–192 (2016)

19. Loyola-González, O., Medina-Pérez, M.A., Martínez-Trinidad, J.F., Carrasco-Ochoa, J.A., Monroy, R., García-Borroto, M.: PBC4cip: a new contrast pattern-based classifier for class imbalance problems. Knowl.-Based Syst. **115**, 100–109 (2016)

20. Moreno-Torres, J.G., Saez, J.A., Herrera, F.: Study on the impact of partition-induced dataset shift on k-fold cross-validation. IEEE Trans. Neural Networks Learn. Syst. **23**(8), 1304–1312 (2012)

21. Orriols-Puig, A., Bernadó-Mansilla, E.: Evolutionary rule-based systems for imbalanced data sets. Soft Comput. **13**(3), 213–225 (2009)

22. Refai, M.H., Yusof, Y.: Partial rule match for filtering rules in associative classification. J. Comput. Sci. **10**(4), 570 (2014)

23. Tan, P.N., Kumar, V., Srivastava, J.: Selecting the right objective measure for association analysis. Inf. Syst. **29**(4), 293–313 (2004)

24. Wang, Y.J., Xin, Q., Coenen, F.: A novel rule weighting approach in classification association rule mining. In: Seventh IEEE International Conference on Data Mining Workshops, pp. 271–276 (2007)

25. Ye, Y., Li, T., Jiang, Q., Wang, Y.: CIMDS: adapting postprocessing techniques of associative classification for malware detection. IEEE Trans. Syst. Man Cybern. Part C Appl. Rev. **40**(3), 298–307 (2010)
26. Yin, X., Han, J.: CPAR: classification based on predictive association rules. In: Proceedings of the Third SIAM International Conference on Data Mining, SDM 2003, pp. 331–335. SIAM (2003)
27. Zhang, X., Dong, G.: Overview and Analysis of Contrast Pattern Based Classification. In: Dong, G., Bailey, J. (eds.) Contrast Data Mining: Concepts, Algorithms, and Applications, Chap. 11. Data Mining and Knowledge Discovery Series, pp. 151–170. Chapman and Hall/CRC (2012)

Efficient Pattern Recognition Using
the Frequency Response of a Spiking Neuron

Sergio Valadez-Godínez, Javier González, and Humberto Sossa$^{(\boxtimes)}$

Centro de Investigación en Computación, Instituto Politécnico Nacional,
Av. Juan de Dios Batiz, S/N, Col. Nva. Industrial Vallejo,
07738 Mexico City, Mexico
svaladezg@gmail.com, xvrgonzalez21@gmail.com, hsossa@cic.ipn.mx

Abstract. In previous works, a successful scheme using a single Spiking Neuron (SN) to solve complex problems in pattern recognition has been proposed. This consists in using the firing frequency response to classify a given input pattern, which is multiplied by a weight vector to produce a constant stimulation current. The weight vector is adjusted by an evolutionary strategy where the objective is to obtain an optimal frequency separation. The problem is that the SN has to be numerically simulated several times when the weight vector is being adjusted. In this work, we propose fitting the SN frequency response curve to a piecewise linear function to be used instead of the costly SN simulation. A high fitting degree was found, but, more importantly, the computational cost of the training and testing phases was drastically reduced.

Keywords: Spiking Neuron · Izhikevich · Pattern recognition · Curve fitting · Frequency Response Curve · Piecewise linear function · Firing Rate · Evolutionary strategy · Differential evolution · Computational cost

1 Introduction

In [1–8], a successful scheme using a single SN to solve complex pattern recognition problems has been proposed. This scheme consists in using the firing rate codification to classify a given input pattern. The SN is stimulated by a constant input current to obtain the firing frequency. The frequency represents the class to which the pattern is associated. The necessary current is obtained by applying the dot product between the input and a weight vectors, as is normally done when using the Perceptron [9]. The weight vector is adjusted by an evolutionary strategy to generate the optimal firing rates maximizing the separation among the classes. The problem is that the SN has to be numerically simulated many times while the weight vector is being adjusted. Therefore, depending on the number of individuals and generations in the evolutionary strategy and on the numerical method, integration step and the time window of the SN simulation, the computational cost during training is highly expensive. Although the authors previously obtained a comparable computational cost to other methodologies, such as the Support Vector Machines or Artificial Neural Networks of

© Springer International Publishing AG 2017
J.A. Carrasco-Ochoa et al. (Eds.): MCPR 2017, LNCS 10267, pp. 53–62, 2017.
DOI: 10.1007/978-3-319-59226-8_6

the second generation, their analysis was performed for the testing phase and not for the training one [8]. In the testing phase, only a single SN simulation is utilized whereas, as previously mentioned, the training stage needs several SN simulations.

As the mentioned methodology uses a constant input current and a firing rate encoding, the firing frequency in function of the current, so-called Frequency Response Curve (FRC), can be obtained. Since the influential work of the Nobel laureate E. D. Adrian, the FRC of a sensory nerve ending to an applied stimulus is well known [10]. In the case of SNs, the FRC can be analytically obtained for the Leaky Integrate-and-Fire (LIF) [11,12]. However, that is more complicated for the Izhikevich (IZH) [13] and Hodgkin-Huxley (HH) [14] models. Moreover, the FRC for the HH can be fitted to a logarithmic function [15]. For the case of the IZH, no fitting is known, but its FRC is highly linear and, hence, this can be accurately fitted to a piecewise linear function using the Least Squares Method (LSM). The IZH is prevalently utilized in the previous works [2–8].

In this paper, we propose fitting the FRC of the IZH to a piecewise linear function to be used instead of the costly SN simulation. A high fitting degree was found, but, more importantly, the computational cost in the training and testing phases was drastically reduced. The Differential Evolution (DE) algorithm was selected as the evolutionary strategy [16,17].

The remainder of the paper is organized as follows. Section 2 focuses on the SN scheme to solve complex problems in pattern recognition. Section 3 describes the IZH model. Section 4 shows the fitting procedure. Section 5, gives the results and Sect. 6 presents the conclusions and future work.

2 Pattern Recognition Using a Single Spiking Neuron

In general, the scheme presented in [1–8] is stated as follows: it is supposed SN groups input patterns producing similar firing rates; once this is made, it allows discriminating input patterns giving different firing rates. So, the SN functions as a pattern classifier. A n-dimension input pattern can be represented as $\mathbf{x} = (x_1, x_2, \cdots, x_n)^T \in \mathbf{X} \subset \mathbb{R}^n$ where $x_1, x_2, x_3, \cdots, x_n$ are the pattern features and $\mathbf{X}$ is the dimensional features domain. A set of K input patterns is defined as $\{\mathbf{x}^k \in \mathbf{X}^n\} \forall k = \{1, 2, \cdots, K\}$ where k is an index and K is the cardinality of the pattern set. A class to which a pattern belongs is defined as $c \in \{1, 2, \cdots, m\}$ where m is the number of classes. Consequently, it is expected that the SN generates m different firing frequencies, each representing a class. For a specific input pattern $\mathbf{x}^k$, there is a class $c^k \forall k = \{1, 2, \cdots, K\}$. Thus, the set A of associations is defined as $A = \{(\mathbf{x}^k, c^k)\} \forall k = \{1, 2, \cdots, K\}$. As the SN is not directly stimulated by an input pattern, the input pattern $\mathbf{x}^k$ is multiplied by a weight vector $\mathbf{w} \in \mathbb{R}^n$ to obtain a constant stimulation current $I = \mathbf{x}^k \cdot \mathbf{w}$ to the SN. For example, the LIF is employed in [1] and the IZH in [2–8].

During training, the weight vector is adjusted by using an evolutionary algorithm with the idea of producing the optimal firing rates that allow maximizing separation between classes. For example, Differential Evolution [16,17] is used in [1–3,6], Cuckoo Search [18] in [4], Particle Swarm Optimization [19] in [5,8] and Artificial Bee Colony [20] in [7]. The firing rate fr^k produced by I for a particular input pattern $\mathbf{x}^k$ is calculated as $fr^k = \text{N}_{\text{sp}}/\text{T}$ where N_{sp} is the number of spikes that occur in the time span T. Later, the Average Firing Rate $AFR \in \mathbb{R}^m$ of all firing rates fr^k produced by the patterns belonging to the same class is estimated. The class to which an input pattern $\mathbf{x}^k$ is classified is $\tilde{c} = arg\ min_{c=1}^m \left(\left| AFR_c - fr^k \right| \right)$. The evolutionary algorithm minimizes the fitness function $f(\mathbf{w}, A) = 1 - {}^{P_{cc}}/_P$ where P_{cc} is the number of input patterns being correctly classified (i.e. $\tilde{c} = c^k$) and P the number of patterns that actually belong to the given class c^k. This way, the evolutionary algorithm automatically determines the optimal frequencies that maximize the separation among the classes. During testing, once the weight vector $\mathbf{w}$ was adjusted and the AFR calculated, the estimulation current to the SN for a unknown pattern $\tilde{\mathbf{x}}$ is determined by $I = \tilde{\mathbf{x}} \cdot \mathbf{w}$, the firing rate by $\widetilde{fr} = \text{N}_{\text{sp}}/\text{T}$ where N_{sp} is the number of spikes that occur in the time span T and the classification by $\tilde{c} = arg\ min_{c=1}^m \left(\left| AFR_c - \widetilde{fr} \right| \right)$.

3   The Izhikevich Model

This SN is a successful dimensional reduction of the HH [14]. In fact, once the HH was reduced to a two-dimensional system [21,22], Izhikevich analyzed the dynamics in the phase plane [23] to propose a new description [13]. This is

$$\begin{aligned} \frac{dV}{dt} &= 0.04V^2 + 5V + 140 - n + I \\ \frac{dn}{dt} &= a\,(bV - n) \end{aligned} \tag{1}$$

where V is the trans-membrane potential, I is the stimulation current, n is a recovery variable representing the potassium activation and sodium inactivation, a is a time scale and b the sensitivity of n to the sub-threshold fluctuations of the trans-membrane potential. When the trans-membrane potential reaches a maximum of 30 mV the voltage V and the recovery variable n are restarted, i.e.

$$If\ V \geq 30\ mV,\ then \begin{cases} V \leftarrow c \\ n \leftarrow n + d \end{cases} \tag{2}$$

where c is the restarting potential and d is the restarting of n.

As in [13], the constants were $a = 0.02$, $b = 0.2$, $c = -65$ and $d = 8$ for generating a regular spiking. Figure 1 shows an example of a spike train produced by a constant input current $I = 31\ \mu\text{A}/\text{cm}^2$.

4   Obtaining and Fitting the Frequency Response Curve

As in [2–5,7,8], the IZH was simulated with the Forward Euler method [24] but with a step size of 0.05 ms to produce an adecuate simulation due to the IZH not being as efficient as it was thought (see [25,26] for details). To obtain the FRC, the IZH was stimulated with various constant input currents I ranging

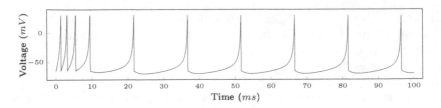

Fig. 1. Spike train generated by the Izhikevich (IZH) model.

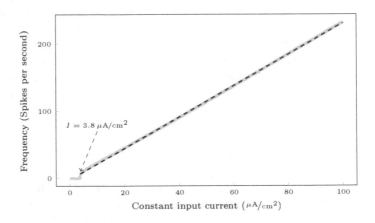

Fig. 2. Linear fitting (dashed line) of the Frequency Response Curve (FRC) (gray continuos line) of the Izhikevich (IZH) model.

from $0\,\mu\mathrm{A/cm^2}$ to $100\,\mu\mathrm{A/cm^2}$ across a current step of $0.1\,\mu\mathrm{A/cm^2}$ and a time span of 1000 ms. The FRC can be seen in Fig. 2 as a gray continuous line.

As can be seen, the FRC is in agreement with a high linear relationship when the IZH produces a firing response for $I \geq 3.8\,\mu\mathrm{A/cm^2}$ (see the arrow in Fig. 2). A strong positive-linear correlation was obtained from the correlation coefficient analysis performed in MATLAB®[27]. The correlation matrix was $\left(\begin{smallmatrix} 1.000 & 0.999 \\ 0.999 & 1.000 \end{smallmatrix}\right)$ for the two variables (current and frequency).

The Curve Fitting Toolbox® software of MATLAB®[27] was employed to fit the FRC from $I \geq 3.8\,\mu\mathrm{A/cm^2}$ to a linear function. As the LSM is sensitive to extreme values in the FRC, the Robust Least Squares was utilized. The Least Absolute Residuals (LAR) method was selected to minimize the influence of such extreme values. With this, a piecewise linear function was obtained

$$fr^k = \begin{cases} 2.324I - 1.898 & I \geq 3.8\,\mu\mathrm{A/cm^2} \\ 0 & otherwise \end{cases} \tag{3}$$

where fr^k is the frequency and $I = \mathbf{x}^k \cdot \mathbf{w}$ the input current that depends on the k-th pattern.

To prove the validity of (3), the Sum of Squares due to Error (SSE), the Root Mean Square Error ($RMSE$) and the ratio (R^2) of the Sum of Regression Squares and the Total Sum of Squares were obtained. With 95% of confidence, the values were $SSE = 735.572$, $RMSE = 0.875$ and $R^2 = 1.000$. In general, the fitting accuracy gives a good level of confidence due to the ratio R^2 indicating that the variance in the frequency is low. Equation (3) for $I \geq 3.8\,\mu A/cm^2$ is depicted in Fig. 2 as a dashed line and the residual errors are shown in Fig. 3. Thus, the frequency to perform the pattern classification task can be obtained from (3) instead of the IZH numerical simulation.

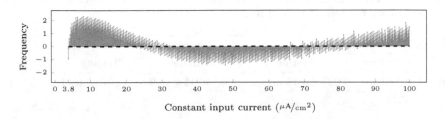

Fig. 3. Residual errors of the linear fitting for the Frequency Response Curve (FRC) of the Izhikevich (IZH) model.

5   Results

Two sets of experiments were carried out to verify that using the piecewise linear function described in (3) is more efficient than using the IZH simulation. See the Appendix for the technical specifications.

The first set of experiments was to obtain a single frequency value. A current $I = 31\,\mu A/cm^2$ was employed in (3) and in (1-2). This experiment was repeated 20 times and the average of the CPU time execution was determined. The IZH was numerically simulated using a time span $T = 1000$ ms and the parameters described in Sects. 3 and 4. The frequencies from (3) and (1-2) were 70.15 and 70 spikes per second. The calculus using (3) consumed on average 2.45 μs (± 0.51) and the IZH simulation 147.15 ms (± 3.26). The frequencies are practically the same, but the difference in the computational cost was four orders of magnitude. The calculus of a single frequency value using (3) is the 0.0017% of the CPU time consumed by the IZH simulation, or, in other words, the computational cost using the IZH simulation is $6,000,000\%$ higher than using the piecewise linear function.

The second set of experiments consisted in solving the well-kown Iris dataset problem [28] by using the methodology described in Sect. 2. To obtain the frequency, in one set of experiments the IZH simulation was employed and in the other, the piecewise linear function in (3). The parameters for the DE [16,17] algorithm were the same as in [2]: $NP = 40$, $MAXGEN = 1000$, $F = 0.9$, $XMAX = 10$, $XMIN = -10$ and $CR = 0.8$. Where NP is the number of

Table 1. Average classification accuracy ($\pm$ Standard deviation) using the simulation and the piecewise linear function of the Izhikevich (IZH) Frequency Response Curve (FRC).

Dataset	IZH simulation		Piecewise FRC	
	Training	Testing	Training	Testing
Iris	0.9933 ($\pm$ 0.0023)	0.9867 ($\pm$ 0.0281)	0.9933 ($\pm$ 0.0023)	0.9800 ($\pm$ 0.0322)

Table 2. Average CPU time ($\pm$ Standard deviation) (h = hours, s = seconds, μs = microseconds) using the simulation and the piecewise linear function of the Izhikevich (IZH) Frequency Response Curve (FRC).

Dataset	IZH simulation		Piecewise FRC	
	Training	Testing	Training	Testing
Iris	1.61 h ($\pm$ 0.02)	93.48 ms ($\pm$ 1.95)	46.41 s ($\pm$ 0.52)	297.60 μs ($\pm$ 155.87)

individuals in the population, $MAXGEN$ is the maximum number of generations, F is a scaling factor that controls the rate at which the population evolves, $XMAX$ is the lower bound of weights, $XMIN$ is the upper bound of weights and CR is the crossover probability [17]. To validate these experiments, 10-Folds Cross-Validation was used [29]. This kind of validation splits the data into ten subsets; each is employed once for the testing and the remaining for the training. The patterns for each subset were arbitrarily selected. Those patterns were presented to both experiments using the IZH simulation and the piecewise linear function. We obtained the average accuracy from the ten experiments of the 10-Folds Cross-Validation. Also, the average of the CPU elapsed time was determined. The classification accuracy and the CPU time execution are shown in Tables 1 and 2, respectively. It can be observed that the classification accuracy in the training stage is the same for both using (3) and using the IZH simulation. Nevertheless, during testing, the accuracy using the piecewise linear function is slightly less. The CPU elapsed time or computational cost using (3) is drastically reduced in both phases. The CPU time execution using the IZH simulation is roughly 12,000% higher than using the piecewise linear function in the training phase. In the testing phase, the percentage is roughly 31,000%.

Figures 4 and 5 show the distribution of a set of training patterns using the IZH simulation and the piecewise linear function of the FRC. As can be seen, the pattern distribution is very similar. The distribution corresponds to the first test in the 10-Folds Cross-Validation. The weight vectors for the IZH simulation and the piecewise linear function were $(-2.79, -2.76\ 10.00, 8.35)^T$ and $(-1.92, -4.10, 10.00, 10.00)^T$.

It can be observed that this scheme is different from that of the traditional Perceptron. The Perceptron distributes the patterns in two regions separated by a hyperplane (see [9]) whereas this proposal distributes the patterns on the

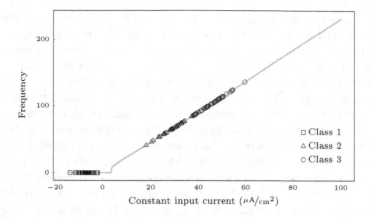

Fig. 4. Pattern distribution using the Frequency Response Curve (FRC) (continuous gray line) obtained from the Izhikevich (IZH) simulation.

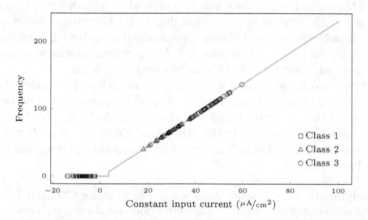

Fig. 5. Pattern distribution using the piecewise linear function (continuous gray line).

FRC. This permits the proposed scheme to solve non-linear problems using a single computational unit. However, as is well-known with a Perceptron, the XOR problem cannot be resolved.

6   Conclusions and Future Work

The single SN scheme consists in using the SN response to group a given input pattern, which is multiplied by a weight vector to produce a constant stimulation current. Depending on the number of individuals and generations in the evolutionary strategy, but also on the numerical method, integration step and the time span of the SN simulation, the computational cost in the training phase is extremely expensive.

The FRC is the relationship between the constant stimulus and the firing frequency emitted by a given SN. The FRC for the IZH model is highly linear. Thus, a piecewise linear function can be fitted to be used instead of the expensive SN simulation. This practically obtains the same accuracy results. Moreover, the computational cost in the training and testing phases was drastically reduced. In the training phase, the CPU time execution using the IZH simulation is roughly 12,000% higher than using the piecewise linear function. In the testing phase, that percentage is approximately 31,000%. The computational cost to obtain a single frequency value using the IZH simulation is 6,000,000% higher than using the piecewise linear function. We drastically reduced the computational cost making the scheme very efficient. As the computational cost to simulate the IZH depends on several factors, further research in this respect is needed.

Simulating a SN to solve pattern recognition problems is not recommended when the two following conditions are satisfied: (1) the stimulation current is constant during the entire simulation, and (2) the output signal encoding is the firing rate. SNs are useful for problems where the input pattern is precise in timing, or the output signal encoding is temporal. It is important to highlight that the single SN scheme is interesting due to its capacity to solve complex problems through the optimal classes (frequencies) distribution over the FRC. We believe that in future this scheme will bring new proposals in the research areas of Artificial Neural Networks and Pattern Recognition.

As a future work, we propose extending the experiments to other datasets, SNs, and training algorithms. A gradient descendant algorithm could be implemented due to the piecewise linear function being derivable. Also, this work could represent a first step in linking Deep and Spiking Neural Networks via the Rectified Linear Unit (ReLU) and the FRC linear fitting. Both functions are piecewise linear.

Acknowledgments. S. Valadez-Godínez would like to thank CONACYT and SIP-IPN for the scholarship granted in pursuit of his doctoral studies. J. González would like to thank CONACYT and SIP-IPN for undertaking his Master studies. H. Sossa would like to thank SIP-IPN and CONACYT under grants 20170693 and 65 (Frontiers of Science) to carry out this research. We are also very grateful to reviewers for their helpful comments.

Appendix

All tests carried out on a computer running the Linux Mint 17.3 64 bits operating system on an Intel Core i7-2600 (3.4 GHz with eight cores) processor. The memory of the computer was 8 GB. All software was implemented in MATLAB®. The *tic* and *toc* functions were utilized to calculate the elapsed time of the experiments described in Sect. 5.

References

1. Vazquez, R.A., Cachón, A.: Integrate and Fire neurons and their application in pattern recognition. In: 7th International Conference on Electrical Engineering Computing Science and Automatic Control, pp. 424–428 (2010)
2. Vazquez, R.: Izhikevich neuron model and its application in pattern recognition. Aust. J. Intell. Inform. Process. Syst. **11**, 35–40 (2010)
3. Vázquez, R.A.: Pattern recognition using spiking neurons and firing rates. In: Kuri-Morales, A., Simari, G.R. (eds.) IBERAMIA 2010. LNCS, vol. 6433, pp. 423–432. Springer, Heidelberg (2010). doi:10.1007/978-3-642-16952-6_43
4. Vazquez, R.A.: Training spiking neural models using cuckoo search algorithm. In: 2011 IEEE Congress of Evolutionary Computation (CEC), pp. 679–686 (2011)
5. Vázquez, R.A., Garro, B.A.: Training spiking neurons by means of particle swarm optimization. In: Tan, Y., Shi, Y., Chai, Y., Wang, G. (eds.) ICSI 2011. LNCS, vol. 6728, pp. 242–249. Springer, Heidelberg (2011). doi:10.1007/978-3-642-21515-5_29
6. Matadamas Ortiz, I.C.: Aplicación de las Redes Neuronales Pulsantes en el reconocimiento de patrones y análisis de imágenes. Master's thesis, Instituto Politécnico Nacional, Centro de Investigación en Computación, México (2014)
7. Vazquez, R.A., Garro, B.A.: Training spiking neural models using artificial bee colony. Comput. Intell. Neurosci. **2015**, 14 (2015). Article ID 947098
8. Carino-Escobar, R.I., Cantillo-Negrete, J., Gutierrez-Martinez, J., Vazquez, R.A.: Classification of motor imagery electroencephalography signals using spiking neurons with different input encoding strategies. Neural Comput. Appl., 1–13 (2016)
9. Minsky, M., Papert, S.: Perceptrons: An Introduction to Computational Geometry. The MIT Press, Cambridge (1969)
10. Adrian, E.D.: The Basis of Sensation. The Action of the Sense Organs. Christophers, London (1928)
11. Lapicque, M.L.: Recherches quantitatives sur l'excitation électrique des nerfs traitée comme une polarisation. J. Physiol. Pathol. Gen. **9**, 620–635 (1907)
12. Stein, R.B.: A theoretical analysis of neuronal variability. Biophys. J. **5**, 173–194 (1965)
13. Izhikevich, E.M.: Simple model of spiking neurons. IEEE Trans. Neural Netw. **14**, 1569–1572 (2003)
14. Hodgkin, A.L., Huxley, A.F.: A quantitative description of membrane current and its application to conduction and excitation in nerve. J. Physiol. **117**, 500–544 (1952)
15. Agin, D.: Hodgkin-Huxley equations: logarithmic relation between membrane current and frequency of repetitive activity. Nature **201**, 625–626 (1964)
16. Storn, R., Price, K.: Differential evolution-a simple and efficient adaptive scheme for global optimization over continuous spaces. J. Global Optim. **11**, 341–359 (1997)
17. Price, K.V., Storn, R.M., Lampinen, J.A.: Differential Evolution: A Practical Approach to Global Optimization. Springer, Heidelberg (2005)
18. Yang, X.S., Deb, S.: Cuckoo search via lévy flights. In: 2009 World Congress on Nature Biologically Inspired Computing (NaBIC), pp. 210–214 (2009)
19. Kennedy, J., Eberhart, R.: Particle swarm optimization. In: IEEE International Conference on Neural Networks Proceedings, vol. 4, pp. 1942–1948 (1995)
20. Karaboga, D.: An idea based on honey bee swarm for numerical optimization. Technical report TR06, Erciyes University, Engineering Faculty, Computer Engineering Department (2005)

21. Krinskii, V.I., Kokoz, Y.M.: Analysis of equations of excitable membranes - I. Reduction of the Hodgkin-Huxley equations to a second order system. Biofizika, pp. 506–511 (1973)
22. Kepler, T.B., Abbott, L.F., Marder, E.: Reduction of conductance-based neuron models. Biol. Cybern. **66**, 381–387 (1992)
23. Izhikevich, E.M.: Dynamical Systems in Neuroscience: The Geometry of Excitability and Bursting. The MIT Press, Cambridge (2007)
24. Euler, L.: Institutionum calculi integralis. Volumen primum. Petropoli: Impenfis Academiae Imperialis Scientiarum (1768)
25. Humphries, M.D., Gurney, K.: Solution methods for a new class of simple model neurons. Neural Comput. **19**, 3216–3225 (2007)
26. Skocik, M.J., Long, L.N.: On the capabilities and computational costs of neuron models. IEEE Trans. Neural Netw. Learn. Syst. **25**, 1474–1483 (2014)
27. MATLAB: Version 8.5.0 (R2015a). The MathWorks Inc., Natick, Massachusetts (2015)
28. Lichman, M.: UCI machine learning repository (2013)
29. Alpayd, E.: Introduction to Machine Learning. Second edn. The MIT Press (2010)

Evolutionary Clustering Using Multi-prototype Representation and Connectivity Criterion

Adán José-García$^{(\boxtimes)}$ and Wilfrido Gómez-Flores

Center for Research and Advanced Studies of the National Polytechnic Institute,
Cinvestav Tamaulipas, Ciudad Victoria, Tamaulipas, Mexico
{ajose,wgomez}@tamps.cinvestav.mx

Abstract. An automatic clustering approach based on differential evolution (DE) algorithm is presented. A clustering solution is represented by a new multi-prototype encoding scheme comprised of three parts: activation thresholds (binary values), cluster centroids (real values), and cluster labels (integer values). In addition, to measure the fitness of potential clustering solutions, an objective function based on a connectivity criterion is used. The performance of the proposed approach is compared with a DE-based automatic clustering technique as well as three conventional clustering algorithms (K-means, Ward, and DBSCAN). Several synthetic and real-life data sets having arbitrary-shaped clusters are considered. The experimental results indicate that the proposed approach outperforms its counterparts because it is capable to discover the actual number of clusters and the appropriate partitioning.

Keywords: Automatic clustering · Differential evolution · Non-linearly separable clusters · Multi-prototype representation · Cluster validity index

1 Introduction

Clustering is an unsupervised learning technique aimed to discover the natural grouping of unlabeled objects according to the similarity of their measured intrinsic characteristics [10]. Formally, let $\mathbf{X} = \{\mathbf{x}_1, \ldots, \mathbf{x}_N\}$ be a set of N objects (or patterns) to be partitioned into K non-overlapping groups (or clusters) $\mathbf{C} = \{\mathbf{c}_1, \ldots, \mathbf{c}_K\}$, such that the following three conditions are satisfied: $\mathbf{c}_i \neq \emptyset$; $\mathbf{c}_1 \cup \ldots \cup \mathbf{c}_K = \mathbf{X}$; and $\mathbf{c}_i \cap \mathbf{c}_j = \emptyset$ for $i, j = 1, \ldots, K$ and $i \neq j$.

In addition, when the number of clusters is unknown *a priori*, the problem is referred to as *automatic clustering* (AC) [3,4,6], which consists in discovering the number of clusters as well as the clustering that best fits the actual data structure.

To find the optimal clustering solution to partition N objects into K clusters is a very difficult combinatorial optimization problem which has been proved to be NP-complete when $K > 3$ [6]. Therefore, the AC problem is frequently formulated as one of numerical optimization, where prototypes (e.g., medoids or

© Springer International Publishing AG 2017
J.A. Carrasco-Ochoa et al. (Eds.): MCPR 2017, LNCS 10267, pp. 63–73, 2017.
DOI: 10.1007/978-3-319-59226-8_7

centroids) are used as representative points of the clusters, that is, a prototype-based representation is considered. This optimization problem consists in finding the optimal locations of the prototypes that best represent the clusters in the dataset. This formulation has been proved to be NP-hard [1]; thus, such a complexity has motivated the use of diverse nature-inspired metaheuristics to address the AC problem [4,6], where the best partition is achieved by optimizing an objective function, which is known as the *cluster validity index* (CVI).

Generally, to evaluate the quality of a clustering solution, the CVI considers both the intracluster dispersion of patterns in every cluster and the intercluster separation among cluster prototypes. Thus, a proper solution representation and an effective evaluation function (or CVI) are important components in the design of an automatic clustering algorithm based on metaheuristics.

On the other hand, data clustering is a difficult problem because the clusters can differ in shape, size, density, overlapping, etc. In addition, the presence of non-linearly separable clusters makes its detection even more difficult. We can stated that a set of patterns is linearly separable if two actual clusters, represented by a couple of prototypes, can be correctly separated by a single hyperplane. This condition is well-estimated by the CVI when the input data is linearly separable. However, the CVI presents a poor performance when the data is non-linearly separable. This behavior is because the criteria of cohesion and separation are unsatisfied simultaneously.

In this work, we propose a multi-prototype clustering representation to encode arbitrary-shaped and non-linearly separable clusters. The proposed representation allows encoding clustering solutions by using a set of prototypes to represent a single cluster. Moreover, a clustering criterion based on connectivity of data is incorporated into the well-known Silhouette index to measure the intracluster dispersion. Finally, both components, the multi-prototype representation and the connectivity-based CVI, are integrated into an automatic clustering algorithm based on differential evolution (MACDE).

The outline of this paper is as follows: Sect. 2 surveys the related work; Sect. 3 presents the proposed approach; Sect. 4 describes the experimental setup; Sect. 5 summarizes the results; and Sect. 6 gives the conclusion.

2   Related Work

In the early years, several researchers performed automatic clustering mainly through modifying K-means and FCM algorithms [6]. Nowadays, diverse nature-inspired metaheuristics have been used to address the AC problem. Indeed, a recent survey [6] reported that the design, development, and application of nature-inspired clustering algorithms to automatic clustering has increased notably during the last decade. In particular, it was reported that evolutionary algorithms are the most used to address this problem.

Regardless of the search mechanisms, the success of nature-inspired clustering algorithms relies mostly on a proper solution representation as well as an effective evaluation function. The related work of these two components is provided in the next.

2.1   Encoding Schemes

The *centroid-based representation of fixed length* [3] is commonly used in automatic clustering algorithms. All the individuals in the population have the same length defined as $K_{max} + K_{max} \times D$, where K_{max} is the maximum number of clusters and D is the dimensionality of the dataset. The first K_{max} entries are real numbers in the range $[0, 1]$ called activation thresholds represented by $\mathbf{T} = \{T_k \mid k = 1, \ldots, K_{max}\}$. The remaining $K_{max} \times D$ entries are reserved for the cluster centroids, $\overline{\mathbf{C}} = \{\bar{\mathbf{c}}_k \mid k = 1, \ldots, K_{max}\}$, where $\bar{\mathbf{c}}_k \in \mathbb{R}^D$. To vary the number of clusters, it is considered the following activation rule: the kth centroid is activated if and only if $T_k > 0.5$; otherwise it is inactivated. Hence, only activated prototypes participate in the clustering process.

2.2   Cluster Validity Indices

A cluster validity index (CVI) is a mathematical function used to quantitatively evaluate a clustering solution by considering the intracluster dispersion and the intercluster separation. Generally, a CVI is used for two purposes: to estimate the number of clusters and to find the corresponding best partition. CVIs are optimization functions by nature, that is, the maximum or minimum values indicate the appropriate partitions. Therefore, CVIs have been used as objective functions by evolutionary clustering algorithms to address the AC problem [3].

The CVIs usually use representative points (e.g., centroids or medoids) to represent the groups in a clustering solution. This approach is suitable for compact and hyperspherical-shaped clusters [10]. However, when clusters are non-linearly separable and present arbitrary shapes, their prototypes could share the same region in the feature space. Therefore, in this case, the CVI presents a poor clustering performance.

In order to discover arbitrary-shaped and non-linearly separable clusters, some approaches based on the connectivity of data have been proposed in the literature. Pan and Biswas [7] incorporated graph theory concepts into some CVIs to compute the intracluster cohesion criteria. Also, Saha and Bandyopadhyay [8] presented a new measure of connectivity based on a relative neighborhood graph, which was incorporated in some conventional CVIs such as Davies–Bouldin, Dunn, and Xie–Beni indices. However, the time complexity of these approaches makes them unsuitable to address the AC problem using evolutionary algorithms.

3   Proposed Approach

3.1   Differential Evolution Algorithm

Differential evolution (DE) is an evolutionary algorithm proposed to solve optimization problems over continuous spaces [9]. Let $\mathbf{P}^g = \{\mathbf{z}_1, \ldots, \mathbf{z}_{NP}\}$ be the current population with NP members at generation g, where the ith individual is a l-dimensional vector denoted by $\mathbf{z}_i = [z_{i,1}, \ldots, z_{i,l}]$.

At the beginning of the algorithm, the variables of the NP individuals are randomly initialized according to a uniform distribution $z_j^{\text{low}} \leq z_{i,j} \leq z_j^{\text{up}}$, for $j = 1, 2, \ldots, l$. After initialization, DE enters into a loop of evolutionary operators (mutation, crossover, and selection) until convergence is reached or the maximum number of generations is attained.

Mutation: At each generation g, mutant vectors $\mathbf{v}_i^g$ are created from the current parent population $\mathbf{P}^g = \{\mathbf{z}_1, \ldots, \mathbf{z}_{NP}\}$. The "DE/rand/1" mutation strategy is frequently used:

$$\mathbf{v}_i^g = \mathbf{z}_{r0}^g + F\left(\mathbf{z}_{r1}^g - \mathbf{z}_{r2}^g\right), \tag{1}$$

where the indices $r0$, $r1$ and $r2$ are distinct integers randomly chosen from the set $\{1, 2, \ldots, NP\} \setminus \{i\}$ and F is a mutation factor which usually ranges within $(0, 1)$.

Crossover: After mutation, a binomial crossover operation creates a trial vector $\mathbf{u}_i^g = \left[u_{i,1}^g, \ldots, u_{i,l}^g\right]$ as

$$u_{i,j}^g = \begin{cases} v_{i,j}^g & \text{if } \text{rand}_j\left(0,1\right) < CR \text{ or } j = j_{\text{rand}}, \\ z_{i,j}^g & \text{otherwise}, \end{cases} \tag{2}$$

where $\text{rand}\left(0,1\right)$ is a uniform random number in the range $[0, 1]$, $j_{\text{rand}} = \text{randint}\left(1, l\right)$ is an integer randomly chosen in the range $[1, l]$, and $CR \in [0, 1]$ is the crossover rate.

Selection: This operator selects the best solution between the target vector $\mathbf{z}_i^g$ and the trial vector $\mathbf{u}_i^g$ according to their fitness value $f(\cdot)$. Without loss of generality, for a minimization problem the selected vector that transcend to the next generation $g + 1$ is given by

$$\mathbf{z}_i^{g+1} = \begin{cases} \mathbf{u}_i^g & \text{if } f\left(\mathbf{u}_i^g\right) < f\left(\mathbf{z}_i^g\right), \\ \mathbf{z}_i^g & \text{otherwise}. \end{cases} \tag{3}$$

3.2 Multi-prototype Representation

In this paper, we introduce a new multi-prototype representation in which every individual in the population, $\mathbf{z}_i$, is a fixed-length vector composed of activation thresholds, cluster centroids, and cluster labels. Thus, every individual is a vector of size $K_{\max} + (K_{\max} \times D) + K_{\max}$. The first $K_{\max}$ entries are real numbers in the range $[0, 1]$, called the activation thresholds denoted by the set $\mathbf{T} = \{T_k \mid k = 1, \ldots, K_{\max}\}$. The next $K_{\max} \times D$ entries correspond to the cluster centroids denoted by the set $\overline{\mathbf{C}} = \{\bar{\mathbf{c}}_k \mid k = 1, \ldots, K_{\max}\}$, where $\bar{\mathbf{c}}_k \in \mathbb{R}^D$. The remaining $K_{\max}$ entries are integer numbers in the set $\{1, \ldots, K_{\max}\}$, called the cluster labels, denoted by $\mathbf{L} = \{L_k \mid k = 1, \ldots, K_{\max}\}$. Then, the ith individual in the DE algorithm is expressed by

$$\mathbf{z}_i = \left[T_{i,1}, \ldots, T_{i,K_{\max}}, \bar{\mathbf{c}}_{i,1}, \ldots, \bar{\mathbf{c}}_{i,K_{\max}}, L_{i,1}, \ldots, L_{i,K_{\max}}\right] = \left\{\mathbf{T}_i, \overline{\mathbf{C}}_i, \mathbf{L}_i\right\}. \tag{4}$$

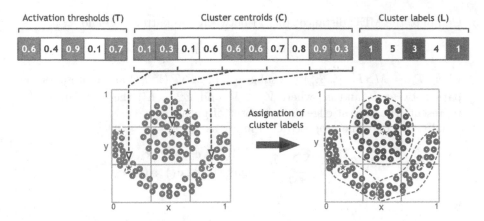

Fig. 1. Example of the multi-prototype representation with $K_{max} = 5$ and $D = 2$. The activated centroids are represented by blue stars and the unactivated ones by red stars. The clustering solution is obtained by assigning to every pattern the cluster label in **L**. (Color figure online)

The kth centroid is activated if and only if $T_{i,k} \geq 0.5$; otherwise, it is inactivated. An activated centroid means that it participates in the clustering process to form "subclusters". Finally, these subclusters are merged according to their corresponding cluster labels in $\mathbf{L}_i$ to form the clusters of a potential solution. It is worth mentioning that a centroid point is approximated to its closest pattern point to represent the cluster by a medoid.

Figure 1 shows an example of the multi-prototype representation, where $K_{max} = 5$ and $D = 2$. First, the positions 1, 3, and 5 satisfy the aforementioned activation rule. Therefore, their centroids $(0.1, 0.3)$, $(0.6, 0.6)$, and $(0.9, 0.3)$ form the corresponding subclusters. Finally, the clusters sharing the same cluster labels are merged, that is, the cluster label "3" is assigned to the third cluster and the label "1" is assigned to the first and fifth subclusters, respectively.

3.3   A Connectivity-Based CVI

The proposed index CSil (Connectivity-Silhouette index) is based on a recent proximity measure called maximum edge distance (MED) [2], which is capable to detect groups of different shape, size, and convexity. The computation of the CSil involves the following steps:

1. Build an undirected complete graph $\mathbf{G}(V, E)$ from the input dataset $\mathbf{X} = \{\mathbf{x}_1, \ldots, \mathbf{x}_N\}$, where $v_i \in V$ corresponds to $\mathbf{x}_i \in \mathbf{X}$, the edge $e_{ij} \in E$ denotes the pairwise Euclidean distance (d_e) between the vertices v_i and v_j, and N is the total number of patterns in the dataset.
2. Construct the minimum spanning tree $MST(V, E^t)$ from $\mathbf{G}$, where $E^t \subset E$ such that $|E^t| = N - 1$.

3. Compute the MED distance[1] (d_{med}) between two patterns $\mathbf{x}_i$ and $\mathbf{x}_j$ as:

$$d_{med}(v_i, v_j) = \{E_p^t \in \mathcal{P}_{ij} \ / \max\left(e_p^t\right)\}. \tag{5}$$

where $\mathcal{P}_{ij} = MST'\left(V_p, E_p^t\right)$ is a subgraph of $MST\left(V, E^t\right)$ and represents the path between v_i and v_j, where $V_p \subset V$ and $E_p^t \subset E^t$; therefore, $\max\left(e_p^t\right)$ represents the longest edge in the subset of paths, e_p^t.

4. Finally, compute the proposed CSil index as

$$\text{CSil}(\mathbf{C}) = \frac{1}{N} \sum_{c_k \in \mathbf{C}} \sum_{x_i \in c_k} \frac{b\left(\mathbf{x}_i, \mathbf{c}_k\right) - a\left(\mathbf{x}_i, \mathbf{c}_k\right)}{\max\left\{b\left(\mathbf{x}_i, \mathbf{c}_k\right), a\left(\mathbf{x}_i, \mathbf{c}_k\right)\right\}}, \tag{6}$$

where

$$a\left(\mathbf{x}_i, \mathbf{c}_k\right) = \frac{1}{n_k} \sum_{x_j \in c_k} d_{\text{med}}\left(\mathbf{x}_i, \mathbf{x}_j\right), \quad b\left(\mathbf{x}_i, \mathbf{c}_k\right) = \min_{c_r \in \mathbf{C} \backslash c_k} \left\{\frac{1}{n_r} \sum_{x_j \in c_r} d_{\text{med}}\left(\mathbf{x}_i, \mathbf{x}_j\right)\right\}.$$

In order to achieve the proper partitioning, the value of CSil is maximized.

3.4   Avoiding Erroneous Clustering Solutions

When a new individual is created by using the DE evolutionary operators, erroneous clustering solutions could be generate; therefore, the following cases must be considered:

– **Minimum number of subclusters**: if all the activation thresholds in $\mathbf{T}_i$ are smaller than 0.5, then, two threshold entries are randomly selected and re-initialized in the range $(0.5, 1)$. Likewise, if all the activated cluster labels in $\mathbf{L}_i$ have the same label, then, a label entry is randomly selected and re-initialized with a different random label chosen from $\{1, \ldots, K_{\max}\}$.
– **Empty clusters**: If any activated cluster has associated less than two patterns, then, all the clusters are re-initialized such that every cluster would have $\frac{N}{K}$ patterns and the corresponding prototypes are recalculated by averaging the patterns of every cluster.
– **Out-of-bound variables**: If any threshold value in $\mathbf{T}_i$ exceeds the unity or becomes negative, then, it is truncated to "1" and "0", respectively. Similarly, if any cluster label in $\mathbf{L}_i$ exceeds $K_{\max}$ or becomes negative, then, it is truncated to $K_{\max}$ and "1", respectively.

3.5   Pseudocode of MACDE Algorithm

The complete pseudocode for the proposed MACDE algorithm is detailed below:

– **Step 1**: Generate an initial random population $\mathbf{P}^0 = \{\mathbf{z}_i^0 \mid i = 1, \ldots, NP\}$ as described in Sect. 3.2.

---

[1] The MED distance is symmetric, always positive, and satisfies the properties of identity and triangle inequality [2].

- **Step 2**: For each $\mathbf{z}_i \in \mathbf{P}^0$, find out the activated centroids in $\mathbf{C}_i$ and cluster labels in $\mathbf{L}_i$ by applying the activation rule to the thresholds values in $\mathbf{T}_i$.
- **Step 3**: For $t = 1$ to $G_{\max}$ do:
  (i) For each $\mathbf{z}_i \in \mathbf{P}^t$ and considering each pattern $\mathbf{x}_p \in \mathbf{X}$ for $p = 1, \ldots, N$, obtain the K_i medoids $\mathbf{M}_i = \{\mathbf{m}_j \mid j = 1, \ldots, K_i\}$ by assigning each activated centroid $\bar{\mathbf{c}}_j \in \bar{\mathbf{C}}_i$ to the closest $\mathbf{x}_p$ such that

$$d_e(\bar{\mathbf{c}}_j, \mathbf{x}_p) = \min_{\mathbf{x}_p} \in \mathbf{X}\{d_e(\bar{\mathbf{c}}_j, \mathbf{x}_p)\}.$$

  (ii) For each $\mathbf{z}_i \in \mathbf{P}^t$ and considering each $\mathbf{x}_p \in \mathbf{X}$, obtain the clustering solution $\mathbf{C}'_i$ by assigning $\mathbf{x}_p$ to the closest medoid $\mathbf{m}_j \in \mathbf{M}_i$ such that

$$d_e(\mathbf{x}_p, \mathbf{m}_j) = \min_{\mathbf{m}_j} \in \mathbf{M}_i\{d_e(\mathbf{x}_p, \mathbf{m}_j)\}.$$

  (iii) For each $\mathbf{z}_i \in \mathbf{P}^t$, check if the number of patterns belonging to any cluster in $\mathbf{C}'_i$ is less than two. If so, update the cluster medoids $\mathbf{M}_i$ using the concept described in Sect. 3.4.
  (iv) For each $\mathbf{z}_i \in \mathbf{P}^t$, create the merged-clustering solution $\mathbf{C}_i$ from $\mathbf{C}'_i$ using the activated cluster labels in $\mathbf{L}_i$.
  (v) Perform the evolutionary operators on each individual $\mathbf{z}_i \in \mathbf{P}^t$ to create a mutant vector $\mathbf{v}_i$ using (1) and then a trial vector $\mathbf{u}_i$ using (2).
  (vi) For each $\mathbf{u}_i$, find out the activated centroids and cluster labels by applying the activation rule. The real values in $\mathbf{L}_i$ must be rounded to its nearest integer in order to generate suitable cluster labels.
  (vii) Repeat steps (i)-(iv) for each trial vector $\mathbf{u}_i$ in order to verify its validity.
  (viii) Evaluate fitness of both the target $\mathbf{z}_i$ and trial $\mathbf{u}_i$ vectors according to the CSil index based on MED distance in (6). Use only the merged-clustering solution of both vectors. Replace the target vector $\mathbf{z}_i^t$ with the trial vector $\mathbf{u}_i^t$ only if the latter yields a better value of the fitness function.
- **Step 4**: Report the final merged-clustering solution obtained by the best individual (the one yielding the highest fitness) at time $t = G_{\max}$.

4   Experimental Setup

The proposed algorithm (MACDE) was compared with four different clustering approaches: an automatic clustering algorithm based on DE, ACDE [3]; a partitional clustering approach, K-means [10]; a hierarchical clustering method, WARD [10]; and a density-based algorithm, DBSCAN [10].

The testing platform used a LINUX-based computer with 8-cores at 2.7 GHz and 8 GB of RAM. All the algorithms were developed in Matlab R2014a (The Mathworks, Boston, Massachusetts, USA).

4.1   Parameter Settings

The settings adopted in MACDE and ACDE are: number of fitness function evaluations, $FE = 1E5$; population size, $NP = D \times 10$; crossover rate, $CR = 0.9$; mutation factor, $F = 0.8$; and maximum number of clusters, $K_{\max} = 20$.

The Silhouette index is used as the selection criterion in K-means (executed for different $K = \{2, \ldots, K_{\max}\}$) and WARD (the dendrogram is cut at different levels); whereas, the best parameter settings were tuned in for DBSCAN.

4.2   Data Sets

Four categories of datasets are studied: (i) linearly separable data having well-separated clusters, $\mathcal{G}_1$; (ii) linearly separable data having overlapping clusters, $\mathcal{G}_2$; (iii) non-linearly separable data having well-separated clusters, $\mathcal{G}_3$; and (iv) real-life datasets, $\mathcal{G}_4$. An example dataset of each category is shown in Fig. 2(a).

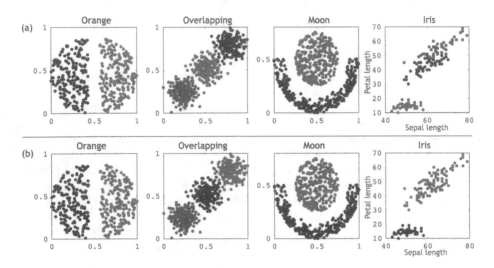

Fig. 2. Example of datasets in categories $\mathcal{G}_1$, $\mathcal{G}_2$, $\mathcal{G}_3$, and $\mathcal{G}_4$ (left to right). (a) Actual cluster labels and (b) clustering solutions obtained by MACDE. Distinct colors are used to represent different clusters. (Color figure online)

4.3   Clustering Quality Evaluation

The adjusted rand index (ARI) [5] takes as input two partitionings and returns a value in the interval $[\sim 0, 1]$, where "1" indicates perfect similarity between them and "~ 0" disagreement. Let $\mathbf{T}$ be the true partitioning and $\mathbf{C}$ be partitioning obtained by a clustering algorithm. Also, let a, b, c and d, denote, respectively, the number of pairs of data points belonging to the same cluster in both $\mathbf{T}$ and $\mathbf{C}$, the number of pairs belonging to the same cluster in $\mathbf{T}$ but to different clusters in $\mathbf{C}$, the number of pairs belonging to different clusters in $\mathbf{T}$ but to the same cluster in $\mathbf{C}$, and the number of pairs belonging to different clusters in both $\mathbf{T}$ and $\mathbf{C}$. The ARI value is then computed as follows

$$ARI\left(\mathbf{T}, \mathbf{C}\right) = \frac{2\left(ad - bc\right)}{\left(a+b\right)\left(b+d\right) + \left(a+c\right)\left(c+d\right)}. \qquad (7)$$

5   Experimental Results

We investigated the effectiveness of the proposed approach focusing on two major issues: (i) quality of the clustering solution and (ii) ability to find the actual number of clusters.

The experimental results in terms of the ARI index and the estimated number of groups, K, are shown in Table 1. Note that ACDE, K-means, and WARD algorithms presented high performance on linearly-separable data (i.e., categories $\mathcal{G}_1$ and $\mathcal{G}_2$), since these datasets contain spherical Gaussian clusters, which is the cluster model assumed by these methods. In contrast, these algorithms presented poor performance in non-linearly separable data having having well-separed clusters (i.e., category $\mathcal{G}_3$). DBSCAN achieved good performance on datasets belonging to category $\mathcal{G}_3$, as these datasets contain dense and spatially well-separated clusters, which is the main assumption made by this algorithm, whereas a lower performance is obtained in datasets of categories $\mathcal{G}_1$ and $\mathcal{G}_2$. In general, ACDE, K-means, WARD, and DBSCAN algorithms obtained poor performance on datasets belonging to $\mathcal{G}_4$, as these contain more complex cluster structures derived from real-life scenarios.

On the other hand, the results for MACDE indicate a good performance across the datasets in categories $\mathcal{G}_1$ and $\mathcal{G}_3$. This is reflected on both the high values of ARI and the slightly difference between the estimated number of clusters

Table 1. Number of clusters and quality of the best solution in terms of the ARI (mean values from 31 runs) generated by MACDE, ACDE, K-means, WARD, and DBSCAN. The statistically best ($\alpha = 0.05$) results are highlighted in bold face. The Kruskal–Wallis test with Bonferroni correction was applied to compare the algorithms.

	Dataset	N	D	K*	MACDE		ACDE		K-means		WARD		DBSCAN	
					K	ARI	K	ARI	K	ARI	K	ARI	K	ARI
$\mathcal{G}_1$	Orange	400	2	2	**2.0**	**1.00**	**2.0**	**1.00**	**2.0**	**1.00**	2	**1.00**	2	**1.00**
	Hepta	212	3	7	**7.0**	**1.00**	**7.0**	**1.00**	7.4	0.96	7	**1.00**	8	0.84
	WingNut	1016	2	2	**2.0**	**1.00**	**2.0**	0.86	2.0	0.86	2	0.79	3	0.38
	Lsun	400	2	3	**3.0**	**0.99**	5.0	0.65	5.0	0.65	5	0.66	4	**1.00**
$\mathcal{G}_2$	Overlapping	600	2	3	3.9	0.92	3.0	0.94	**3.0**	**0.95**	3	**0.95**	4	0.92
	TwoDiamonds	800	2	2	2.0	0.98	**2.0**	**1.00**	**2.0**	**1.00**	2	**1.00**	2	0.99
	Tetra	400	3	4	**4.0**	**1.00**	**4.0**	**1.00**	**4.0**	**1.00**	4	0.97	5	0.75
	Data_5_2	250	2	5	4.7	0.72	5.0	0.86	**5.0**	**0.89**	5	0.88	5	0.59
$\mathcal{G}_3$	Moon	600	2	2	**2.0**	**0.94**	3.8	0.51	11.5	0.19	9	0.24	2	**1.00**
	Inside	600	2	2	**2.0**	**0.99**	6.4	0.59	6.4	0.59	7	0.58	2	**1.00**
	Arcs	600	2	2	**2.0**	**0.98**	4.4	0.28	9.4	0.22	12	0.17	2	**1.00**
	Atom	800	3	2	**2.0**	**0.95**	11.3	0.56	10.2	0.56	11	0.56	2	**1.00**
$\mathcal{G}_4$	Iris	150	4	3	2.0	0.57	2.0	0.54	2.0	0.54	2	**0.57**	3	0.55
	Breast	676	9	2	2.5	0.70	2.0	0.80	**2.0**	**0.84**	2	0.83	2	**0.86**
	Seeds	199	7	3	3.1	0.42	2.0	0.46	**2.0**	**0.49**	2	**0.47**	2	0.44
	Voting	232	16	2	3.2	0.59	2.0	0.60	2.0	0.63	2	**0.67**	13	0.10

K and the actual number of clusters K^*. It is notable that the performance of MACDE is affected by the overlap degree between clusters, that is, as the overlap increases the performance diminishes. This disadvantage is because the minimum separation between clusters is less than the maximum first neighbor distance (this restriction is related to the use of the MST in the CSil index). However, as expected, the multi-prototype representation and the cluster validity index based on the connectivity criteria (CSil) allows discovering arbitrary-shaped cluster regardless the property of linear separability of data. Finally, Fig. 2(b) presents some clustering solutions for the different studied categories of datasets.

6 Conclusions

In this paper, an automatic clustering approach based on DE algorithm was proposed. A new multi-prototype representation and a cluster validity index based on connectivity were proposed to represent and evaluate, respectively, clustering solutions having non-linearly separable clusters. The proposed algorithm (MACDE) has been shown to outperform some well-know clustering techniques across a diverse range of datasets separated by categories. The proposed approach has two main advantages which are: (i) the automatic discovering of the number of clusters and (ii) the data clustering independently of its linear separability (arbitrary-shaped clusters). Finally, it should be noted that data having overlapping clusters may decrease the performance of MACDE.

In the future, it is interesting to investigate and overcome the performance of MACDE for data having strong-overlapping clusters.

Acknowledgments. The authors would like to thank the support from CONA-CyT Mexico through a scholarship to pursue doctoral studies at Unidad Cinvestav Tamaulipas.

References

1. Aloise, D., Deshpande, A., Hansen, P., Popat, P.: NP-hardness of euclidean sum-of-squares clustering. Mach. Learn. **75**(2), 245–248 (2009)
2. Bayá, A.E., Granitto, P.M.: How many clusters: a validation index for arbitrary-shaped clusters. IEEE/ACM Trans. Comput. Biol. Bioinform. **10**(2), 401–414 (2013)
3. Das, S., Abraham, A., Konar, A.: Automatic clustering using an improved differential evolution algorithm. IEEE Trans. Syst. Man Cybern. **38**(1), 218–237 (2008)
4. Hruschka, E.R., Campello, R.J.G.B., Freitas, A.A., de Carvalho, A.C.P.L.: A survey of evolutionary algorithms for clustering. IEEE Trans. Syst. Man Cybern. Part C **39**(2), 133–155 (2009)
5. Hubert, L., Arabie, P.: Comparing partitions. J. Classification **2**(1), 193–218 (1985)
6. José-García, A., Gómez-Flores, W.: Automatic clustering using nature-inspired metaheuristics: a survey. Appl. Soft Comput. **41**, 192–213 (2016)

7. Pal, N., Biswas, J.: Cluster validation using graph theoretic concepts. Pattern Recognit. **30**(6), 847–857 (1997)
8. Saha, S., Bandyopadhyay, S.: Some connectivity based cluster validity indices. Appl. Soft Comput. **12**(5), 1555–1565 (2012)
9. Storn, R., Price, K.: Differential evolution - a simple and efficient heuristic for global optimization over continuous spaces. J. Global Optim. **11**(4), 341–359 (1997)
10. Theodoridis, S., Koutrumbas, K.: Pattern Recognition, 4th edn. Elsevier Inc., Burlington (2009)

Fixed Height Queries Tree Permutation Index for Proximity Searching

Karina Figueroa[1(✉)], Rodrigo Paredes[2], J. Antonio Camarena-Ibarrola[1],
and Nora Reyes[3]

[1] Universidad Michoacana, Morelia, Mexico
karina@fismat.umich.mx, camarena@umich.mx
[2] Universidad de Talca, Talca, Chile
raparede@utalca.cl
[3] Universidad Nacional de San Luis, San Luis, Argentina
nreyes@unsl.edu.ar

Abstract. Similarity searching consists in retrieving from a database
the objects, also known as nearest neighbors, that are most similar to
a given query, it is a crucial task to several applications of the pattern
recognition problem. In this paper we propose a new technique to reduce
the number of comparisons needed to locate the nearest neighbors of a
query. This new index takes advantage of two known algorithms: FHQT
(Fixed Height Queries Tree) and PBA (Permutation-Based Algorithm),
one for low dimension and the second for high dimension. Our results
show that this combination brings out the best of both algorithms, this
winner combination of FHQT and PBA locates nearest neighbors up to
four times faster in high dimensions leaving the known well performance
of FHQT in low dimensions unaffected.

1 Introduction

Similarity searching consists in retrieving the most similar objects from a database to a given query. This problem is also known as nearest neighbor searching,
which is a crucial task to several areas such as multimedia retrieval (i.e. images),
computational biology, pattern recognition, etc. The similarity between objects
can be measured with a distance function, usually considered expensive to compute, defined by experts in a specific data domain. Thus, the main objective of
several proposed indexes is to reduce the number of distance evaluations to get
the most similar objects in a database with respect to a given query.

Similarity searching can be mapped into a metric space problem. It can be
seen as a pair $(\mathbb{X}, d)$, where $\mathbb{X}$ is the universe of objects and d is the distance
function $d : \mathbb{X} \times \mathbb{X} \to \mathbb{R}^+ \cup \{0\}$. The *distance* satisfies, for all $x, y, z \in \mathbb{X}$, the
following properties: reflexivity $d(x, y) = 0$ iff $x = y$, symmetry $d(x, y) = d(y, x)$,
and triangle inequality $d(x, y) \leq d(x, z) + d(z, y)$. In practical applications, we
have a working database $|\mathbb{U}| = n$, $\mathbb{U} \subseteq \mathbb{X}$.

Basically, there are two kinds of queries: range query (q, r) and k-nearest
neighbor query $kNN(q)$. The first one retrieves all the objects within a given

© Springer International Publishing AG 2017
J.A. Carrasco-Ochoa et al. (Eds.): MCPR 2017, LNCS 10267, pp. 74–83, 2017.
DOI: 10.1007/978-3-319-59226-8_8

radius r measured from q, that is, $(q, r) = \{u \in \mathbb{U}, d(q, u) \leq r\}$. The other one retrieves the k objects in $\mathbb{U}$ that are the closest to q. Formally, $|kNN(q)| = k$, and $\forall\, u \in kNN(q), v \in \mathbb{U},\, d(u, q) \leq d(v, q)$.

Several algorithms [8,10,11] have been proposed in order to answer these kind of queries. Examples of indexes that are relevant to this work are Fixed Queries Tree (FQT) [2] and the Permutation-Based Algorithm (PBA) [5]. Some indexes suffer the well known *curse of dimensionality* (that is, the searching effort increases as the dataset intrinsic dimensionality grows) [8]. In practice, there are indexes more adequate to certain dimensionality. For instance, the FQT is well suited for low dimensionality and the PBA for medium to high dimensionality. As the distance is considered expensive to compute, in this work we measure any cost in terms of the number of distance computations needed.

In this paper, we introduce an improvement on top of these two classic metric spaces searching algorithms: a variant of FQT (Fixed Height Queries Tree - FHQT) and PBA. This new index works well in both low and high dimensionality, because adequately combines the best of them.

The organization of this paper is as follows. Section 2 presents the basic concepts about metric spaces algorithms. Section 3 describes in detail the indexes FHQT and PBA, that are the base of our work. Section 4 introduces our proposal in details (indexing and searching). In Sect. 5, we experimentally prove our claims using both synthetic and real world datasets, the first one help us to known the performance of the parameters. In the last section we arrive to some conclusions and a discussion of future work.

2   Basic Concepts

Firstly, an algorithm aims to establish some structure or index over the database $\mathbb{U}$; then, when a query is given, the algorithm uses this structure to speed up the response time. Of course, in order to traverse through the index some distances computations are needed. This process obtains a set of non-discarded objects, then the query is compared with all these objects to answer the similarity query.

The answer to a similarity query can be *exact* or *approximate*. An approximate similarity searching is appealing when we require efficiency and instead we accept to lose some accuracy. This is specially relevant when we work on high intrinsic dimensionality spaces. Algorithms that obtain exact answers can be classified in two groups, namely pivots-based algorithms (PB) and compact-partitions algorithms (CP). While PB algorithms work well in low dimensions, CP algorithms work better in high dimensions. On the other hand, a good approach to solve similarity queries in an approximated fashion is to use the PBA, that are unbeatable in high and very high dimensions.

Pivot-Based Algorithms. A pivot-based algorithm chooses a set of *pivots* $\mathbb{P} = \{p_1, p_2, \ldots, p_j\} \subseteq \mathbb{U}, j = |\mathbb{P}|$. For each database element $u \in \mathbb{U}$, the PB algorithm computes and stores all the distances between u and the members of $\mathbb{P}$. The resulting set of distances $\{d(p_1, u), d(p_2, u), \ldots, d(p_j, u)\}, \forall u \in \mathbb{U}$ is used for building the index.

For a range query (q, r), the distances $d(p_i, q) \; \forall p_i \in \mathbb{P}$ are computed. By virtue of the triangle inequality property, for each $u \in \mathbb{U}$, it holds that $\max_{1 \leq i \leq j} |d(q, p_i) - d(p_i, u)| \leq d(q, u)$. Therefore, if an element u satisfies that $r < \max_{1 \leq i \leq j} |d(q, p_i) - d(p_i, u)|$, then u can be safely discarded. Finally, every non-discarded element is directly compared with q and reported if it fulfills the range criterion.

Compact Partition Algorithms. A compact partition algorithm exploits the idea of dividing the space in compact zones, usually in a recursive manner, and storing a representative object (a "center") c_i for each zone plus a few extra data that permits quickly discarding the zone at query time. During search, entire zones can be discarded depending on the distance from their cluster center c_i to the query q. Two criteria can be used to delimit a zone, namely hyperplane and covering radius.

3   Related Works

In this section we describe the two algorithms that we use as base of our work: Fixed Height Queries Tree and Permutation-based Algorithm.

3.1   Fixed Height Queries Tree

The FHQT belongs to a family of indexes, all of them with about the same efficiency in terms of number of distances computed. These indexes are known as Fixed-Queries Tree (FQT) [2], Fixed Height FQT (FHQT) [1,2], Fixed-Queries Array (FQA) [7], and Fixed Queries Trie (FQTrie) [4]. All of them partition $\mathbb{U}$ in classes according to a distance, or range of distances, to a pivot p. In particular, the FHQT is built as follows. Firstly, a set of j pivots is chosen, the ranges of distances are stablished, and we associate a label l to each range. We start with pivot p_1 and for every range of distance l we pick the objects whose distance to p_1 belongs to the range labeled as l. For every non-empty subset, a tree branch with label l is generated. Next, we repeat the process recursively using the next pivot. Each pivot is used for all subtrees in the same level; therefore, pivot p_2 is used for all subtrees in the second level, p_3 for the third level and so on. The height of the tree is j.

For a given query q and radius r, $d(q, p_1)$ is computed and all branches whose range of distance do not intersect with $[d(q, p_1) - r, d(q, p_1) + r]$ are discarded. The process is repeated recursively for those branches not yet discarded using the next pivot. Hence, a set of candidates elements is obtained. Finally, all the candidates are directly compared with q to answer the similarity query.

3.2   Permutation-Based Algorithm (PBA)

A permutation-based algorithm can be described as follows: Let $\mathbb{P} \subset \mathbb{U}$ be a set of permutants with m members. Each element $u \in \mathbb{U}$ induces a preorder $\leq_u$ given by the distance from u towards each permutant, defined as $y \leq_u z \Leftrightarrow d(u, y) \leq d(u, z)$, for any pair $y, z \in \mathbb{P}$.

Let $\Pi_u = i_1, i_2, \ldots, i_m$ be the permutation of u, where permutant $p_{i_j} \leq_u p_{i_{j+1}}$. Permutants at the same distance take an arbitrary but consistent order. For every object in $\mathbb{U}$, its preorder of $\mathbb{P}$ is computed and associated to a permutation. The resulting set of permutations conforms the index, since a PBA does not store any distance.

Given the query $q \in \mathbb{X}$, the PBA search algorithm computes its permutation Π_q and compares it with all the permutations stored in the index. Then, the dataset $\mathbb{U}$ is traversed in increasing dissimilarity of permutations, comparing directly the objects in $\mathbb{U}$ with the query using the distance d of the particular metric space. There are many similarity measures between permutations. One of them is the L_s family of distances, that obeys Eq. (1), where $\Pi^{-1}(i_j)$ denotes the position of permutant p_{i_j} within permutation Π.

$$L_s(\Pi_u, \Pi_q) = \sum_{1 \leq j \leq |\mathbb{P}|} |\Pi_u^{-1}(i_j) - \Pi_q^{-1}(i_j)|^s \tag{1}$$

There are some special values for s. If $s = 2$ the distance is known as *Spearman Rho* (S_ρ); and for $s = 1$ it, is called *Spearman Footrule* (S_f).

4   Fixed Height Queries Tree Permutation (FHQTP)

We propose a new efficient metric index as result of a clever combination of the best features of the two aforementioned indexes. This way, at search time we can take advantage of both the search pruning of the FHQT and the prediction capability of the PBA. We use the same pivots of the FHQT as the permutants of our PBA, so that we produce the PBA index with no extra distance computation.

4.1   Index Construction

The first stage of the FHQTP consists in building the classic FHQT, maintaining all the distances computed during the process. As we have computed the real distances between every object $u \in \mathbb{U}$ and each pivot $p \in \mathbb{P}$, we use them to compute the permutation for each u. The FHQTP stores all these permutations in the index, and after computing them, it discards the distances. Finally, we have a complete FHQT and each element has its permutation. The construction cost of FHQTP is the same as that of the classic FHQT.

If we have memory restrictions, we can save some space by not storing the central part of the permutations, because these permutants have a lesser contribution to the PBA prediction power. This was previously shown in [9].

4.2 Searching

To solve a range query (q, r), we divide the process in two steps. In the first one, we use the FHQTP as a standard FHQT in order to discard non-relevant objects. However, once a leaf is reached, instead of computing the distance d between all the objects stored in the leaf with the query, we put them in a candidate list (since they were not discarded by the FHQT). This process continues adding to the candidate list every non discarded object.

In the second step, we compute Π_q and we sort the candidate list in increasing order of permutation distance (e.g., Spearman Rho or Footrule), in a similar way as the standard PBA, but only considering the resulting list of candidates. Thanks to PBAs prediction property, it suffices to review just a small percentage of the top of the list.

4.3 Example

Figure 1 depicts an example of our technique. This database consists of the small circles and we use the Euclidean distance. Pivots/permutants are the black filled circles. The circle in bold line centered at q represents a range query. Each circle in dash line, centered around pivots/permutants, represents a branch; and the elements inside it are part of the same branch. Note that for all pivots/permutants, we consider that every element beyond the third branch belongs to branch four. For example, object u_1 is placed in the first root branch as it is located at the first concentric circle from p_1 (notice that u_1 is the only one), and u_1 is in the third ring of both p_2 and p_3. The remaining elements are subject to the same process. Finally, the resulting FHQTP is composed by the FHQT shown in Fig. 1(b) and the corresponding permutations in Fig. 1(c).

In Fig. 1(a), the query q is located inside the rings 2, 4, and 2 of pivots/permutants p_1, p_2, and p_3, respectively; its permutation is $\Pi_q = 1\ 3\ 2$. Thin lines in Fig. 1(b) show the excluded branches. This means that elements u_1, u_5, u_6, u_7, u_8, and u_{11} are not discarded when solving the range query (q, r). However, according to their permutations (Fig. 1(c)), the permutations distances are $S_f(q, u_1) = 2, S_f(q, u_5) = 0, S_f(q, u_6) = 0, S_f(q, u_7) = 2, S_f(q, u_8) = 2$, and $S_f(q, u_{11}) = 4$. Therefore, the possible relevant objects can be reviewed in the following order: u_5, u_6, u_1, u_7, u_8, and u_{11}. Note that if we review a fraction (as authors describe in [6]) from the top of this list we have a good chance of retrieving the complete answer, and the chances increase as long as we review more non discarded elements.

4.4 Partial vs Full Permutation

Since we are processing the candidate list with PBA, we propose just to evaluate (and to keep) just a small part of permutations. According to [9], the most important part of permutations are the first and the last permutants. For example, if we divide each permutation in 3 parts and dismiss the central part, we use just 2/3 part of the permutations and we can get a good order to review

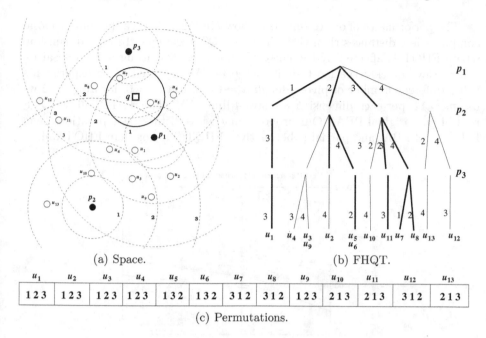

(a) Space.   (b) FHQT.

u_1	u_2	u_3	u_4	u_5	u_6	u_7	u_8	u_9	u_{10}	u_{11}	u_{12}	u_{13}
1 2 3	1 2 3	1 2 3	1 2 3	1 3 2	1 3 2	3 1 2	3 1 2	1 2 3	2 1 3	2 1 3	3 1 2	2 1 3

(c) Permutations.

Fig. 1. Sketch of our proposal.

the candidate list. In our example, in Fig. 1(c), we have $u_1 = 1\ 3, u_2 = 1\ 3, u_3 = 1\ 3, u_4 = 1\ 3, u_5 = 1\ 2$, and so on.

5   Experiment Results

We tested our proposal using two kinds of metric databases. The first one is composed by several synthetic datasets that consists of uniformly distributed vectors in the unitary hypercube of dimensions 6 to 16 using the Euclidean distance. The second one consists of three real world datasets, namely, a set of English words using the Levenshtein's distance, a set of images from NASA, and a set of images from CoPhIR (Content-based Photo Image Retrieval). For the two image datasets we also use the Euclidean distance.

Since kNN queries are more appealing than range queries in practical applications, during the experimental evaluation, we simulate kNN queries with range queries using a radius that retrieves exactly the number of neighbors we need.

5.1   Synthetic Datasets

Each synthetic datasets is composed by 100,000 vectors uniformly distributed. These datasets allow us to control the intrinsic dimensionality of the space and analyze how the *curse of dimensionality* affects our proposal. We tested $kNN(q)$ queries with a query set of size 100 in dimensions that vary from 6 to 16.

The performance of our technique is shown in Fig. 2. Notice that our proposal computes less distances than the original FHQT technique. In high dimension, where FHQT is affected by the curse of dimensionality, using our proposal we can retrieve quickly the most similar objects. We repeated these experiments using different number of pivots, for $dim = 6$, we use $m = 6$, for $dim = 12$ we use $m = 12$, per each dimension we have 4 lines (FHQT, FHQTP full, FHQTP partial and original PBA). Our proposal made less distance computations than FHQT and PBA, and partial is better than FHQT but not than FHQT full.

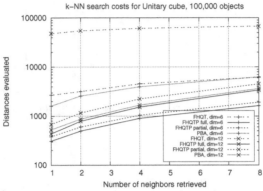

(a) Performance of FHQTP over k in the unitary cube.

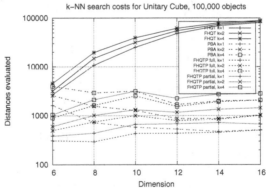

(b) Performance of FHQTP over dimension in the unitary cube.

Fig. 2. Comparison of search performance in the unitary cube synthetic spaces, 100,000 objects.

5.2   Real World Datasets

We use three real datasets. The first one is an English dictionary consisting of 69,069 words using the Levenshtein's distance, also known as edit distance. This

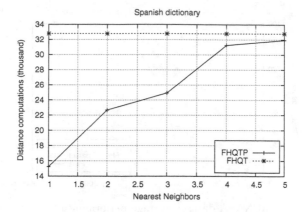

Fig. 3. Performance of FQTP using an English dictionary.

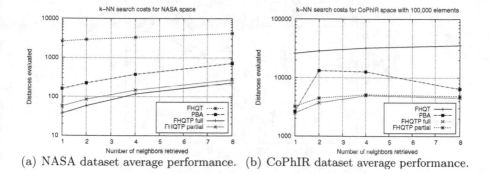

(a) NASA dataset average performance. (b) CoPhIR dataset average performance.

Fig. 4. Performance of FHQTP on real database.

distance is equivalent to the minimum number of single character edit operations (character insertion, deletion or substitution) needed to convert one word into the other.

The second dataset contains 40,150 20-dimensional feature vectors, generated from images downloaded from NASA[1], where duplicated vectors were eliminated. Any quadratic form can be used as a distance, so we chose the Euclidean distance as the simplest, meaningful alternative.

Finally, the third dataset is a subset of 100,000 images from CoPhIR [3]. For each image, the standard MPEG-7 image feature have been extracted. For the aforementioned reason, in this space we also use the Euclidean distance.

In Fig. 3, we show the performance of our technique for the English dictionary. Notice that our technique has an excellent performance for 1, 2, and 3 *NN*. Figure 4 shows how FHQT is not competitive in real databases. In the left side, Fig. 4(a) is for NASA database and, in the right side, Fig. 4(b) is for the CoPhIR

---

[1] At http://www.dimacs.rutgers.edu/Challenges/Sixth/software.html.

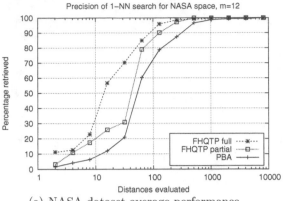

(a) NASA dataset average performance.

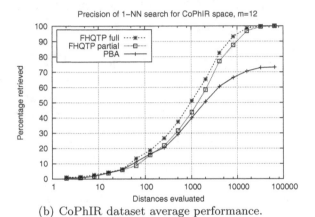

(b) CoPhIR dataset average performance.

Fig. 5. Performance of FHQTP on real databases, $1NN$ queries.

dataset. Both figures are showing the performance of our proposal for different values for k in kNN searches. We can get the answers up to 10x times faster. In this cases we use $m = 12$.

Finally, as shown in Fig. 5, we get the nearest neighbors faster than PBA when we review an incremental fraction, both for NASA database (Fig. 5(a)) and the CoPhIR dataset (Fig. 5(b)). Notice that we have a better performance than PBA in both databases.

6   Conclusions and Future Work

In this paper a new index is proposed. It consists in merging the best features of two known algorithms for metric spaces FHQT (Fixed Height Query Tree) and the Permutation based algorithm (PBA). Following the FHQT, when a query decides which branches are used and arrives at a leaf, all the objects in this leaf must be compared with the query. Instead, we can reuse the distance comparison performed, and to make the permutation per each object. At quering time, we use the FHQT to make a candidate list, then using the PBA technique, we review in a new order this list. This combination is a winner approach, it has an excellent performance because it works well in all dimensions.

For future work, we are interested in designing an algorithm for solving the k nearest neighbor query using our proposal, and extend this technique for other algorithms like FQA (Fixed Query Array).

References

1. Baeza-Yates, R.: Searching: an algorithmic tour. In: Kent, A., Williams, J. (eds.) Encyclopedia of Computer Science and Technology, vol. 37, pp. 331–359. Marcel Dekker Inc., New York (1997)
2. Baeza-Yates, R., Cunto, W., Manber, U., Wu, S.: Proximity matching using fixed-queries trees. In: Crochemore, M., Gusfield, D. (eds.) CPM 1994. LNCS, vol. 807, pp. 198–212. Springer, Heidelberg (1994). doi:10.1007/3-540-58094-8_18
3. Bolettieri, P., Esuli, A., Falchi, F., Lucchese, C., Perego, R., Piccioli, T., Rabitti, F.: CoPhIR: a test collection for content-based image retrieval. CoRR abs/0905.4627v2 (2009). http://cophir.isti.cnr.it
4. Chávez, E., Figueroa, K.: Faster proximity searching in metric data. In: Monroy, R., Arroyo-Figueroa, G., Sucar, L.E., Sossa, H. (eds.) MICAI 2004. LNCS (LNAI), vol. 2972, pp. 222–231. Springer, Heidelberg (2004). doi:10.1007/978-3-540-24694-7_23
5. Chávez, E., Figueroa, K., Navarro, G.: Proximity searching in high dimensional spaces with a proximity preserving order. In: Gelbukh, A., Albornoz, Á., Terashima-Marín, H. (eds.) MICAI 2005. LNCS (LNAI), vol. 3789, pp. 405–414. Springer, Heidelberg (2005). doi:10.1007/11579427_41
6. Chávez, E., Figueroa, K., Navarro, G.: Effective proximity retrieval by ordering permutations. IEEE Trans. Pattern Anal. Mach. Intell. (TPAMI) **30**(9), 1647–1658 (2009)
7. Chávez, E., Marroquín, J., Navarro, G.: Fixed queries array: a fast and economical data structure for proximity searching. Multimed. Tools Appl. (MTAP) **14**(2), 113–135 (2001)
8. Chávez, E., Navarro, G., Baeza-Yates, R., Marroquín, J.: Proximity searching in metric spaces. ACM Comput. Surv. **33**(3), 273–321 (2001)
9. Figueroa, K., Paredes, R.: An effective permutant selection heuristic for proximity searching in metric spaces. In: Martínez-Trinidad, J.F., Carrasco-Ochoa, J.A., Olvera-Lopez, J.A., Salas-Rodríguez, J., Suen, C.Y. (eds.) MCPR 2014. LNCS, vol. 8495, pp. 102–111. Springer, Cham (2014). doi:10.1007/978-3-319-07491-7_11
10. Samet, H.: Foundations of Multidimensional and Metric Data Structures. The Morgan Kaufmann Series in Computer Graphics and Geometric Modeling. Morgan Kaufmann Publishers Inc., San Francisco (2005)
11. Zezula, P., Amato, G., Dohnal, V., Batko, M.: Similarity Search: The Metric Space Approach. Advances in Database Systems, vol. 32. Springer, Heidelberg (2006)

A Projection Method for Optimization Problems on the Stiefel Manifold

Oscar Dalmau-Cedeño and Harry Oviedo$^{(\boxtimes)}$

Mathematics Research Center, CIMAT A.C., Guanajuato, Mexico
{dalmau,harry.oviedo}@cimat.mx

Abstract. In this paper we propose a feasible method based on projections using a curvilinear search for solving optimization problems with orthogonality constraints. Our algorithm computes the SVD decomposition in each iteration in order to preserve feasibility. Additionally, we present some convergence results. Finally, we perform numerical experiments with simulated problems; and analyze the performance of the proposed methods compared with state-of-the-art algorithms.

Keywords: Constrained optimization · Orthogonality constraints · Non-monotone algorithm · Stiefel manifold · Optimization on manifolds

1 Introduction

In this paper we consider the following optimization problem with orthogonality constraints:

$$\min_{X \in \mathbb{R}^{n \times p}} \mathcal{F}(X) \quad \text{s.t.} \quad X^\top X = I_p, \tag{1}$$

where $\mathcal{F} : \mathbb{R}^{n \times p} \to \mathbb{R}$ is a differentiable function and $I_p \in \mathbb{R}^{p \times p}$ represents the identity matrix. The feasible set $Stf(n,p) := \{X \in \mathbb{R}^{n \times p} | X^\top X = I\}$ is known as the "Stiefel Manifold". This manifold is simplified to the unit sphere when $p = 1$ and in the case $p = n$ is called "Orthogonal group". The Stiefel manifold can be seen as an embedded sub-manifold of $\mathbb{R}^{n \times p}$ with dimension equals to $np - \frac{1}{2}p(p + 1)$, see [1].

Problem (1) admits many applications such as, linear eigenvalue problem [14], sparse principal component analysis [4], Kohn-Sham total energy minimization [16], orthogonal procrustes problem [5], weighted orthogonal procrustes problem [6], nearest low-rank correlation matrix problem [7,12], joint diagonalization (blind source separation) [8], among others. In addition, some problems such as PCA, LDA, multidimensional scaling, orthogonal neighborhood preserving projection can be formulated as problem (1) [9].

On the other hand, the Stiefel manifold is a compact set, which ensures that (1) has a global optimum at least. However, this manifold is not a convex set, which transforms (1) in a hard optimization problem. For example, the *quadratic assignment problem* (QAP) and the *leakage interference minimization* are NP-hard [10].

© Springer International Publishing AG 2017
J.A. Carrasco-Ochoa et al. (Eds.): MCPR 2017, LNCS 10267, pp. 84–93, 2017.
DOI: 10.1007/978-3-319-59226-8_9

In this paper we propose a new method based on projections onto the Stiefel manifold. In particular, we study two algorithms to solve problem (1). At each iteration of the algorithms, we project the corresponding update onto the Stiefel manifold using the singular value decomposition (SVD) which guarantees to obtain a feasible sequence. Although, the SVD decomposition is computationally expensive, this is less expensive than building a geodesic. In the literature, we can find other feasible methods that solve problem (1), for example, the ones based on retractions methods use projections that involve QR factorization, polar decomposition, Gram-Schmidt process or SVD decomposition [1].

This paper is organized as follows. In Subsect. 2.1 we present some standard notation and in Subsect. 2.2 we give the optimality conditions of the problem (1), Subsect. 2.3 describes the proposed update scheme, where we present a linear search monotone algorithm and a globally convergent non-monotone algorithm for solving problem (1), Subsect. 2.4 shows different strategies to choose the step size according to Armijo-Wolfe like condition, and a non-monotone search using the Barzilai Borwein step size. Some theoretical results are presented in Sect. 3. Section 4 is dedicated to numerical experiments in order to demonstrate the efficiency and robustness of the proposed algorithms.

2 Algorithms

In the first two subsections, we introduce some standard notation and the optimality conditions of problem (1) respectively. Next subsections are devoted to introduce our proposed method.

2.1 Notation

We say that a matrix $W \in \mathbb{R}^{n \times n}$ is skew-symmetric if $W = -W^\top$. The trace of X is defined as the sum its diagonal elements, and we will denote by $Tr[X]$. The Euclidean inner product of two matrices $A, B \in \mathbb{R}^{m \times n}$ is defined as $\langle A, B \rangle := \sum_{i,j} A_{i,j} B_{i,j} = Tr[A^\top B]$. The Frobenius norm is defined using the previous inner product, i.e., $\|A\|_F = \sqrt{\langle A, A \rangle}$. Let $\mathcal{F} : \mathbb{R}^{n \times p} \to \mathbb{R}$ be a differentiable function, then the derivative of $\mathcal{F}$ with respect to X is denoted as $G := \mathcal{D}\mathcal{F}(X) := (\frac{\partial \mathcal{F}(X)}{\partial X_{ij}})$ and the derivative of the function $\mathcal{F}$ in X in the direction Z is defined as:

$$\mathcal{D}\mathcal{F}(X)[Z] := \frac{\partial \mathcal{F}(X+tZ)}{\partial \tau}\bigg|_{t=0} = \lim_{t \to 0} \frac{\mathcal{F}(X+tZ) - \mathcal{F}(X)}{t} = \langle \mathcal{D}\mathcal{F}(X), Z \rangle. \ (2)$$

2.2 Optimality Conditions

The Lagrangian function associated to the optimization problem (1) is given by:

$$\mathcal{L}(X, \Lambda) = \mathcal{F}(X) - \frac{1}{2} Tr[\Lambda(X^\top X - I_p)], \tag{3}$$

where I_p is the identity matrix and Λ is the Lagrange multipliers matrix, which is symmetric due to the matrix $X^\top X$ is also symmetric. The Lagrangian function leads to the first order optimality conditions for problem (1):

$$G - X\Lambda = 0 \tag{4a}$$
$$X^\top X - I_p = 0. \tag{4b}$$

Lemma 1 *(cf. Wen and Yin [15]). Suppose that X is a local minimizer of problem (1). Then X satisfies the first order optimality conditions (4a) and (4b) with the associated Lagrangian multiplier $\Lambda = G^\top X$. Defining $\nabla\mathcal{F}(X) := G - XG^\top X$ and $A := GX^\top - XG^\top$. Then $\nabla\mathcal{F}(X) = AX$. Moreover, $\nabla\mathcal{F} = 0$ if and only if $A = 0$.*

Proof. See [15].

The Lemma 1 establishes an equivalence to the (4a) and (4b) conditions, i.e., if $X \in Stf(n, p)$ satisfies that $\nabla\mathcal{F}(X) = 0$ then X also satisfies (4a) and (4b), so we can use this result as a stopping criterion for our algorithms.

2.3   Update Schemes

In this subsection we present a linear combination based algorithm. As the new iterated of our proposals does not necessarily belong to the Stiefel Manifold, we use a projection operator, in order to force the feasibility of the new iterated. Specifically, we use the classical projection operator which is defined as $\pi(X) := \arg\min_{Q \in Stf(n,p)} \|X - Q\|_F^2$, it is known that the solution of this problem is given by $\pi(X) = U I_{n,p} V^\top$ where $X = U\Sigma V^\top$ is the SVD decomposition of X, for details of the demonstration of this result see [11].

In our updating formula, we use the previous result for obtaining a new point that satisfies the constraints of the problem (1). For example, if $Y_k(\tau)$ is obtained from our proposal, i.e., the linear combination scheme, then the new test point is:

$$X_{k+1} := Z_k(\tau) := \pi(Y_k(\tau)). \tag{5}$$

In the next subsections we explain in more detail our updating formula $Y_k(\tau)$.

A Scheme Based on a Linear Combination. Our proposal uses the following update formula:

$$Y_k^{CL}(\tau) := X_k - \tau\left(\lambda B_k L + \mu C_k R\right), \tag{6}$$

where $G_k = \mathcal{D}\mathcal{F}(X_k)$, $B_k = G_k L^\top - LG_k^\top$, $C_k = G_k R^\top - RG_k^\top$, $L, R \in \mathbb{R}^{n \times p}$, τ is the step size and (λ, μ) are any two scalars satisfying:

$$\lambda\|B_k\|_F^2 + \mu\|C_k\|_F^2 > 0.$$

The following lemma shows that the curve $Y_k^{CL}(\cdot)$ defined by Eq. (6) is a descent curve at $\tau = 0$.

Lemma 2. *Let $Y_k^{CL}(\tau)$ be defined by Eq. (6), then $Y_k^{CL}(\tau)$ is a descent curve at $\tau = 0$, i.e.,*

$$\mathcal{D}\mathcal{F}(X_k)[\dot{Y}_k^{CL}(0)] = -\frac{\lambda}{2}||B_k||_F^2 - \frac{\mu}{2}||C_k||_F^2 < 0. \tag{7}$$

Proof. The proof is straightforward, and it can be obtained by using trace properties and using Eq. (2).

Remark 1. Note that in the updating formula (6), we can select any matrix L or R, in particular one can use matrices L, R with random entries. The parameters (λ, μ) can appropriately selected, for example, we can choose both positive. This ensures that the method will descent and may eventually converge to a local minimum. In our implementation, we select $L = X_k$, $R = X_{k-1}$ and $(\lambda, \mu) = (2/3, 1/3)$.

2.4 Strategies to Select the Step Size

From now on, $Y_k(\tau)$ represents our proposal, i.e., the based on the linear combination method.

A Descent Condition. In our method, we will choose the biggest step size τ that satisfies the following condition:

$$\mathcal{F}(Z_k(\tau)) \leq \mathcal{F}(X_k) + \sigma\tau Tr[G_k^\top \dot{Y}_k(0)], \tag{8}$$

with $0 < \sigma < 1$.

Note that Eq. (8) is not exactly the classic *"Armijo condition"*, since we use $\dot{Y}_k(0)$ instead of $\dot{Z}_k(0)$. However, if we only use the condition (8) for computing the step size, it ensures the descent of the objective function as long as the directional derivative $Tr[\mathcal{D}\mathcal{F}(X_k)^\top \dot{Y}_k(0)]$ is negative. In this work, we also study the behavior of our algorithms calculating the step size as satisfying (8).

Nonmonotone Search with Barzilai Borwein Step Size. It is known that the *Barzilai-Borwein* (BB) step size, see [2], can sometimes improve the performance of linear search algorithms such as the steepest descent method without adding too much computational cost. This technique considers the classic *steepest descent method* and proposes to use any of the following step sizes:

$$\alpha_k^{BB1} = \frac{||S_k||_F^2}{Tr[S_k^\top R_k]} \quad \text{and} \quad \alpha_k^{BB2} = \frac{Tr[S_k^\top R_k]}{||R_k||_F^2}. \tag{9}$$

where $S_k = X_{k+1} - X_k$, $R_k = \mathcal{D}\mathcal{F}(X_{k+1}) - \mathcal{D}\mathcal{F}(X_k)$ and the matrix $B(\alpha) = (\alpha I)^{-1}$, is considered an approximation of the Hessian of the objective function. For more details see [2, 13].

Since the quantities α_k^{BB1}, α_k^{BB2} could be negatives, the absolute value of these step sizes is usually considered. On the other hand, the BB-steps do not necessarily guarantee the descent of the objective function at each iteration, this may imply that the method does not converge. In order to solve this problematic, we use a technique that guarantees global convergence, see Refs. [3, 13] for details. In particular, we use a non-monotone line search algorithm, see [17], combined with the BB-step in order to select the step size, see Algorithm 1.

Algorithm 1. Non-monotone linear search algorithm for solve optimization problems on Stiefel manifold

---

Require: $X_0 \in Stf(n, p)$, $\tau > 0$, $0 < \tau_m \ll \tau_M$, $\sigma, \epsilon, \eta, \delta \in (0, 1)$, $X_{-1} = X_0$, $C_0 = \mathcal{F}(X_0)$, $Q_0 = 1$, $k = 0$.
Ensure: X^* a local minimizer.
1: **while** $\|\nabla \mathcal{F}(X_k)\|_F > \epsilon$ **do**
2:    **while** $\mathcal{F}(Z_k(\tau)) \geq C_k + \sigma \tau D\mathcal{F}(X_k)[\dot{Y}_k(0)]$ **do**
3:        $\tau = \delta \tau$,
4:    **end while**
5:    $X_{k+1} = Z_k(\tau) := \pi(Y_k(\tau))$, with $Y_k(\tau)$ using (6).
6:    Calculate $Q_{k+1} = \eta Q_k + 1$ and $C_{k+1} = (\eta Q_k C_k + \mathcal{F}(X_{k+1}))/Q_{k+1}$.
7:    Choose $\tau = |\alpha_k^{BB1}|$ or well $\tau = |\alpha_k^{BB2}|$, where α_k^{BB1} and α_k^{BB2} are defined as in (9).
8:    Set, $\tau = \max(\min(\tau, \tau_M), \tau_m)$.
9:    $k = k + 1$.
10: **end while**
11: $X^* = X_k$.

Note that when $\eta = 0$, Algorithm 1 is reduced to a monotonous algorithm which generates points satisfying the descent condition (8).

3   Theoretical Results

In this section we prove some convergence results of our Algorithm 1 when it's use with $\eta = 0$.

Lemma 3. *Let $\{X_k\}$ be an infinite sequence generated by Algorithm 1. Then $\{\mathcal{F}(X_k)\}$ is a convergent sequence. Moreover any accumulation point X_* of $\{X_k\}$ is feasible, i.e., $X_*^\top X_* = I$.*

Proof. By construction of the Algorithm 1 we have,

$$\mathcal{F}(X_{k+1}) \leq \mathcal{F}(X_k) + \sigma \tau_k Tr[G_k^\top \dot{Y}_k(0)], \qquad \forall k \qquad (10)$$

or equivalently,

$$\mathcal{F}(X_k) - \mathcal{F}(X_{k+1}) \geq -\sigma \tau_k Tr[G_k^\top \dot{Y}_k(0)], \qquad \forall k$$
$$> 0 \qquad \text{(due } Y_k(\tau) \text{ is a descent curve at } \tau = 0 \text{)},$$

so, $\{\mathcal{F}(X_k)\}$ is a monotonically decreasing sequence. Now, since Stiefel manifold is a compact set and $\mathcal{F}$ is a continuous function, we obtain that $\mathcal{F}$ has maximum and minimum on $Stf(n,p)$. Therefore, $\{\mathcal{F}(X_k)\}$ is bounded, and then $\{\mathcal{F}(X_k)\}$ is a convergent sequence.

On the other hand, let $\{X_k\}_{k\in\mathcal{K}}$ be a convergent subsequence of $\{X_k\}$ and suppose that this subsequence converges to X_*, that is $\lim_{k\in\mathcal{K}} X_k = X_*$, since X_k is a feasible point for all $k \in \mathcal{K}$ and $Stf(n,p)$ is a compact set, then we have $X_* \in Stf(n,p)$, i.e.,

$$X_*^\top X_* = I,$$

therefore every accumulation point is feasible.

Theorem 1. *Let $\{X_k\}$ be an infinite sequence generated by Algorithm 1. Then any accumulation point X_* of $\{X_k\}$ satisfies the the first order optimality conditions.*

The proof of Theorem 1 is obtained by following the ideas of the demonstration of Theorem 4.3.1 in [1] except for slight adaptations.

4 Numerical Experiments

In this section we analyze the performance of our method by solving several simulated experiments with the format of the problem (1), for different objective functions and different sizes of problems. We also make comparisons between some state of the art methods and our proposal, in order to measure the performance and efficiency of our algorithms.

4.1 Implementation Details

All our experiments were performed using Matlab R2013a on an Intel processor i3-380M, 2.53 GHz CPU with 500 Gb HD and 8 Gb of Ram. For the different parameters of our two algorithms, we use the following values: initial step size $\tau = 1e{-}2$, $\sigma = 1e{-}4$, $\eta = 0.85$, $\delta = 0.1$. Moreover, as the convergence of the first-order methods (methods using the first derivative of the objective function) can be very slow we will use several stop criteria:

$$\|\nabla\mathcal{F}(X_k)\|_F < \epsilon, \quad \text{and} \quad (tol_k^x < xtol \wedge tol_k^f < ftol), \tag{11}$$

and a maximum of K iterations, where

$$tol_k^x := \frac{\|X_{k+1} - X_k\|_F}{\sqrt{n}}, \quad \text{and} \quad tol_k^f := \frac{\mathcal{F}(X_k) - \mathcal{F}(X_{k+1})}{|\mathcal{F}(X_k)| + 1}.$$

In the experiments, we used the following default values: $xtol = 1e{-}6$, $ftol = 1e{-}12$, $T = 5$ and $\epsilon = 1e{-}4$.

In all experiments presented in the following subsections we use the following notation:

- *Nfe*: The number of evaluations of the objective function.
- *Nitr*: The number of iterations performed by the algorithm to convergence.
- *Time*: The time (in seconds) used by the algorithm to converge.
- *NrmG*: The gradient norm of the Lagrangian function with respect to primal variables evaluated at the estimated "optimal".
- *Fval*: Evaluation of the objective function at the estimated "optimal".
- *Feasi*: Corresponds to the following error $||\hat{X}^\top \hat{X} - I_p||_F$, where $\hat{X}$ denotes the "optimal" estimated by the algorithm.

In addition, we denote by the Steepest Descent *Steep-Dest*, the Trust-Region method *Trust-Reg* and the Conjugate Gradient method *Conj-Grad* from "*manopt*" toolbox[1], and *PGST* the algorithm presented in [6]. On the other hand, *Linear-Co* denote our Algorithm 1.

4.2   Weighted Orthogonal Procrustes Problem (WOPP)

Let $X \in \mathbb{R}^{m \times n}$, $A \in \mathbb{R}^{p \times m}$, $B \in \mathbb{R}^{p \times q}$ and $C \in \mathbb{R}^{n \times q}$. The *Weighted Orthogonal Procrustes Problem* (WOPP) consists in solving the following constrained optimization problem:

$$\min_{X \in \mathbb{R}^{m \times n}} \tfrac{1}{2}||AXC - B||_F^2 \tag{12}$$
$$\text{s. t.} \quad X^\top X = I_n.$$

When C is the identity matrix with appropriate dimensions, this problem is known as *Unbalanced Orthogonal Procrustes Problem* (UOPP), for more details see [1].

Experiments with WOPP Problems. The problems in this subsection were taken from [18]. In particular, we considered $n = q$, $p = m$, $A = PSR^\top$ and $C = Q\Lambda Q^\top$, where P, Q and R are orthogonal matrices generated randomly with $Q \in \mathbb{R}^{n \times n}$, $R, P \in \mathbb{R}^{m \times m}$, $\Lambda \in \mathbb{R}^{n \times n}$ is a diagonal matrix with entries generated from a uniform distribution in the range $[\tfrac{1}{2}, 2]$ and S is a diagonal matrix defined for each type of problem, see below for details. As a starting point $X_0 \in \mathbb{R}^{m \times n}$, we generated random matrices on the Stiefel manifold. When not specified, the entries of the matrix were generated using a *standard Gaussian* distribution.

For comparison purposes, we created problems with a known solution $Q_* \in \mathbb{R}^{m \times n}$ randomly selected on the Stiefel manifold. Then, we built the matrix B as $B = AQ_*C$. Finally, for the different tested problems the diagonal matrix S is described below.

Problem 1: The diagonal elements of S were generated by a normal distribution in the interval [10,12].

---

[1] The tool-box manopt is available in http://www.manopt.org/.

Problem 2: The diagonal of S is given by $S_{ii} = i + 2r_i$, where r_i was a random number uniformly distributed in the interval $[0, 1]$.

For each experiment, a total of 300 WOOP's problems were built with the matrix S generated according to problems **Problem 1** and **Problem 2** respectively. The maximum number of iterations, for all methods, was $K = 8000$.

The results of the previous experiments are presented in Tables 1 and 2. We denote by *Error* to the standard error with respect to the global solution Q_*, i.e., $\|\hat{X} - Q_*\|_F$ where $\hat{X}$ is the optimum estimated by the algorithms. Furthermore, *min, mean, max* denote the minimum, maximum and average obtained by each algorithm in the 300 runs.

According to Table 1 for well-conditioned problems, i.e., **Problem 1**, all the algorithms present similar results. Note that **PGST** obtained a lower number of iterations. In general, all the methods presented a similar performance for this type of problems. On the other hand, for ill-conditioned problems, i.e., **Problem 2**, we observe that all the method arrived to the solution Q_*, according to **NrmG**, **Fval** and **Error** measures. Moreover, our **Linear-Co** procedure obtained similar results compared with the **PGST** algorithm when $n < m$, and when $m = n$ **Linear-Co** method achieved better results that the **PGST**, see Table 2.

Table 1. Performance of the methods for well conditioned WOPP problems (**Problem 1**)

Method		Nitr	Nfe	Time	NrmG	Fval	Error
		\multicolumn{6}{l}{**Problem 1** with m = 500 and n = 70}
Linear-Co	Min	48	49	2.60	1.33e−05	7.13e−13	1.44e−07
	Mean	59.7	60.7	3.72	6.10e−05	3.63e−11	1.27e−06
	Max	71	72	5.06	9.95e−05	1.34e−10	2.92e−06
PGST	Min	36	35	1.87	9.63e−06	7.96e−13	1.75e−07
	Mean	41.6	40.0	2.37	7.85e−05	3.68e−11	1.12e−06
	Max	49	42	3.22	2.23e−04	1.35e−10	2.98e−06
Method		Nitr	Nfe	Time	NrmG	Fval	Error
		\multicolumn{6}{l}{**Problem 1** with m = 200 and n = 200}
Linear-Co	Min	46	47	1.77	1.35e−05	9.35e−14	4.51e−08
	Mean	53.0	54.1	2.64	6.16e−05	1.08e−11	6.40e−07
	Max	63	65	3.81	9.97e−05	3.94e−11	1.58e−06
PGST	Min	33	36	1.86	1.64e−04	2.25e−11	5.48e−07
	Mean	38.2	42.0	2.75	6.56e−04	9.62e−10	5.95e−06
	Max	43	45	3.73	9.99e−04	3.83e−09	1.55e−05

Table 2. Performance of the methods for ill-conditioned WOPP problems (**Problem 2**)

Method		Problem 2 with m = 300 and n = 20					
		Nitr	Nfe	Time	NrmG	Fval	Error
Linear-Co	Min	2078	2133	16.36	5.20e−04	4.17e−09	2.38e−05
	Mean	4732.2	4861.3	40.25	1.01e−02	9.57e−02	8.04e−02
	Max	8000	8229	72.37	3.40e−01	9.91e−01	4.89e−01
PGST	Min	3118	2080	18.15	6.52e−05	1.59e−13	1.46e−07
	Mean	6373.1	4142.3	37.38	4.67e−01	8.66e−02	8.14e−02
	Max	8000	8478	53.75	2.62e+01	1.22	4.96e−01
Method		Problem 2 with m = 150 and n = 150					
		Nitr	Nfe	Time	NrmG	Fval	Error
Linear-Co	Min	576	775	13.48	1.20e−04	6.74e−11	2.66e−06
	Mean	1164.1	1210.6	20.80	1.30e−03	1.06e−08	3.71e−05
	Max	1881	1945	33.56	1.29e−02	2.56e−07	2.53e−04
PGST	Min	1125	962	27.16	1.67e−04	3.66e−12	6.59e−08
	Mean	2039.6	1921.1	50.62	8.52e−04	5.50e−09	2.85e−05
	Max	3521	3558	116.36	1.00e−03	1.98e−08	8.50e−05

5   Conclusions

In this paper we proposed a feasible method for solving optimization problems with orthogonality constraints. This method is very general and was based on a linear combination of descent directions and using the same manifold framework. We are currently exploring several variants of this procedure. In order to preserve feasibility, our proposal requires to project onto the Stiefel manifold. In particular, we used the SVD decomposition in each iteration. In this work, we also presented some convergence results. Finally, in numerical experiments, the proposed algorithms obtained a competitive performance compared with some state of the art algorithms.

Acknowledgments. This work was supported in part by CONACYT (Mexico), Grant 258033.

References

1. Absil, P.A., Mahony, R., Sepulchre, R.: Optimization Algorithms on Matrix Manifolds. Princeton University Press, Princeton (2009)
2. Barzilai, J., Borwein, J.M.: Two-point step size gradient methods. IMA J. Numer. Anal. **8**(1), 141–148 (1988)
3. Dai, Y.H., Fletcher, R.: Projected barzilai-borwein methods for large-scale box-constrained quadratic programming. Numerische Mathematik **100**(1), 21–47 (2005)

4. d'Aspremont, A., Ghaoui, L., Jordan, M.I., Lanckriet, G.R.: A direct formulation for sparse PCA using semidefinite programming. SIAM Rev. **49**(3), 434–448 (2007)
5. Eldén, L., Park, H.: A procrustes problem on the stiefel manifold. Numerische Mathematik **82**(4), 599–619 (1999)
6. Francisco, J., Martini, T.: Spectral projected gradient method for the procrustes problem. TEMA (São Carlos) **15**(1), 83–96 (2014)
7. Grubisi, I., Pietersz, R.: Efficient rank reduction of correlation matrices. Linear Algebra Appl. **422**(2), 629–653 (2007)
8. Joho, M., Mathis, H.: Joint diagonalization of correlation matrices by using gradient methods with application to blind signal separation. In: Sensor Array and Multichannel Signal Processing Workshop Proceedings, pp. 273–277. IEEE (2002)
9. Kokiopoulou, E., Chen, J., Saad, Y.: Trace optimization and eigenproblems in dimension reduction methods. Numer. Linear Algebra Appl. **18**(3), 565–602 (2011)
10. Liu, Y.F., Dai, Y.H., Luo, Z.Q.: On the complexity of leakage interference minimization for interference alignment. In: 2011 IEEE 12th International Workshop on Signal Processing Advances in Wireless Communications (SPAWC), pp. 471–475. IEEE (2011)
11. Manton, J.H.: Optimization algorithms exploiting unitary constraints. IEEE Trans. Signal Process. **50**(3), 635–650 (2002)
12. Pietersz, R., Groenen, P.J.: Rank reduction of correlation matrices by majorization. Quant. Fin. **4**(6), 649–662 (2004)
13. Raydan, M.: The Barzilai and Borwein gradient method for the large scale unconstrained minimization problem. SIAM J. Optim. **7**(1), 26–33 (1997)
14. Saad, Y.: Numerical Methods for Large Eigenvalue Problems, vol. 158. SIAM, Manchester (1992)
15. Wen, Z., Yin, W.: A feasible method for optimization with orthogonality constraints. Math. Program. **142**(1–2), 397–434 (2013)
16. Yang, C., Meza, J.C., Lee, B., Wang, L.W.: KSSOLVoa MATLAB toolbox for solving the Kohn-Sham equations. ACM Trans. Math. Softw. (TOMS) **36**(2), 10 (2009)
17. Zhang, H., Hager, W.W.: A nonmonotone line search technique and its application to unconstrained optimization. SIAM J. Optim. **14**(4), 1043–1056 (2004)
18. Zhang, Z., Du, K.: Successive projection method for solving the unbalanced procrustes problem. Sci. China Ser. A **49**(7), 971–986 (2006)

An Alternating Genetic Algorithm for Selecting SVM Model and Training Set

Michal Kawulok[1,2](✉), Jakub Nalepa[1,2](✉), and Wojciech Dudzik[2]

[1] Silesian University of Technology, Gliwice, Poland
{michal.kawulok,jakub.nalepa}@polsl.pl
[2] Future Processing, Gliwice, Poland
wdudzik@future-processing.com

Abstract. Support vector machines (SVMs) have been found highly helpful in solving numerous pattern recognition tasks. Although it is challenging to train SVMs from large data sets, this obstacle may be mitigated by selecting a small, yet representative, subset of the entire training set. Another crucial and deeply-investigated problem consists in selecting the SVM model. There have been a plethora of methods proposed to effectively deal with these two problems treated independently, however to the best of our knowledge, it was not explored how to effectively combine these two processes. It is a noteworthy observation that depending on the subset selected for training, a different SVM model may be optimal, hence performing these two operations simultaneously is potentially beneficial. In this paper, we propose a new method to select both the training set and the SVM model, using a genetic algorithm which alternately optimizes two different populations. We demonstrate that our approach is competitive with sequential optimization of the hyperparameters followed by selecting the training set. We report the results obtained for several benchmark data sets and we visualize the results elaborated for artificial sets of 2D points.

Keywords: Support vector machines · Model selection · Training set selection · Genetic algorithms

1 Introduction

Support vector machines (SVMs) are a supervised classifier that has been successfully deployed to solve a variety of pattern recognition and computer vision tasks. SVM training consists in determining a hyperplane to separate the training data belonging to two classes. Position of this hyperplane is defined with a (usually small) subset of all the vectors from the training set (T)—the selected ones are termed *support vectors* (SVs). Though the decision hyperplane separates the data linearly, the input vectors could be mapped into higher-dimensional spaces, in which they become linearly separable—this mapping is achieved with *kernel functions*. The most frequently used kernel, which we also consider in this paper, is the radial basis function (RBF): $\mathcal{K}(\boldsymbol{u}, \boldsymbol{v}) = \exp\left(-\gamma\|\boldsymbol{u} - \boldsymbol{v}\|^2\right)$, where $\boldsymbol{u}$ and $\boldsymbol{v}$ are the input vectors, and γ is the kernel width.

© Springer International Publishing AG 2017
J.A. Carrasco-Ochoa et al. (Eds.): MCPR 2017, LNCS 10267, pp. 94–104, 2017.
DOI: 10.1007/978-3-319-59226-8_10

Selecting the SVM model ($\mathcal{M}$), namely the kernel function along with its parameters and the slack penalty coefficient (C) stays among the main difficulties to be faced while applying SVMs in practice. Failing to properly tune these hyperparameters leads to poor performance of SVMs and it is hardly possible to estimate the optimal values *a priori*. This problem has been extensively studied and a number of solutions were proposed, including improvements to the standard grid search [19], trial-and-error approaches [6] and a variety of automated methods, which often involve evolutionary computation [3].

Another important problem that is becoming increasingly important nowadays, in the era of big data, is concerned with high $O(t^3)$ time and $O(t^2)$ memory complexity of training SVMs, where t is the cardinality of $\boldsymbol{T}$. Furthermore, the number of obtained SVs (s) depends in practice on t, and the classification time depends linearly on s. Overall, there are two principal problems here: (i) some data sets are too large to train SVM, and (ii) large training sets may result in slow classification, even if the SVM can be trained. Apart from enhancing the SVM training [12], this obstacle can be mitigated by selecting a small subset of the training set ($\boldsymbol{T'}$), which contains the potential SVs capable of determining a proper decision hyperplane. This makes SVM training feasible for very large sets, and usually reduces the resulting s without affecting the classification score. Selecting a "good" training set is not trivial, though, and this problem has attracted considerable research attention, including our earlier successful attempts to exploit genetic [11] and memetic [16] algorithms.

1.1 Contribution

Tuning the SVM model is tricky, when the training set is to be reduced, because of the mutual dependence—the model is usually required to select $\boldsymbol{T'}$, while depending on the subset used for training, a different SVM model may be optimal. In most of the existing approaches, the model is selected either for the entire $\boldsymbol{T}$ (when the goal is to reduce s rather than enable the SVM training at all), or for a randomly selected $\boldsymbol{T'}$, prior to its proper refinement. In this paper, we propose a new approach towards solving these two optimization problems in an alternating manner, using a genetic algorithm (GA). We focus on the RBF kernel, hence the SVM model $\mathcal{M} = (\gamma, C)$ in this case, and we exploit our earlier GASVM algorithm [10] for training set selection. The key aspects of our contribution are as follows: (i) we make it possible to run a single optimization process to select both $\mathcal{M}$ and $\boldsymbol{T'}$, (ii) we alternately evolve two different populations to solve two optimization problems having a common fitness function, (iii) we establish a new scheme that can easily embrace other techniques for selecting $\boldsymbol{T'}$ and may be enhanced beyond relying on the RBF kernel.

1.2 Paper Structure

In Sect. 2, we outline the state-of-the-art on selecting $\mathcal{M}$ and training SVMs from large sets. Our approach is described in Sect. 3 and the experimental results are reported in Sect. 4. The paper is concluded in Sect. 5.

2   Related Literature

To the best of our knowledge, there are no methods for selecting $\mathcal{M}$ and T' in a simultaneous manner, however there are many approaches to solve these problems independently from each other. They are briefly outlined in this section.

Model selection for SVMs is a computationally expensive task, especially if the trial-and-error approaches are utilized [6]. It may consist in selecting parameters of the predefined kernels [23], but also the desired kernel can be determined— in [5], this is achieved using an evolution strategy. GAs are also applied for this purpose, as in [25] for the smooth twin parametric-margin SVMs. In another recent algorithm [3], the SVM parameters are optimized using a fast messy GA.

Other interesting approaches include tabu searches [13], genetic programming [22], and compression-based techniques [14] (the coding precision is related to the geometrical characteristics of the data). A dynamic model adaptation strategy, which combines the swarm intelligence with a standard grid search, was proposed in [9]. A promising research direction is to construct new kernels tailored for a problem at hand, including the use of neuro-fuzzy systems [17,21].

The algorithms to deal with large training sets can be divided into those which: (i) enhance the SVM training [4,8,12], and (ii) decrease the size of training sets by retrieving the most valuable vectors. Importantly, the approaches from the first group still induce the problem of high memory complexity of the training which must be endured in big data problems.

Decreasing the size of T makes SVM training feasible for large data sets, but it also allows for reducing s, which accelerates the classification. The methods which exploit the information on the layout of T, encompass clustering-based algorithms [20,24] and those utilizing the T's geometry without grouping the data [1]. Significant research effort has been put into proposing algorithms exploiting statistical properties of the T vectors [7]. Other techniques include various random sampling algorithms [18] as well as the induction trees [2].

In our recent research, we pioneered the use of evolutionary algorithms for this task. In our initial approach (GASVM) [10], a population of individuals (chromosomes) representing refined sets of a fixed size (t'), evolves in time using standard genetic operators—selection, crossover, and mutation. GASVM was enhanced to dynamically adapt the crucial parameters, including t', during the evolution [11]. We also exploited the knowledge concerning T, attained during the evolution or extracted beforehand in our memetic algorithms [15,16].

3   Proposed Method

In the work reported here, we introduce ALGA—**AL**ternating **G**enetic **A**lgorithm for selecting the SVM model and refining the training set. ALGA alternates between two main phases—one is aimed at optimizing T', while the other optimizes $\mathcal{M}$, as illustrated in Fig. 1. The model selection phase is inspired by [3], while for selecting the training set, we exploit our relatively simple GASVM algorithm [10] to verify the very foundations of the new alternating scheme. The pseudocode of ALGA is given in Algorithm 1. In the first generation (G_0), two populations representing (i) refined training sets T' ($\{p_i\}$) and (ii) SVM models ($\{q_i\}$) are initialized (lines 1–2). For a randomly chosen SVM model (q_{init}), we evaluate every individual p_i to select the best one (p^B) (lines 4–7). Afterwards, the algorithm enters the model optimization phase (lines 9–15), in which $\mathcal{M}$ is optimized for the currently best refined training set p^B. If the average fitness of all individuals in the population does not grow in two subsequent generations, then the local stop condition is met (line 15) and the algorithm enters the training set optimization phase (lines 17–23)—T' is optimized using the currently best hyperparameters (q^B), again until the same local stop condition is met (line 23). This alternating process is repeated as long as at least one of two subsequent phases manages to improve the average fitness. Otherwise, the global stop condition (line 25) is reached and the SVM trained with the best individuals ($p^B; q^B$) is retrieved.

3.1 Individuals and Their Fitness

The process of computing the fitness is shown in Fig. 2. A chromosome p_i defines a single $T' \subset T$, containing K vectors from each class, hence the length of the chromosome is $2K$. SVM model is represented with q_j—as we limit the search to the RBF kernel, the chromosome contains two elements (γ_j, C_j). Two individuals from two populations are required to train the SVM, which is subsequently used to classify the validation set V. Based on the ground-truth labels from V, the classification accuracy is evaluated and used as the fitness $\eta(p_i; q_j)$. Importantly, the test set Ψ used for final evaluation, is not seen during the optimization.

3.2 Selecting Training Set

Each individual p in the initial population (of a size N) is created by choosing a randomly selected subset of K vectors from each class of T. For selection, we exploit the high-low fit scheme. The population is sorted by the fitness, evaluated as outlined in Fig. 2, and divided into two equally-sized parts. The parent p_A is selected from the more fitted part, while the parent p_B is drawn from the less-fitted part of the population. The offspring solutions are appended to the population, forming a new population of a size $2N$. The N individuals with the highest fitness survive to maintain the constant population size.

Crossover of two individuals p_a and p_b is done by creating a sum of two training sets defined by these individuals, from which $2K$ unique samples are

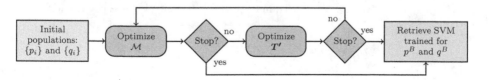

Fig. 1. Flowchart of the proposed method.

Algorithm 1. Alternating Genetic Algorithm (ALGA).

1: Initialize population of T''s ($\{p_i\}$) of size N;
2: Initialize population of $\mathcal{M}$'s ($\{q_i\}$) of size M;
3: $q_{\text{init}} \leftarrow$ GETRANDOM($\{q_i\}$);
4: **for all** $\{p_i\}$ **do**
5:     $\eta_i \leftarrow$ COMPUTEFITNESS($p_i;q_{\text{init}}$);
6: **end for**
7: Select p^B—individual with the highest fitness in $\{p_i\}$;
8: **repeat**
9:     **repeat** $\triangleright$ $\mathcal{M}$ optimization phase
10:         **for all** $\{p_i\}$ **do**
11:             $\eta_i \leftarrow$ COMPUTEFITNESS($p^B;q_i$);
12:         **end for**
13:         Select q^B—individual with the highest fitness in $\{q_i\}$;
14:         Create new population of $\{q_i\}$;
15:     **until** LOCALSTOP;
16:     **if** $\neg$ GLOBALSTOP **then**
17:         **repeat** $\triangleright$ T' optimization phase
18:             **for all** $\{q_i\}$ **do**
19:                 $\eta_i \leftarrow$ COMPUTEFITNESS($q_i;q^B$);
20:             **end for**
21:             Select p^B—individual with the highest fitness in $\{p_i^B\}$;
22:             Create new population of $\{p_i\}$;
23:         **until** LOCALSTOP;
24:     **end if**
25: **until** GLOBALSTOP;
26: **return** $(p^B;q^B)$;

---

selected randomly to form p_{a+b}. Then, p_{a+b} is subject to mutation with the probability $\mathcal{P}_m^{T'}$. Finally, $\lfloor 2K \cdot f_m \rfloor$ randomly-chosen vectors are substituted with others from T (it is assured that T' contains unique elements).

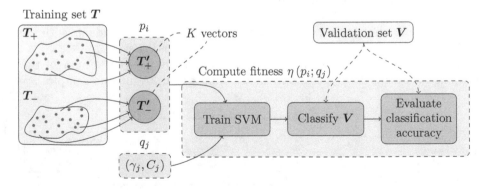

Fig. 2. The process of computing the fitness.

3.3   SVM Model Optimization

The process of optimizing the hyperparameters is similar to the training set optimization with three main differences concerning (i) initialization, (ii) crossover
and (iii) mutation. The first population is initialized deterministically to cover
a large range of the values with a logarithmic step ($\gamma \in \{0.01, 0.1, 1, 10, 100\}$
and $C \in \{0.1, 1, 10, 100\}$, hence $M = 20$). Two individuals $q_a = (\gamma_a, C_a)$ and
$q_b = (\gamma_b, C_b)$ are crossed over (with the probability $\mathcal{P}_c^{\mathcal{M}}$) to form a new individual $q_{a+b} = (\gamma_{a+b}, C_{a+b})$. The child parameter values x_{a+b} (x_{a+b} is either γ or
C) become $x_{a+b} = x_a + \alpha_{\mathcal{M}} \cdot (x_a - x_b)$, where $\alpha_{\mathcal{M}}$ is the crossover weight randomly drawn from the interval $[-0.5, 1.5]$ to diversify the search (and $x_{a+b} > 0$).
Mutation is proceeded with the probability $\mathcal{P}_m^{\mathcal{M}}$, and it consists in modifying a
value x within a range $x \in [x - \delta_m \cdot x, x + \delta_m \cdot x]$, where x is γ or C.

4   Experimental Validation

The algorithms were implemented in C++ (using LIBSVM) and run on a computer equipped with an Intel Xeon 3.2 GHz (16 GB RAM) processor. The parameters of the GA were set experimentally to $N = 20$, $\mathcal{P}_m^{T'} = 0.3$, $f_m = 0.2$,
$\mathcal{P}_m^{\mathcal{M}} = 0.2$, $\delta_m = 0.1$ and $\mathcal{P}_c^{\mathcal{M}} = 0.7$ (such values are commonly used in GAs,
including our earlier works [10,11]). We tested our algorithms using three sets of
2D points[1], for which we visualize the results: *2d-random-dots* ($1.7 \cdot 10^4$ samples),
2d-random-points (1552 samples) and *2d-chessboard-dots* ($2.7 \cdot 10^4$ samples). In
the *dots* variants, groups of the vectors from each class form clusters on the 2D
plane, while in the *points* variant the vectors are isolated. Each set is divided
into: a training set T (from which T''s are selected), a validation set V (for
which the fitness is evaluated) and a test set Ψ. Furthermore, ALGA was validated for three benchmark sets from the UCI repository: *German*, *Ionosphere*
and *Wisconsin breast cancer*, using 5-fold cross-validation. ALGA was run 30×
for each 2D set and 50× for the benchmarks (10× for every fold), and we verified
the statistical significance of the differences using two-tailed Wilcoxon test.

   We compared ALGA against the SVM trained with whole T, whose model is
optimized using grid search (GS) with a logarithmic step (we start with a step
of 10, subsequently decreased to 2 for the best range) as well as with our GA
working only in the $\mathcal{M}$ optimization phase (termed GA-model), trained with the
whole T. Furthermore, we report the scores obtained using GASVM with the
model selected with GS, performed using T. ALGA and GASVM were tested
for different values of K (we start with K equal to the data dimensionality and
then we increase it logarithmically with a step of 4, until $2K$ reaches $t/2$).

   In Fig. 3, we report the results obtained for the 2D sets—the Pareto fronts
present the accuracy obtained using different K's for V (i.e., the fitness) vs.
the number of SVs (the smaller, the better). It can be seen that ALGA renders
better or comparable scores to GASVM, taking into account both criteria (the

---

[1] Available at http://sun.aei.polsl.pl/~mkawulok/mcpr2017 (along with visual
  results).

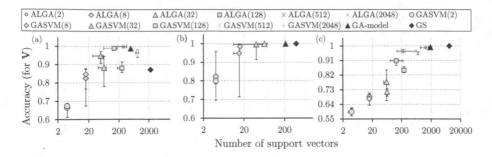

Fig. 3. Pareto fronts (the accuracy for V vs. s) for (a) 2d-random-dots, (b) 2d-random-points and (c) 2d-chessboard-dots (for GASVM and ALGA, K is given in parentheses).

Table 1. Results for the 2D data sets—accuracy and evolution time (in seconds).

Set	Alg.	Accuracy for Ψ		Evolution time	
		ALGA	GASVM	ALGA	GASVM
2d-dots	$K = 2$	$.619 \pm .024$	$.653 \pm .026$	5.79 ± 2.42	$2.10 \pm .71$
	$K = 8$	$.816 \pm .087$	$.869 \pm .021$	11.63 ± 2.85	4.98 ± 1.73
	$K = 32$	$.958 \pm .011$	$.909 \pm .042$	18.58 ± 7.11	5.71 ± 3.82
	$K = 128$	$.989 \pm .003$	$.868 \pm .018$	33.83 ± 10.05	7.38 ± 2.07
	$K = 512$	$.994 \pm .001$	$.967 \pm .011$	79.48 ± 15.23	37.33 ± 5.42
	GA-model	$.993 \pm .000$		180.92 ± 29.40	
	GS	$.871$		2523.39	
2d-pts	$K = 2$	$.783 \pm .054$	$.807 \pm .014$	2.27 ± 1.67	$.81 \pm 1.08$
	$K = 8$	$.901 \pm .108$	$.954 \pm .017$	3.60 ± 1.56	$.41 \pm .38$
	$K = 32$	$.987 \pm .029$	$.997 \pm .002$	1.49 ± 1.47	$.98 \pm .69$
	GA-model	$.992 \pm .001$		$.825 \pm .149$	
	GS	1.000		16.79	
2d-chess.	$K = 2$	$.585 \pm .020$	$.581 \pm .020$	9.75 ± 3.81	3.84 ± 1.32
	$K = 8$	$.680 \pm .032$	$.673 \pm .027$	19.63 ± 3.81	10.16 ± 3.39
	$K = 32$	$.791 \pm .058$	$.717 \pm .012$	42.50 ± 16.47	8.38 ± 2.46
	$K = 128$	$.918 \pm .011$	$.865 \pm .007$	80.00 ± 36.18	16.23 ± 5.09
	$K = 512$	$.971 \pm .004$	$.959 \pm .004$	291.48 ± 168.13	53.98 ± 25.52
	$K = 2048$	$.986 \pm .001$	$.987 \pm .001$	622.80 ± 168.46	181.89 ± 54.71
	GA-model	$.989 \pm .000$		2153.78 ± 288.15	
	GS	$.909$		20723.34	

differences are statistically significant at p $= .05$, except for 2d-random-points for $K = 8$ and 2d-chessboard-dots for $K \in \{2, 8\}$). GA-model and GS usually allow for high accuracy at the cost of very high s, which in fact leads to the overfitting (this is the reason for low accuracy of GS for 2d-random-dots). This

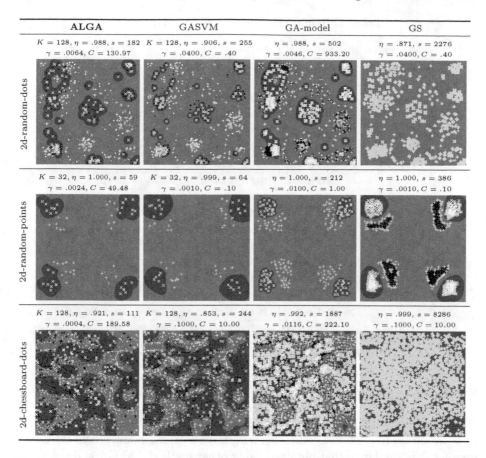

Fig. 4. Examples of the results retrieved using various methods for 2D data sets. (Color figure online)

problem can also be seen in Fig. 4, where we visualize the results for selected K's and compare them with GS and GA-model. Black and white points indicate the vectors from V, and those marked with white and black crosses (the colors are swapped for better visualization) show the data selected to T' (yellow crosses indicate the SVs). It can be seen that when the model is selected with GS or GA-model, there are lots of SVs and the kernel width is small (the same happens for GASVM, as it relies on the model obtained with GS). The model selected with ALGA better discovers the data structure—this is because the limited size of $T' = 2K$ requires such a model, for which the SVM must generalize well to classify V (this is most evident for 2d-random-dots, where GASVM fails to generalize).

The scores obtained for the 2D test sets are reported in Table 1. The tendencies are the same as those observed for V's. Although the evolution times for ALGA are longer than for GASVM, the latter requires the model to be selected

Table 2. Results obtained for the UCI data sets.

Set	Algorithm	K	Acc. for V	Acc. for Ψ	s	Time (in sec.)
German	**ALGA**	24	$.742 \pm .009$	$.694 \pm .028$	$48.00 \pm .00$	$2.28 \pm .38$
		96	$.825 \pm .008$	$.696 \pm .029$	$192.00 \pm .00$	$2.92 \pm .57$
	GASVM	24	$.726 \pm .012$	$.714 \pm .022$	45.20 ± 2.32	$2.29 \pm .89$
		96	$.769 \pm .018$	$.723 \pm .012$	173.30 ± 14.72	$3.56 \pm .34$
	GA-model	–	$1.000 \pm .000$	$.728 \pm .014$	775.44 ± 22.90	$1.36 \pm .31$
	GS	–	$.864 \pm .053$	$.734 \pm .025$	545.00 ± 46.39	12.14 ± 2.33
Ionosphere	**ALGA**	34	$.981 \pm .002$	$.947 \pm .020$	47.04 ± 4.07	$1.90 \pm .90$
		68	$.992 \pm .003$	$.948 \pm .018$	67.06 ± 7.55	$1.90 \pm .22$
	GASVM	34	$.979 \pm .005$	$.962 \pm .021$	45.12 ± 7.99	$1.13 \pm .42$
		68	$.988 \pm .008$	$.965 \pm .022$	67.88 ± 17.18	$1.41 \pm .35$
	GA-model	–	$1.000 \pm .000$	$.949 \pm .035$	213.68 ± 42.19	$.30 \pm .01$
	GS	–	$.987 \pm .010$	$.952 \pm .032$	92.80 ± 25.44	$1.20 \pm .04$
Wisconsin	**ALGA**	9	$.979 \pm .004$	$.965 \pm .019$	10.40 ± 1.79	$1.59 \pm .55$
		36	$.981 \pm .002$	$.961 \pm .022$	22.04 ± 4.54	$1.92 \pm .74$
	GASVM	9	$.978 \pm .003$	$.968 \pm .020$	12.14 ± 1.16	$.66 \pm .31$
		36	$.981 \pm .004$	$.968 \pm .020$	25.88 ± 7.84	$.81 \pm .20$
	GA-model	–	$1.000 \pm .000$	$.961 \pm .022$	342.42 ± 15.96	$.31 \pm .02$
	GS	–	$.977 \pm .006$	$.966 \pm .020$	77.20 ± 16.30	$1.14 \pm .06$

beforehand—here, we used GS, which is very time consuming, especially for 2d-chessboard-dots—the largest set used in our experimental study.

For the UCI sets (the scores are shown in Table 2), GA-model is overfitted to V, resulting in perfect classification with large s values, however still the accuracy for Ψ is similar to that obtained using other methods. ALGA and GASVM render similar accuracies (though statistically different)—ALGA behaves slightly better for V, while GASVM delivers slightly higher accuracy for Ψ. Although GASVM is better here in terms of the classification performance, ALGA remains competitive, without the need for selecting $\mathcal{M}$ prior to the training set selection.

5   Conclusions and Outlook

In this paper, we introduced ALGA—a new approach to select both the SVM model, as well as the training set, within a single optimization process that alternates between two phases. We incorporated relatively simple GAs to solve these two optimization problems and we compared our method with the same GAs applied in a sequential manner. This allowed us to verify the very foundations of the new scheme and based on extensive experiments we showed that ALGA is capable of selecting the training set without the necessity to tune the SVM hyperparameters beforehand (which is non-trivial for the sets that are too large to train SVM from them).

The main shortcoming of the new method is the same as that of GASVM, which we use to refine the training set—it is necessary to select a proper value of K prior to starting the optimization. Importantly, we addressed this problem recently [11] by increasing the chromosome's length during the evolution, and we developed a memetic algorithm [16] which outperforms the state-of-the-art methods, including those based on data structure analysis [24]. Therefore, our ongoing work is aimed at enhancing ALGA with these adaptive and memetic algorithms, which we expect to increase its competitiveness substantially. Furthermore, we aim at improving the model selection phase as well—instead of tuning the parameters of predefined kernel functions, we plan to allow for selecting them or constructing from scratch. This will be an important step towards parameter-less evolutionary SVMs, easily applicable to deal with large data sets.

Acknowledgments. This work was supported by the National Centre for Research and Development under the grant: POIR.01.02.00-00-0030/15.

References

1. Angiulli, F., Astorino, A.: Scaling up support vector machines using nearest neighbor condensation. IEEE Trans. Neural Netw. **21**(2), 351–357 (2010)
2. Cervantes, J., Lamont, F.G., López-Chau, A., Mazahua, L.R., Ruíz, J.S.: Data selection based on decision tree for SVM classification on large data sets. Appl. Soft Comput. **37**, 787–798 (2015)
3. Chou, J.S., Cheng, M.Y., Wu, Y.W., Pham, A.D.: Optimizing parameters of SVM using fast messy genetic algorithm for dispute classification. Expert Syst. Appl. **41**(8), 3955–3964 (2014)
4. Ferragut, E., Laska, J.: Randomized sampling for large data applications of SVM. In: Proceedings of the ICMLA, vol. 1, pp. 350–355 (2012)
5. Friedrichs, F., Igel, C.: Evolutionary tuning of multiple SVM parameters. Neurocomputing **64**, 107–117 (2005)
6. Gold, C., Sollich, P.: Model selection for support vector machine classification. Neurocomputing **55**(1–2), 221–249 (2003)
7. Guo, L., Boukir, S.: Fast data selection for SVM training using ensemble margin. Pattern Recognit. Lett. **51**, 112–119 (2015)
8. Joachims, T.: Making large-scale SVM learning practical. In: Advances in Kernel Methods, pp. 169–184. MIT Press, Cambridge (1999)
9. Kapp, M.N., Sabourin, R., Maupin, P.: A dynamic model selection strategy for support vector machine classifiers. Appl. Soft Comput. **12**(8), 2550–2565 (2012)
10. Kawulok, M., Nalepa, J.: Support vector machines training data selection using a genetic algorithm. In: Gimel'farb, G.L., et al. (eds.) SSPR & SPR 2012. LNCS, vol. 7626, pp. 557–565. Springer, Heidelberg (2012). doi:10.1007/978-3-642-34166-3_61
11. Kawulok, M., Nalepa, J.: Dynamically adaptive genetic algorithm to select training data for SVMs. In: Bazzan, A.L.C., Pichara, K. (eds.) IBERAMIA 2014. LNCS (LNAI), vol. 8864, pp. 242–254. Springer, Cham (2014). doi:10.1007/978-3-319-12027-0_20
12. Le, Q., Sarlos, T., Smola, A.: Fastfood - approximating kernel expansions in loglinear time. In: Proceedings of the ICML, pp. 1–9 (2013)
13. Lebrun, G., Charrier, C., Lezoray, O., Cardot, H.: Tabu search model selection for SVM. Int. J. Neural Syst. **18**(01), 19–31 (2008)
14. von Luxburg, U., Bousquet, O., Schölkopf, B.: A compression approach to support vector model selection. J. Mach. Learn. Res. **5**, 293–323 (2004)

15. Nalepa, J., Kawulok, M.: A memetic algorithm to select training data for support vector machines. In: Proceedings of the GECCO, pp. 573–580. ACM (2014)
16. Nalepa, J., Kawulok, M.: Adaptive memetic algorithm enhanced with data geometry analysis to select training data for SVMs. Neurocomputing **185**, 113–132 (2016)
17. Nalepa, J., Siminski, K., Kawulok, M.: Towards parameter-less support vector machines. In: Proceedings of the ACPR, pp. 211–215 (2015)
18. Nishida, K., Kurita, T.: RANSAC-SVM for large-scale datasets. In: Proceedings of the IEEE ICPR, pp. 1–4 (2008)
19. Ripepi, G., Clematis, A., DAgostino, D.: A hybrid parallel implementation of model selection for support vector machines. In: Proceedings of the PDP, pp. 145–149 (2015)
20. Shen, X.J., Mu, L., Li, Z., Wu, H.X., Gou, J.P., Chen, X.: Large-scale SVM classification with redundant data reduction. Neurocomputing **172**, 189–197 (2016)
21. Simiński, K.: Neuro-fuzzy system based kernel for classification with support vector machines. In: Gruca, D.A., Czachórski, T., Kozielski, S. (eds.) Man-Machine Interactions 3. AISC, vol. 242, pp. 415–422. Springer, Cham (2014). doi:10.1007/978-3-319-02309-0_45
22. Sullivan, K.M., Luke, S.: Evolving kernels for support vector machine classification. In: Proceedings of the GECCO, pp. 1702–1707. ACM, New York (2007)
23. Tang, Y., Guo, W., Gao, J.: Efficient model selection for support vector machine with Gaussian kernel function. In: Proceedings of the IEEE CIDM, pp. 40–45 (2009)
24. Wang, D., Shi, L.: Selecting valuable training samples for SVMs via data structure analysis. Neurocomputing **71**, 2772–2781 (2008)
25. Wang, Z., Shao, Y.H., Wu, T.R.: A GA-based model selection for smooth twin parametric-margin SVM. Pattern Recognit. **46**(8), 2267–2277 (2013)

An Exploration Strategy for RL with Considerations of Budget and Risk

Jonathan Serrano Cuevas$^{(\boxtimes)}$ and Eduardo Morales Manzanares$^{(\boxtimes)}$

Department of Computer Science, Instituto Nacional de Astrofísica,
Óptica y Electrónica, 72840 Puebla, Mexico
{jonathan.serrano,emorales}@inaoep.mx

Abstract. Reinforcement Learning (RL) algorithms create a mapping from states to actions, in order to maximize an expected reward and derive an optimal policy. However, traditional learning algorithms rarely consider that learning has an associated cost and that the available resources to learn may be limited. Therefore, we can think of learning over a limited budget. If we are developing a learning algorithm for an agent i.e. a robot, we should consider that it may have a limited amount of battery; if we do the same for a finance broker, it will have a limited amount of money. Both examples require planning according to a limited budget. Another important concept, related to budget-aware reinforcement learning, is called risk profile, and it relates to how risk-averse the agent is. The risk profile can be used as an input to the learning algorithm so that different policies can be learned according to how much risk the agent is willing to expose itself to. This paper describes a new strategy to incorporate the agent's risk profile as an input to the learning framework by using reward shaping. The paper also studies the effect of a constrained budget on RL and shows that, under such restrictions, RL algorithms can be forced to make a more efficient use of the available resources. The experiments show that as the even if it is possible to learn on a constrained budget with low budgets the learning process becomes slow. They also show that the reward shaping process is able to guide the agent to learn a less risky policy.

Keywords: Reinforcement learning · Risk · Budget · Reward shaping

1 Introduction

The purpose of executing a reinforcement learning algorithm is to generate a mapping from situations to actions so as to maximize a function reward or reinforcement signal. The agent must discover by itself which actions yield the best reward by executing an experimentation process, considering as well the impact of the current decision over future rewards. These two characteristics, trial-and-error and delayed reward, are the two most important features of reinforcement learning [1]. The agent task is to develop a knowledge of its environment by using an experimentation process. This knowledge is to be exploited afterwards by the

© Springer International Publishing AG 2017
J.A. Carrasco-Ochoa et al. (Eds.): MCPR 2017, LNCS 10267, pp. 105–116, 2017.
DOI: 10.1007/978-3-319-59226-8_11

agent to obtain a reward. However, most RL algorithms learn an (near) optimal policy without considering a learning cost. The process of learning has a cost because of the exploration process [2], since deciding to explore an unknown area implies an expense of some sort, and visiting certain areas in the environment might lead to large costs. If the agent has a limited amount of resources, or a budget, to pay for these costs, the learning process has to be optimized accounting for it. This optimization becomes critical within certain applications.

An example of such applications might appear in robotics [3], where a robot has to learn a task before running out of batteries but, at the same time, avoid certain actions that might yield a catastrophic outcome, such as the destruction of the robot. Another example might occur in a finance application [4], where a policy tries to maximize the utility at a certain time horizon, but at the same time, avoiding any chance of running out of money. The concept of risk arises naturally on the latter example, and also the concept of risk aversion, if the problem is stated as to learn a policy which yields the largest reward but at the same time minimizing the risk of running out of money. In general terms our problem is to optimize a certain parameter observing, at the same time, a safety margin over another parameter. To deal with these type of problems some techniques and algorithms have been developed under the concept of Safe Reinforcement Learning or SRL [5]. SRL can be defined as the process of learning policies that maximize the expectation of the reward in problems where it is important to ensure reasonable system performance and/or respect safety constraints during the learning and/or deployment processes [6]. Therefore one could say that SRL studies the process of reinforcement learning accounting for the safety of the agent.

So far we have the problem of learning a policy to optimize a resource, while reducing the probability (or risk) of running out of such a resource during the learning process. Since the traditional learning algorithms only aim to optimize a reward, this work proposes the use of reward shaping to model the concept of risk profile β_p and learn policies accounting for it.

The remainder of this paper is organized as follows. Section 2 describes the most closely related work. In Sect. 3 the proposed approach is described in detail. Section 4 describes the learning environment and in Sect. 5 the experimental results are given. Finally conclusions and future research work are given in Sect. 6.

2   Related Work

SRL is a requirement in many scenarios where the safety of the agent is particularly important and, for this reason, researchers are paying increasing attention not only to maximize the long-term reward, but also to damage and risk avoidance [7,8]. SRL is a relatively new topic, therefore there is still some debate on how to classify the different techniques used to accomplish it. However, García and Fernández [6] proposed a SRL taxonomy which classifies the SRL techniques in two broad groups. The first group includes techniques which modify the optimality criterion, the second group includes techniques which modify the exploration process through the incorporation of external knowledge or the guidance of a risk metric. This work fits into the first group, since it proposes and modifies the reward function in order to consider the agent's risk profile.

Risk metrics are considered in several forms in the RL literature, but in most of them the risk is related to the stochasticity of the environment, therefore it is related to the inherent uncertainty of the environment [9]. Dealing with environment uncertainty is not easy because in those environments, even an optimal policy (with respect to the return) may perform poorly in some cases. For this reason, and in order to be able to test our exploration strategy without the randomness which the environment's inherent uncertainty might generate, our work will deal only with stationary rewards, and our risk metric will be tightly related to the amount of resources the agent has at any given time and to the probability of running out of these resources.

In RL, techniques for selecting actions during the learning phase are called exploration/exploitation strategies. Most exploration methods are based on heuristics, rely on statistics collected from sampling the environment, or have a random exploratory component, i.e. $\epsilon - greedy$, which aim to explore the state space efficiently. To avoid risky situations, the exploration process is often modified by including prior knowledge of the task. This prior knowledge can be used to provide initial information to the RL algorithm biasing the subsequent exploratory process [10,11], to provide a finite set of demonstrations on the task [12], or to provide guidance [13]. It is important to mention that most of these exploration methods are blind to the risk of actions, and all of them are blind to the notion of a budget. It is left as future work to make the exploration dependent on the budget.

Finally some background on reward shaping will be given now. The practice of reward shaping in reinforcement learning consists of supplying additional rewards to a learning agent to guide its learning process, beyond those supplied by the underlying MDP [14], thus shaping its behavior. This shaping process results on a faster convergence time to an optimal policy because the additional rewards provided to the agent makes the exploration process more efficient. Therefore, reward shaping has the potential to be a very powerful technique for scaling up reinforcement learning methods to handle complex problems [15], and it can be used with any reinforcement learning algorithm such as Q-Learning [16].

3   Learning Framework Description

Our learning framework was inspired by the real-life perception of danger and the concept of risk averseness. As shown in Fig. 1, the same danger can be perceived as larger or smaller according to the agent who observes it. For a risk-seeking agent, the danger perception is diminished, but for a risk-averse agent, the danger is amplified. The same idea applies for budget; any investment is seen as less risky as the amount of budget increases. Now we will explain our learning framework.

Fig. 1. Risk-averse agent experiment.

Formally the agent starts with a given amount of resources B_0, called budget, then at each time $t = i$ it receives a reward r_i, therefore the accumulated reward at time, or step, t is:

$$R_t = \sum_{i=1}^{t} r_i \tag{1}$$

And the accumulated budget at time t is shown in Eq. 2. In this research, each step represents a cost so the budget is reduced accordingly to the length of the path.

$$B_t = B_0 + R_t \tag{2}$$

In order to guarantee that a policy can be found with a given budget we must ensure that $B_0 > E(R_t)$, where $E(Rt)$ corresponds to the expected value of the reward's sum assuming no discount rate. For this simple gridworld scenario with only two possible paths this value is simple to calculate as it will be shown later.

3.1 Shaping Rewards

Intuitively we are trying to learn a policy for some Markov Decision Process (MDP) $M = (S, A, T, \gamma, R)$[1], and we wish to help our learning algorithm by giving it additional shaping rewards which will hopefully guide it towards learning a policy which accounts for β_p. To formalize this, we assume that, rather than running our reinforcement learning algorithm on $M = (S, A, T, \gamma, R)$, we will run it on some transformed MDP $M' = (S, A, T, \gamma, R')$, where $R' = f(R)$ is the reward function in the transformed MDP, and and f can take several forms.

---

[1] Where S is a set of states, A is a set of actions, T is a transition function, γ is a discount factor and R is a reward function.

In traditional reward shaping research, it has been an additive function, in this research we use a simple, yet effective form to represent the reward provided to the learning algorithm. So, if in the original MDP M we would have received a reward $R(s, a, s')$ for moving from s to s' by executing action a, then in the new MDP M' we would receive reward $\beta_p R(s, a, s')$ on the same event. Now our job is to select the value of β_p to properly shape the reward and derive a risk-aware policy.

3.2 Mapping the Risk Profile Using Reward Shaping

We used a simple approach to map risk profile by using reward shaping: by carefully selecting the value of β_p. If we pick $\beta_p > 1$ the agent will have a risk averse profile, and if $0 < \beta_p \leq {<}1$, the agent will have a risk seeking profile. Effectively, with this simple mechanism, a risk averse agent will consider the rewards (costs) in a pessimistic way and will take lower risks, while a risk seeking agent will consider the rewards in an optimistic way and will take higher risks.

In this work we only change the original value of R when the rewards are negative, thus changing the agent's perception of the investment or effort required to earn any given final reward B_t. As a sidenote, in an analogous manner the rewards could be shaped as well by affecting only the positive rewards, however, in order to be consistent with the explanation given on Sect. 3, only the negative rewards were shaped.

4 Scenario Description

In order to test the ideas about reinforcement learning considering budget and risk we used Q-Learning as the reinforcement learning algorithm, and a grid world with some considerations which will be described on this section.

The strategy was tested in a 10×5 grid world, as shown in Fig. 2. The bottom left square has the coordinate $(0, 0)$, while the upper right square has the coordinate $(10, 5)$. The grid world includes a wall which the agent cannot cross (marked in black) and two special squares marked with an E and a \$ sign. The E shows the position of an exit, while the \$ square provides the agent with a special reward. As a convention the reward provided by any given square where $x = i$ and $y = j$ will be named as $r_{i,j}$, and the reward provided by the \$ square will be named as $r_\$$ regardless of its position. The reward $r_\$ \geq 1$, while the rest of the squares provide a reward equal to -1.

The task that the agent has to perform is to find its way to the E square in order to maximize its final reward R. The agent can as well decide to get the coin first and then head to the exit. Since the reward of every grid world square equals -1, the reward the agent will receive after reaching the exit will depend on the Manhattan distance [17] of the route it chooses and on wether it decides to pick up the coin or not. Let's call the shortest route which picks up the coin as $Route_1$ and the shortest route which does not pick up the coin as $Route_2$. The Manhattan distance of $Route_1$ is $MD(Route_1) = 17$, while

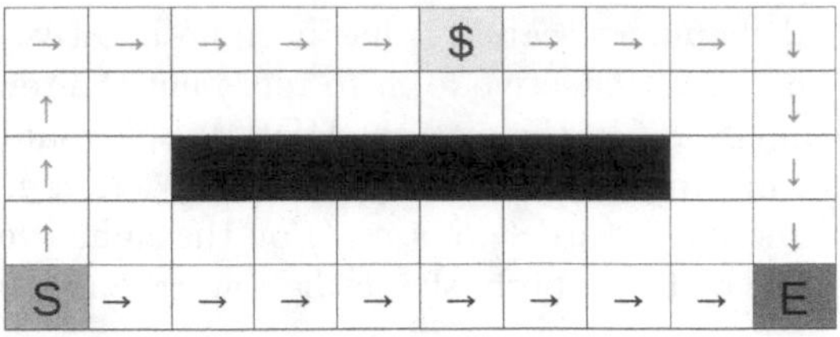

Fig. 2. Simple grid world used to show the effect on the learned policy of using reward shaping to modelate the user's risk profile.

$MD(Route_2) = 9$, therefore the reward for $Route_1$ is $R(Route_1) = -17 + r_\$$ and $R(Route_2) = -9$. Note that $R(Route_1)$ considers the reward provided by the $\$$ square. The accumulated reward and the accumulated budget are calculated as stated by Eqs. 1 and 2. Considering that the agent requires one time step to move from one square to another one, the time required to complete the task following $Route_1$ is $T(Route_1) = 17$ and $T(Route_2) = 9$. One could anticipate that the agents decision to choose $Route_1$ will depend on $r_\$$, however it will be shown that it depends as well on the agents risk profile β_p.

As mentioned before the initial budget B_0 is the amount of resources that the agent has to complete its task. If we think in terms of a robot and fuel then the budget is the amount of fuel the agent has, while $r_\$$ corresponds to a fuel tank that the robot can find and use. The variable $r_{i,j}$ is the reward received at any given (i, j) square, so following on with the robot example, it represents the fuel the agent has to spend in order to move. The agent aims to find the exit with as much fuel remaining as possible. If the agent runs out of fuel before finding the exit then the task is considered as failed and the game is over. With this in mind a game start when the agent receives its budget and ends if any of the following conditions occur:

- Condition 1 (Tn_1). The agent completed the task and found the exit.
- Condition 2 (Tn_2). The agent ran out of budget.

Note that finding the $\$$ square is not part of the task, therefore the agent has to decide wether it is convenient to visit it to receive $r_\$$ or not.

5   Experiments

To test our experiments we used the simulation software Burlap [18] and the RL algorithm Q-Learning shown in Eq. 3 with $\gamma = 0.90$ and the values for α shown in Table 1. The experiments aim to prove that (i) it is possible to learn a policy which completes a task with a constrained budget and (ii) that reward shaping is a good alternative to learn policies which account for the agent's risk profile.

$$Q(s_t, a_t) \leftarrow (1 - \alpha_t)Q(s_t, a_t) + \alpha_t[R(s) + \gamma \max_a Q(s_{t+1}, a)] \tag{3}$$

Table 1. Experiment values.

Variable	Value(s)	Comments
α	$\{0.3, 0.9\}$	The learning rate
B_0	$\{20, 30, 40, 50\}$	The budget at the beginning of each game
$r_\$$	$\{9\}$	The reward for visiting the square $
β_p	$\{0.5, 2\}$	The agent's risk profile

In order to prove these ideas we executed experiments changing the values of β_p and B_0. The used values are shown in Table 1.

To refer to any particular combination of these variables a shorthand is used. For instance the shorthand *COIN9-0.9-2-20* is used to label an experiment with $r_\$ = 9$, $\alpha = 0.9$, $B_0 = 25$ of a risk-averse agent (the number 2 represents a risk-averse agent while the number 1 represents a risk-seeking agent).

5.1   Experiment Settings

The series of experiments is divided in learning episodes (LE), each one of these started when the agent received its initial budget and ended when either Tn_1 or Tn_2 was accomplished. One experiment consisted of 500 learning episodes and each experiment was repeated for $n = 25$ times. The metrics used to evaluate the results are described in Table 1.

Table 2. Experiment metrics.

Variable	Description
EXIT	For any given LE it represents the percentage of n where the agent reached the exit
C1	For any given LE it represents the percentage of n where the agent visited square $
FR	Average final reward that the agent received at the end of any given learning episode

Table 3 shows that the agent can only receive three different rewards: any movement on any direction gives the agent a reward equal to -1, if it reaches square $ it receives a reward of 9, and if it runs out of budget it receives a reward of -10. Considering this, and assuming that the agent does not run out of budget, then $R(Route_1) = -17 + 9 = -8$ and $R(Route_2) = -9$; therefore, if the agent is willing to tolerate the risk of running out of budget it should try to learn a path similar to $Route_1$, otherwise its safer choice is a path like $Route_2$.

To determine the value of $E(R_t)$ we assumed that there is an equal probability for the agent to pick either $Route_1$ or $Route_2$. Therefore $E(R_t) = 0.5 \cdot -8 + 0.5 \cdot$

Table 3. The possible rewards the agent can receive.

State change	Reward
Transition to any square	−1
Transition to square $	9
Transition to $B_0 = 0$	−10

$-9 = -8.5$. From this simple calculation we can tell that B_0 has to be larger than $|-8.5|$ so that it is sufficient to complete the task.

5.2   Experimental Results

Each plot shown in this section includes three subplots, and each one with one of the metrics described on Table 2. At any given subplot four series are plotted, and these are labeled as described in Sect. 4. The subplots share the horizontal axis which represents the learning episode LE.

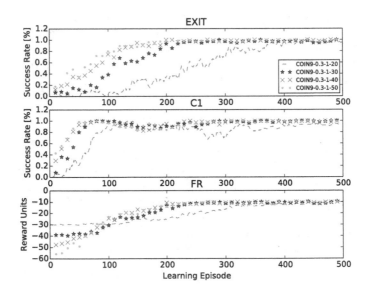

Fig. 3. Risk-seeking agent experiment.

Figure 3 shows the learning process of a risk-seeking agent ($\beta_p = 0.5$), and Fig. 4 shows the learning process of a risk-averse agent ($\beta_p = 2$). Both risk profiles were modeled by using reward shaping and the difference between them is clear: the risk-seeking agent learns to follow $Route_1$ while the risk-averse agent learns that it is better to follow $Route_2$ while both learn to find the exit located at the bottom right square of the gridworld. This can be told by observing the subplots labeled as $EXIT$ and $C1$ of each plot. On the $EXIT$ subplot after learning episode 400 all the plotted series have a success rate larger than 75%. On the $C1$ subplot is where the differences between the risk-seeking and risk-averse agent are: after learning episode 400 the risk-seeking success rate for this subplot is more than 80%, while for the risk-averse agent is less than 40%, therefore the latter is developing a policy which ignores the square \$ and instead heads directly towards the exit square.

Each experiment was tested with 4 different values of B_0: 20, 30, 40 and 50. Regardless the value of α and the agent's risk profile, the more budget the faster the agent learns its final policy, however it also requires more money to learn the same policy.

Now lets analyze the impact of modifying the budget by observing Fig. 3. As mentioned before 4 different values of B_0. Also if the agent runs out of budget (condition Tn_2), it receives a reward of -10. These two facts support the agents behavior near learning episode 0 on subplot FR, which plots the variable B_t at the moment t when either condition Tn_1 or Tn_2 is reached: the minimum value B_t that the agent can reach corresponds to $-B_0$ minus the penalization for running out of budget. For this reason the minimum value for B_t of series $COIN9\text{-}0.3\text{-}1\text{-}20$ is -30. Another important observation on FR subplots is that as B_0 increases, it allows the agent to learn its policy faster, however the tradeoff is that the learning process becomes more expensive since the area below the x-axis and the FR curve increases proportionally to B_0. Table 4 shows this area from $t = 0$ to time $t = 100$ and from time $t = 0$ to $t = 200$ for the FR subplot shown on Fig. 3.

Our final analysis is related to the impact of learning rate α. A high learning rate value was used on the experiments reported previously, therefore, just to prove that this decision has no relevant impact on the learned policy we will report an experiment with a low learning rate value. Figure 5 shows the results of the same experiment plotted on Fig. 3, which corresponds to a risk-seeking agent. The only difference is the value of $alpha$: the first one uses Q-Learning with $\alpha = 0.9$ and the latter uses $\alpha = 0.3$. Both agents learn a policy which follows $Route_1$, however the agent which uses $\alpha = 0.9$ develops its policy faster. Since the rewards are stationary we expected this behavior. Therefore we can state that the value of α makes no impact on our process of modeling risk profile by using reward shaping.

Table 4. Learning cost for different values of B_0

B_0	FR @ 100	FR @ 200
20	44,896	84,799
30	55,937	88,879
40	59,931	86,766
50	63,647	91,013

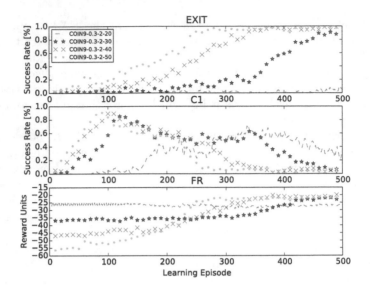

Fig. 4. Risk-averse agent experiment.

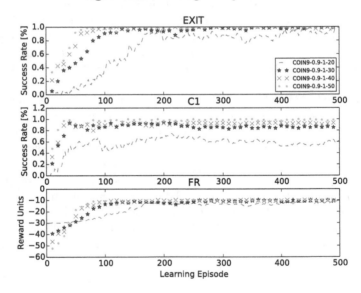

Fig. 5. Risk-seeking experiment with low learning rate value.

So far our experiments showed that: it is possible to learn an optimal policy with a constrained budget, given that this budget is large enough to cover a minimum cost[2]; and that with a high budget the agent learns the optimal policy

---

[2] In this case a transportation cost set by the distance between the agent's starting point and the exit square.

faster, but since it is allowed to explore more, the learning cost increases as shown in Table 4.

6 Conclusions and Future Work

This work provided some references about the lack of consideration of budget in RL, and how relevant the concept of budget is within several RL applications. As mentioned before, RL techniques attempt to optimize a reward no matter the learning cost, and this situation is not suitable when the learning process requires resources which are limited, as occurs in many real scenarios. Furthermore, not all agents share the same risk tolerance, however RL techniques do not account for this issue either.

For these reasons we aimed to work on a SRL technique which consider budget and risk profile. We provided the agent a negative reward for running out of resources and weighted this reward by using reward shaping. The negative reward received by the agent if it runs out of budget forces it to consider the risk of trying to improve its final reward, while the shaping process allows us to quantify a qualitative agent's attribute, such as how risk tolerant it is, and use it as an input for the learning process.

Our results showed that it is possible to learn a policy even on a constrained budget. They also showed that the reward shaping process helps learning algorithm to understand the agent's risk preferences, and allows it to learn different policies according to its risk profile, i.e. sometimes the agent decided that, in order to maintain a safety level, it was not an optimal policy to visit square $.

Our future work will be focused on developing strategies to determine the minimum budget required by a given scenario in order to determine if a solution is reachable with a certain budget or at a given time horizon for any given profile. This task is not hard to do in our simple grid world, however it becomes a very important issue in non-trivial scenarios. We also aim to develop exploration strategies which consider a budget input; the idea that justifies to have this target is that an agent should be less prone to exploring new actions as it resources run low. There is also work to be done related to improving the exploration process accounting again for budget and risk profile.

References

1. Sutton, R.S., Barto, A.G.: Reinforcement Learning: An Introduction. MIT Press, Cambridge (1998)
2. Thrun, S.B.: Efficient Exploration in Reinforcement Learning. Springer, New York (1992)
3. Mahadevan, S., Connell, J.: Automatic programming of behavior-based robots using reinforcement learning. Artif. Intell. **55**(2), 311–365 (1992)
4. Nevmyvaka, Y., Feng, Y., Kearns, M.: Reinforcement learning for optimized trade execution. In: Proceedings of the 23rd International Conference on Machine Learning, pp. 673–680. ACM (2006)

5. Thomas, P.S.: Safe reinforcement learning (2015)
6. García, J., Fernández, F.: A comprehensive survey on safe reinforcement learning. J. Mach. Learn. Res. **16**, 1437–1480 (2015)
7. Mihatsch, O., Neuneier, R.: Risk-sensitive reinforcement learning. Mach. Learn. **49**(2–3), 267–290 (2002)
8. Heger, M.: Consideration of risk in reinforcement learning. In: Proceedings of the Eleventh International Conference on Machine Learning, pp. 105–111 (1994)
9. Coraluppi, S.P., Marcus, S.I.: Risk-sensitive and minimax control of discrete-time, finite-state Markov decision processes. Automatica **35**(2), 301–309 (1999)
10. Driessens, K., Džeroski, S.: Integrating guidance into relational reinforcement learning. Mach. Learn. **57**(3), 271–304 (2004)
11. Martín H., J.A., Lope, J.: Learning autonomous helicopter flight with evolutionary reinforcement learning. In: Moreno-Díaz, R., Pichler, F., Quesada-Arencibia, A. (eds.) EUROCAST 2009. LNCS, vol. 5717, pp. 75–82. Springer, Heidelberg (2009). doi:10.1007/978-3-642-04772-5_11
12. Abbeel, P.: Apprenticeship learning and reinforcement learning with application to robotic control. In: ProQuest (2008)
13. Garcia, J., Fernández, F.: Safe exploration of state and action spaces in reinforcement learning. J. Artif. Intell. Res. **45**, 515–564 (2012)
14. Ng, A.Y., Harada, D., Russell, S.: Policy invariance under reward transformations: theory and application to reward shaping. In: ICML, vol. 99, pp. 278–287 (1999)
15. Dorigo, M., Colombetti, M.: Robot shaping: developing autonomous agents through learning. Artif. Intell. **71**(2), 321–370 (1994)
16. Devlin, S., Kudenko, D.: Dynamic potential-based reward shaping. In: Proceedings of the 11th International Conference on Autonomous Agents and Multiagent Systems, vol. 1, pp. 433–440 (2012)
17. Black, P.E.: Manhattan distance. Dict. Algorithms Data Struct. **18**, 2012 (2006)
18. MacGlashan, J.: Brown UMBC reinforcement learning and planning BURLAP. http://burlap.cs.brown.edu/. Accessed 5 Jan 2017

Modeling Dependencies
in Supervised Classification

Rogelio Salinas-Gutiérrez[1](✉), Angélica Hernández-Quintero[1],
Oscar Dalmau-Cedeño[2], and Ángela Paulina Pérez-Díaz[1]

[1] Universidad Autónoma de Aguascalientes, Aguascalientes, Mexico
rsalinas@correo.uaa.mx, angelica.hernandez.q@gmail.com,
alegna_287@hotmail.com
[2] Centro de Investigación en Matemáticas, Guanajuato, Mexico
osdalmau@gmail.com

Abstract. In this paper we show the advantage of modeling dependencies in supervised classification. The dependencies among variables in a multivariate data set can be linear or non linear. For this reason, it is important to consider flexible tools for modeling such dependencies. Copula functions are able to model different kinds of dependence structures. These copulas were studied and applied in classification of pixels. The results show that the performance of classifiers is improved when using copula functions.

Keywords: Copula function · Graphical model · Likelihood function

1 Introduction

Classification is an important task in Pattern Recognition. The goal in supervised classification is to assign a new object to a category based on its features [1]. Applications in this subject use training data in order to model the distribution of features for each class. In this work we propose the use of bivariate copula functions in order to design a probabilistic model. The copula function allows us to properly model dependencies, not necessarily linear dependencies, among the object features.

By using copula theory, a joint distribution can be built with a copula function and, possibly, several different marginal distributions. Copula theory has been used for modeling multivariate distributions in *unsupervised learning* problems [3,5,9,13] as well as in *supervised classification* [4,6,7,10,12,14,15]. For instance, in [4], a challenging classification problem is solved by means of copula functions and vine graphical models. However, all marginal distributions are modelled with gaussian distributions and the copula parameter is calculated by inverting Kendall's tau. In [10,15], simulated and real data are used to solve classification problems within the framework of copula theory. No graphical models are employed and marginal distributions are based on parametric models. In this paper, we employed flexible marginal distributions such as Gaussian kernels and

J.A. Carrasco-Ochoa et al. (Eds.): MCPR 2017, LNCS 10267, pp. 117–126, 2017.
DOI: 10.1007/978-3-319-59226-8_12

the copula parameter is estimated by using the maximum likelihood method. Moreover, the proposed classifier takes into account the most important dependencies by means of a graphical model. The reader interested in applications of copula theory in supervised classification is referred to [6,7,12,14].

The content of the paper is the following: Sect. 2 is a short introduction to copula functions, Sect. 3 presents a copula based probabilistic model for classification. Section 4 presents the experimental setting to classify an image database, and Sect. 5 summarizes the results.

2   Copula Functions

The copula theory was introduced by [11] to separate the effect of dependence from the effect of marginal distributions in a joint distribution. Although copula functions can model linear and nonlinear dependencies, they have rarely been used in supervised classification where nonlinear dependencies are common and need to be represented.

Definition 1. *A copula function is a joint distribution function of standard uniform random variables. That is,*

$$C(u_1, \ldots, u_d) = Pr[U_1 \leq u_1, \ldots, U_d \leq u_d],$$

where $U_i \sim U(0,1)$ for $i = 1, \ldots, d$.

Due to the Sklar's Theorem, *any* d-dimensional density f can be represented as

$$f(x_1, \ldots, x_d) = c(F_1(x_1), \ldots, F_d(x_d)) \cdot \prod_{i=1}^{d} f_i(x_i), \tag{1}$$

where c is the density of the copula C, $F_i(x_i)$ is the marginal distribution function of random variable x_i, and $f_i(x_i)$ is the marginal density of variable x_i. Equation (1) shows that the dependence structure is modeled by the copula function. This expression separates any joint density function into the product of copula density and marginal densities. This is contrasted with the usual way to model multivariate distributions, which suffers from the restriction that the marginal distributions are usually of the same type. The separation between marginal distributions and a dependence structure explains the modeling flexibility given by copula functions.

In this paper we use two-dimensional parametric copula functions to model the dependence structure of random variables associated by a joint distribution function. The densities of these copula functions are shown in Table 1. We consider the Farlie-Gumbel-Morgenstern (FGM) copula function, elliptical copulas (Gaussian) and archimedean copulas (Independent, Ali-Mikhail-Haq (AMH), Clayton, Frank, Gumbel). These copula functions have been chosen because they cover a wide range of dependencies. For instance, the AMH, Clayton,

Table 1. Bivariate copula densities.

Copula	Description
Independent	$c(u_1, u_2) = 1$
AMH	$c(u_1, u_2; \theta) =$ $\dfrac{1 + \theta(u_1 + u_2 + u_1 u_2 - 2) - \theta^2(u_1 + u_2 - u_1 u_2 - 1)}{(1 - \theta(1 - u_1)(1 - u_2))^3}$
Clayton	$c(u_1, u_2; \theta) =$ $(1 + \theta)(u_1 u_2)^{-\theta - 1}\left(u_1^{-\theta} + u_2^{-\theta} - 1\right)^{-2 - 1/\theta}$
FGM	$c(u_1, u_2; \theta) = 1 + \theta(1 - 2u_1)(1 - 2u_2)$
Frank	$c(u_1, u_2; \theta) = \dfrac{-\theta(e^{-\theta} - 1)e^{-\theta(u_1 + u_2)}}{((e^{-\theta u_1} - 1)(e^{-\theta u_2} - 1) + (e^{-\theta} - 1))^2}$
Gaussian	$c(u_1, u_2; \theta) = \left(1 - \theta^2\right)^{-1/2} \exp\left(-\dfrac{(x_1^2 + x_2^2 - 2\theta x_1 x_2)}{2(1 - \theta^2)} + \dfrac{(x_1^2 + x_2^2)}{2}\right)$ where $x_1 = \Phi^{-1}(u_1)$ and $x_2 = \Phi^{-1}(u_2)$
Gumbel	$c(u_1, u_2; \theta) = \dfrac{C(u_1, u_2)}{u_1 u_2} \dfrac{(\tilde{u}_1 \tilde{u}_2)^{\theta - 1}}{(\tilde{u}_1^\theta + \tilde{u}_2^\theta)^{2 - 1/\theta}}\left((\tilde{u}_1^\theta + \tilde{u}_2^\theta)^{1/\theta} + \theta - 1\right)$ where $\tilde{u}_1 = -\ln(u_1)$ and $\tilde{u}_2 = -\ln(u_2)$

FGM, Frank and Gaussian copula functions can model negative and positive dependences between the marginals. One exception is the Gumbel copula, which does not model negative dependence. The AMH and FGM copula functions are adequate for marginals with modest dependence. When dependence is strong between extremes values, the Clayton and Gumbel copula functions can model left and right tail association respectively. The Frank copula is appropriate for data that exhibit weak dependence between extreme values and strong dependence between centered values, while the Gaussian copula is adequate for data that exhibit weak dependence between centered values and strong dependence between extreme values. In general, when the Gaussian copula is used with standard Gaussian marginals, then the joint probabilistic model is equivalent to a multivariate normal distribution.

The dependence parameter θ of a bivariate copula function can be estimated using the maximum likelihood method (ML). To do so, the one-dimensional log-likelihood function

$$\ell\left(\theta; \{(u_{1i}, u_{2i})\}_{i=1}^n\right) = \sum_{i=1}^n \log\left(c(u_{1i}, u_{2i}; \theta)\right), \tag{2}$$

is maximized. Assuming the marginal distributions are known, the pseudo copula observations $\{(u_{1i}, u_{2i})\}_{i=1}^n$ in Eq. (2) are obtained by using the marginal distribution functions of variables X_1 and X_2. Once the maximum likelihood estimator of θ has been found, it is represented by the notation $\hat{\theta}$. It has been shown in [16] that the ML estimator $\hat{\theta}$ has better properties than other estimators.

3   The Probabilistic Model for Classification

The proposed classifier explicitly considers dependencies among variables. The dependence structure for the design of the probabilistic classifier is based on a chain graphical model. Such model, for a d-dimensional continuous random vector $\mathbf{X}$, represents a probabilistic model with the following density:

$$f_{\text{chain}}(\mathbf{x}) = f(x_{\alpha_1}) \prod_{i=2}^{d} f\left(x_{\alpha_i}|x_{\alpha_{(i-1)}}\right), \tag{3}$$

where $\boldsymbol{\alpha} = (\alpha_1, \ldots, \alpha_d)$ is a permutation of the integers between 1 and d. Figure 1 shows an example of a chain graphical model for a three dimensional vector. Notice that a permutation could not be unique, in the sense that different permutations could yield the same density values in (3).

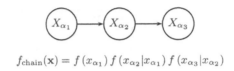

$$f_{\text{chain}}(\mathbf{x}) = f(x_{\alpha_1}) f(x_{\alpha_2}|x_{\alpha_1}) f(x_{\alpha_3}|x_{\alpha_2})$$

Fig. 1. Joint distribution over three variables represented by a chain graphical model.

In practice the permutation $\boldsymbol{\alpha}$ is unknown and the chain graphical model must be learnt from data. A way of choosing the permutation $\boldsymbol{\alpha}$ is based on the Kullback-Leibler divergence (D_{KL}). This divergence is an information measure between two distributions. It is always non-negative for any two distributions, and is zero if and only if the distributions are identical. Hence, the Kullback-Leibler divergence can be interpreted as a measure of the dissimilarity between two distributions. Then, the goal is to choose a permutation $\boldsymbol{\alpha}$ that minimizes the Kullback-Leibler divergence between the true distribution $f(\mathbf{x})$ of the data set and the distribution associated to a chain model, $f_{\text{chain}}(\mathbf{x})$. For instance, the Kullback-Leibler divergence between joint densities f and f_{chain} for a continuous random vector $\mathbf{X} = (X_1, X_2, X_3)$ is given by:

$$D_{KL}(f\|f_{\text{chain}}) = E_f\left[\log \frac{f(\mathbf{x})}{f_{\text{chain}}(\mathbf{x})}\right]$$
$$= -H(\mathbf{X}) + \int \log\left(f(x_{\alpha_1}) f(x_{\alpha_2}|x_{\alpha_1}) f(x_{\alpha_3}|x_{\alpha_2})\right) f d\mathbf{x}. \tag{4}$$

The first term in Eq. (4), $H(\mathbf{X})$, is the entropy of the joint distribution $f(\mathbf{x})$ and does not depend on the permutation $\boldsymbol{\alpha}$. By using copula theory and Eq. (1), the second term can be decomposed into the product of marginal distributions and bivariate copula functions.

$$D_{KL}\left(f\|f_{\text{chain}}\right) = -H(\mathbf{X}) + \sum_{i=1}^{d} H(X_i)$$

$$- \int \log\left(c\left(u_{\alpha_1}, u_{\alpha_2}; \hat{\theta}_{\alpha_1,\alpha_2}\right)\right) f dx$$

$$- \int \log\left(c\left(u_{\alpha_2}, u_{\alpha_3}; \hat{\theta}_{\alpha_2,\alpha_3}\right)\right) f dx. \tag{5}$$

The second term of Eq. (5), the sum of marginal entropies, also does not depend on the permutation $\boldsymbol{\alpha}$. Therefore, minimizing Eq. (5) is equivalent to maximize the sum of the last two terms. Once a sample of size n is obtained from the joint density f, the last two terms can be approximated by a Monte Carlo approach:

$$\int \log\left(c\left(u_{\alpha_1}, u_{\alpha_2}; \hat{\theta}_{\alpha_1,\alpha_2}\right)\right) f dx \approx \frac{1}{n} \sum_{i=1}^{n} \log\left(c\left(u_{1i}, u_{2i}; \hat{\theta}_{\alpha_1,\alpha_2}\right)\right). \tag{6}$$

Through Eq. (6), the D_{KL} is minimized by maximizing the sum of the log-likelihood for the copula parameters. It is worth to noting that the log-likelihood allows us to estimate the copula parameter and to select the appropriate permutation $\boldsymbol{\alpha}$. Finally, by means of copula theory, a chain graphical model for a three dimensional vector has the density

$$f_{\text{chain}}(\mathbf{x}) = f(x_{\alpha_1}) f(x_{\alpha_2}) f(x_{\alpha_3}) c(u_{\alpha_1}, u_{\alpha_2}) c(u_{\alpha_2}, u_{\alpha_3}) \tag{7}$$

3.1   The Probabilistic Classifier

Here, we present the incorporation of bivariate copula functions and a chain graphical model in order to design a probabilistic classifier.

The Bayes' theorem states the following:

$$P(K = k|\mathbf{X} = \mathbf{x}) = \frac{P(\mathbf{X} = \mathbf{x}|K = k) \times P(K = k)}{P(\mathbf{X} = \mathbf{x})}, \tag{8}$$

where $P(K = k|\mathbf{X} = \mathbf{x})$ is the posterior probability, $P(\mathbf{X} = \mathbf{x}|K = k)$ is the likelihood function, $P(K = k)$ is the prior probability and $P(\mathbf{X} = \mathbf{x})$ is the data probability.

Equation (8) has been used as a tool in supervised classification. A probabilistic classifier can be designed comparing the posterior probability that an object belongs to the class K given its features $\mathbf{X}$. The object is then assigned to the class with the highest posterior probability. For practical reasons, the data probability $P(\mathbf{X})$ does not need to be evaluated for comparing posterior probabilities. Furthermore, the prior probability $P(K)$ can be substituted by a uniform distribution if the user does not have an informative distribution.

The joint density in Eq. (7) can be used for modeling the likelihood function in Eq. (8). In this case, the Bayes' theorem can be written as:

$$P(K = k|\mathbf{x}) = \frac{\prod_{j=1}^{2} c(F_{\alpha_j}, F_{\alpha_{(j+1)}}|k; \hat{\theta}_{\alpha_j,\alpha_{(j+1)}}) \cdot \prod_{i=1}^{3} f_i(x_i|k) \cdot P(K = k)}{f(x_1, x_2, x_3)} \tag{9}$$

where F_i are the marginal distribution functions and f_i are the marginal densities for each feature. The function c is a bivariate copula density taken from Table 1. As can be seen in Eq. (9), each class determines a likelihood function.

4   Experiments

We use Eq. (9) and copula functions from Table 1 in order to classify pixels of 50 test images. Hence, we prove seven probabilistic classifiers. The image database was used in [2] and is available online [8]. This image database provides information about two classes: the foreground and the background. The training data and the test data are contained in the labelling-lasso files [8], whereas the correct classification is contained in the segmentation files. Figure 2 shows the description of one image from the database. Although the database is used for segmentation purposes, the aim of this work is to model dependencies in supervised classification. Only color features are considered for classifying pixels.

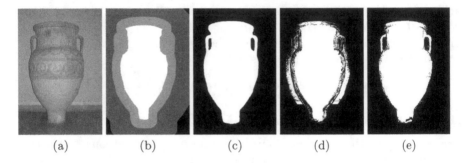

(a)   (b)   (c)   (d)   (e)

Fig. 2. (a) The color image. (b) The labelling-lasso image with the training data for background (dark gray), for foreground (white) and the test data (gray). (c) The correct classification with foreground (white) and background (black). (d) Classification made by independence. (e) Classification made by Frank Copula.

Three evaluation measures are used in this work: *accuracy, sensitivity* and *specificity*. These measures are described in Fig. 3. The sensitivity and specificity measures explain the percentage of well classified pixels for each class, foreground and background, respectively. We define the positive class as the foreground and the negative class as the background.

4.1   Numerical Results

In Table 2 we summarize the measure values reached by the classifiers according to the copula function used to model the dependencies.

To properly compare the performance of the probabilistic classifiers, we conducted an ANOVA test for comparing the accuracy mean among the classifiers. The test reports a statistical difference between Clayton, Frank, Gaussian and

	Truth	
	Positive	Negative
Model Positive	tp	fp
Model Negative	fn	tn

$$accuracy = \frac{tp + tn}{tp + fp + fn + tn}$$

$$sensitivity = \frac{tp}{tp + fn}$$

$$specificity = \frac{tn}{tn + fp}$$

(a)   (b)

Fig. 3. (a) A confusion matrix for binary classification, where tp are true positive, fp false positive, fn false negative, and tn true negative counts. (b) Definitions of accuracy, sensitivity and specificity used in this work.

Table 2. Descriptive results for all evaluation measures. The results are presented in percentages.

Copula	Accuracy		Sensitivity		Specificity	
Model	Mean	Std. dev.	Mean	Std. dev.	Mean	Std. dev.
Independent	79.4	10.8	77.3	16.6	81.3	13.6
AMH	82.9	9.5	80.7	15.9	84.7	11.9
Clayton	86.0	8.5	81.6	16.4	89.5	9.2
FGM	80.9	9.8	78.9	16.5	82.5	13.2
Frank	87.7	7.1	87.1	12.2	88.1	9.0
Gaussian	86.0	10.6	87.1	11.0	85.0	18.6
Gumbel	86.7	8.2	87.0	10.9	86.5	13.2

Gumbel copula functions with respect to the Independent copula (p-value < 0.05). The major difference of accuracy with respect to the independent copula is given by the Frank copula.

4.2   Discussion

According to Table 2, the classifier based on the Frank copula shows the best behavior for accuracy. For sensitivity, Frank and Gaussian copulas provide the best results. The best mean specificity is reached by the classifier based on the Clayton copula.

As can be seen, the average performance of a classifier is improved by the incorporation of the copula functions. The lowest average performance corresponds to the classifier that uses the independence assumption. Figure 4 shows how the accuracy is increased when dependencies are taken into account by the probabilistic classifier. The line of Fig. 4(a) represents the identity function, so the points above this line correspond to a better accuracy than the

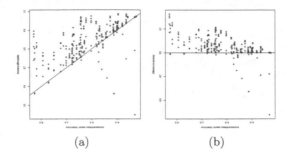

(a)   (b)

Fig. 4. (a) Scatterplot of the accuracy values between classifier based on independence assumption (horizontal axis) and classifiers based on copula functions (vertical axis). (b) The gain of accuracy by using copula functions.

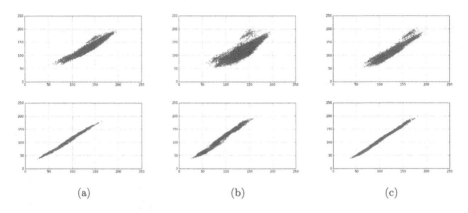

(a)   (b)   (c)

Fig. 5. The first line shows the scatterplot among (a) red and green, (b) red and blue, and (c) green and blue colors for the foreground class. The second line similarly shows the scatterplots for the background class. (Color figure online)

accuracy achieved by the classifier based on the independent copula. To get a better insight, Fig. 4(b) shows the difference in accuracy between using copula functions respect to the naive classifier (independent copula).

Table 2 also shows information about the standard deviations for each evaluation measure. For accuracy, the standard deviation indicates that using a Frank copula in pixel classification is more consistent than the other classifiers.

Figure 2 shows the results of one of the 50 images mentioned before, once we worked on them. In (d), we can see the resultant image when it is classified by independence, (e) shows the same image classified by Frank copula. It is possible to visually perceive the improvement that the use of Frank copula provides to the classifier. For this image, the color data for each class is shown in Fig. 5. In this case, it can be seen that the dependence structure does not correspond to the dependence structure of a bivariate Gaussian distribution. According to the numerical results, the copula Frank is the best model for this kind of dependence.

5   Conclusions

In this paper we have compared the performance of several copula based probabilistic classifiers. The results show that the dependence among features provides important information for supervised classifying. For the images used in this work, the Gumbel copula performs very well in most of the cases. One advantage of using a chain graphical model consists in detecting the most important dependencies among variables. This can be valuable for different applications where associations among variables gives additional knowledge of the problem. Though accuracy is increased by the classifiers based on copula functions, the selection of the copula function has relevant consequences for the performance of the classifier. For instance, in Fig. 4, a few classifiers do not improve the performance achieved by the classifier based on the independent copula. It suggests more experiments are needed in order to select the adequate copula function for a given problem. Moreover, as future work, the classifier based on copula functions must be proved in other datasets and compared with other classifiers in order to achieve a better insight of its benefits and limitations.

Acknowledgments. The authors acknowledge the financial support from the National Council of Science and Technology of México (CONACyT, grant number 258033) and from the Universidad Autónoma de Aguascalientes (project number PIM17-3). The student Ángela Paulina also acknowledges to the CONACYT for the financial support given through the scholarship number 628293.

References

1. Bishop, C.: Pattern Recognition and Machine Learning. Information Science and Statistics. Springer, New York (2007)
2. Blake, A., Rother, C., Brown, M., Perez, P., Torr, P.: Interactive Image Segmentation Using an Adaptive GMMRF Model. In: Pajdla, T., Matas, J. (eds.) ECCV 2004. LNCS, vol. 3021, pp. 428–441. Springer, Heidelberg (2004). doi:10.1007/978-3-540-24670-1_33
3. Brunel, N., Pieczynski, W., Derrode, S.: Copulas in vectorial hidden Markov chains for multicomponent image segmentation. In: Proceedings of the 2005 IEEE International Conference on Acoustics, Speech and Signal Processing (ICASSP 2005), pp. 717–720 (2005). doi:10.1109/ICASSP.2005.1415505
4. Carrera, D., Santana, R., Lozano, J.: Vine copula classifiers for the mind reading problem. Prog. Artif. Intell. (2016). doi:10.1007/s13748-016-0095-z
5. Mercier, G., Bouchemakh, L., Smara, Y.: The use of multidimensional Copulas to describe amplitude distribution of polarimetric SAR Data. In: IGARSS 2007 (2007). doi:10.1109/IGARSS.2007.4423284
6. Ouhbi, N., Voivret, C., Perrin, G., Roux, J.: Real grain shape analysis: characterization and generation of representative virtual grains. application to railway ballast. In: Oñate, E., Bischoff, M., Owen, D., Wriggers, P., Zohdi, T. (eds.) Proceedings of the IV International Conference on Particle-based Methods Fundamentals and Applications (2015)
7. Resti, Y.: Dependence in classification of aluminium waste. J. Phys. Conf. Ser. **622**(012052), 1–6 (2015). doi:10.1088/1742-6596/622/1/012052

8. Rother, C., Kolmogorov, V., Blake, A., Brown, M.: Image and video editing. http://research.microsoft.com/en-us/um/cambridge/projects/visionimagevideoediting/segmentation/grabcut.htm
9. Sakji-Nsibi, S., Benazza-Benyahia, A.: Multivariate indexing of multichannel images based on the copula theory. In: IPTA08 (2008)
10. Sen, S., Diawara, N., Iftekharuddin, K.: Statistical pattern recognition using Gaussian Copula. J. Stat. Theor. Pract. **9**(4), 768–777 (2015). doi:10.1080/15598608.2015.1008607
11. Sklar, A.: Fonctions de répartition à n dimensions et leurs marges. Publications de l'Institut de Statistique de l'Université de Paris **8**, 229–231 (1959)
12. Slechan, L., Górecki, J.: On the accuracy of Copula-based Bayesian classifiers: an experimental comparison with neural networks. In: Núñez, M., Nguyen, N.T., Camacho, D., Trawiński, B. (eds.) ICCCI 2015. LNCS, vol. 9329, pp. 485–493. Springer, Cham (2015). doi:10.1007/978-3-319-24069-5_46
13. Stitou, Y., Lasmar, N., Berthoumieu, Y.: Copulas based multivariate gamma modeling for texture classification. In: Proceedings of the 2009 IEEE International Conference on Acoustics, Speech and Signal Processing (ICASSP 2009), pp. 1045–1048. IEEE Computer Society, Washington, DC (2009). doi:10.1109/ICASSP.2009.4959766
14. Voisin, A., Krylov, V., Moser, G., Serpico, S., Zerubia, J.: Classification of very high resolution SAR images of urban areas using copulas and texture in a hierarchical Markov random field model. IEEE Geosci. Remote Sens. Lett. **10**(1), 96–100 (2013). doi:10.1109/LGRS.2012.2193869
15. Ščavnický, M.: A study of applying copulas in data mining. Master's thesis, Charles University in Prague, Prague (2013)
16. Weiß, G.: Copula parameter estimation by maximum-likelihood and minimum-distance estimators: a simulation study. Comput. Stat. **26**(1), 31–54 (2011). doi:10.1007/s00180-010-0203-7

Fast-BR vs. Fast-CT_EXT:
An Empirical Performance Study

Vladímir Rodríguez-Diez[1,2]([⊠]), José Fco. Martínez-Trinidad[1],
J. Ariel Carrasco-Ochoa[1], and Manuel S. Lazo-Cortés[1]

[1] Coordinación de Ciencias Computacionales, Instituto Nacional de Astrofísica,
Óptica y Electrónica, Luis Enrique Erro # 1, Tonantzintla, Puebla, Mexico
vladimir.rdguez@gmail.com, mlazo@inaoep.mx
[2] Universidad de Camagüey, Circunvalación Nte. km 5 1/2, Camagüey, Cuba

Abstract. Testor Theory allows performing feature selection in supervised classification problems through typical testors. Typical testors are
irreducible subsets of features preserving the object discernibility ability
of the original set of features. However, finding the complete set of typical
testors for a dataset requires a high computational effort. In this paper,
we make an empirical study about the performance of two of the most
recent and fastest algorithms of the state of the art for computing typical testors, regarding the density of the basic matrix. For our study we
use synthetic basic matrices to control their characteristics, but we also
include public standard datasets taken from the UCI machine learning
repository. Finally, we discuss our conclusions drawn from this study.

Keywords: Testor theory · Algorithms · Basic matrix

1 Introduction

Feature selection is an important task for supervised classification. It consists
in identifying those features that provide relevant information for the classification process. Feature selection may improve the efficiency of pattern recognition
and machine learning tools without significantly degrading their efficacy. In the
Logical Combinatorial Pattern Recognition [13], Testor Theory emerges as a
solution to feature selection [10,17]. A testor is a subset of features which allows
discerning between objects from different classes by using only its features. A
Typical Testor (TT) is defined as a testor which is minimal with respect to inclusion. The main practical limitation of Testor Theory is that finding all TTs has
exponential complexity regarding the number of attributes in the dataset [7].

The algorithms BT and TB [16] were the first approaches to computing all
TTs using the basic matrix. The basic matrix is a reduced representation of the
comparison between all pairs of objects in a sample belonging to different classes.
These algorithms codify a feature subset as a binary word with as many bits as
features in the dataset. In this codification, a 1 represents the inclusion of the
corresponding feature in the subset. In this way, the candidate feature subsets

© Springer International Publishing AG 2017
J.A. Carrasco-Ochoa et al. (Eds.): MCPR 2017, LNCS 10267, pp. 127–136, 2017.
DOI: 10.1007/978-3-319-59226-8_13

are evaluated following the order of the binary numbers, and some candidates are skipped based on the result of previous evaluations over the basic matrix. Several years later, a new algorithm named REC was introduced [15]. The main drawback of REC is that it operates directly over the dataset (instead of the basic matrix), handling a huge amount of superfluous information. With the purpose of solving this issue, the CER algorithm was proposed [4], which also introduced a different traversing order for the candidate subsets.

A new traversing order for the candidate subsets was introduced in [20] for the LEX algorithm. This new traversing order (which resembles the lexicographical order in which strings of characters are compared) is used in the most recent and fastest algorithms for computing all TTs. This algorithm also introduced the concept of gap into its pruning process. Once obtained a TT (or a non testor) candidate which includes the last feature in the dataset, the concept of gap allows avoiding the evaluation of any subset of this candidate. In LEX, the typical condition is verified first and the testor condition is evaluated only for those potentially TTs. Later, the CT_EXT algorithm for computing all TTs [18] was proposed. This algorithm searches for testors without verifying the typical condition, following the traversing order of LEX. This approach, usually leads to a higher number of candidate evaluations, in comparison to LEX. However, the cost of each candidate evaluation is lower. The authors of CT_EXT show that it is faster than the previous existing algorithms for most datasets. Then, the BR algorithm [11] was presented. This new Recursive algorithm based on Binary operations is similar to LEX but its recursive nature encloses a reduction in the number of evaluated candidates. Given a candidate subset, the remaining features (according to the lexicographical order) are tested, and those being rejected are excluded from subsequent evaluations in the supersets of the current candidate.

In [19] a cumulative procedure for the CT_EXT algorithm was presented. This fast-CT_EXT implementation drastically reduces the runtime for most datasets at no extra cost. Then, in [12] gap elimination and column reduction are added to BR, also using binary cumulative operations. The main drawback of fast-BR and BR is, as in LEX, the high cost of evaluating the typical condition for each contributing candidate.

We have succinctly described above the evolution of the so called *external scale* algorithms for typical testor computation. In addition, a different kind of algorithms such as CT [6], CC [1] and YYC [3] have been developed. Algorithms of this kind are called *internal scale* algorithms, and they analyze the distribution of 1's into the basic matrix to find out some conditions to guarantee that a subset of attributes is a typical testor. Internal scale algorithms usually evaluate less candidates than external scale algorithms but each candidate evaluation has a higher computational cost. Therefore, the search for fast algorithms for computing typical testors has been biased to external scale algorithms [3].

Recently, a thorough study presented in [2] concluded that no single typical testor finding algorithm has the best performance for any given problem. Other studies [12,14] performed experiments by categorizing the basic matrices according to the density of 1's they have; i.e. the number of 1's divided by the total number of cells of the matrix. Furthermore, in [9], the authors highlighted that the number of rows, the density of 1's and the number of typical testors of the basic matrix impact the performance of the algorithms for computing typical testors. With these precedents, in this paper, we present an empirical study to identify a relationship between the density of the basic matrix and the performance of the most successful algorithms: fast-CT_EXT [19] and fast-BR [12]. From this empirical study, we obtained a simple rule to determine a priori the fastest algorithm for a given dataset. Finally we validate our rule on standard datasets from the UCI machine learning repository [5].

We have organized the rest of this paper in the following way. In Sect. 2, the theoretical basis for this work are introduced. In Sect. 3, we present our comparative study over synthetic basic matrices and discuss the results. Then, in Sect. 4, the conclusions drawn from the empirical study are validated over standard datasets from the UCI machine learning repository. Finally, in Sect. 5, our conclusions are presented.

2 Basic Concepts

In this section, we introduce the main concepts, definitions and propositions supporting the pruning strategies of fast-CT_EXT and fast-BR. Here, we aim to provide the key elements to understand the differences between these two algorithms.

Let DS be a dataset with k objects described by n attributes (features) and grouped in r classes. Each attribute in the set of attributes $R = \{x_1, ..., x_n\}$, has a predefined Boolean comparison criterion. Let DM be the binary comparison matrix obtained from feature wise comparing every pair of objects in DS belonging to different classes. Each comparison of a pair of objects adds a row to DM with $0 -$ equal, $1 =$ different in the corresponding attribute position (column). DM has m rows and n columns. Comparisons generating a row with only 0's, hereinafter referred to as *empty row*, imply that two objects from different classes are equal according to their attribute values.

Definition 1. *Let $T \subseteq R$ be a subset of attributes from DS. We say that T is a testor if in the sub-matrix of DM formed by the columns corresponding to attributes in T, there is no an empty row.*

Usually the number of rows in DM (m) is large. In [10] a reduction of DM without loss of relevant information is proposed, and it was proved that this reduced matrix, called *basic matrix* (BM) has the same set of testors as DM. Then, we can substitute DM by BM in the Definition 1.

Definition 2. *A subset of attributes $T \subseteq R$ is a typical testor in BM iff T is a testor and $\forall x_i \in T, T \backslash x_i$ is not a testor.*

2.1 Concepts for Fast-CT_EXT

Definition 3. *Given $T \subseteq R$ and $x_i \in R$ such that $x_i \notin T$. We say that x_i contributes to T iff the sub-matrix of BM formed with only those attributes in T has more empty rows than the sub-matrix formed by the attributes in $T \cup \{x_i\}$.*

The core of the CT_EXT algorithm is supported by Propositions 1 and 2; which are stated and proved in [19] (Theorems 1 and 2 respectively).

Proposition 1. *Given $T \subseteq R$ and $x_i \in R$ such that $x_i \notin T$. If x_i does not contribute to T, then $T \cup \{x_i\}$ cannot be a subset of any typical testor.*

Proposition 2. *Given $T \subseteq R$ and $Z \subseteq R$ such that $Z \cap T = \emptyset$. If T is a testor, then $T \cup Z$ is a testor too, but it is not a typical testor.*

2.2 Concepts for Fast-BR

In addition to the propositions exposed above, fast-BR is supported by the following propositions; which are stated and proved in [12].

Definition 4. *Given $T \subseteq R$. We call compatibility mask of T, denoted as cm_T, to the binary word in which its jth bit is 1 if the jth row of BM has a 1 in only one of the columns corresponding to attributes in T, otherwise the jth bit in cm_T is 0.*

Proposition 3. *Given $T \subseteq R$ and $x_i \in R$ such that $x_i \notin T$. We denote c_{x_k} to the binary word in which its jth bit is 1 if the jth row of BM has a 1 in the column corresponding to x_k. If $\exists x_k \in T$ such that $cm_{T \cup \{x_i\}} \wedge c_{x_k} = (0,...,0)$, then, $T \cup \{x_i\}$ cannot be a subset of any typical testor, and we will say that x_i is exclusionary with T.*

We will refer to Proposition 3 as exclusion evaluation. Proposition 4 expresses how to apply the exclusion evaluation for determining whether a subset T of features is a typical testor.

Proposition 4. *Given $T \subseteq R$ and $x_i \in R$ such that $x_i \notin T$. The subset $T \cup \{x_i\}$ is a typical testor iff it is a testor and x_i is not exclusionary with T.*

3 Comparative Study

From our literature review, we found two families of external scale algorithms: those evaluating first the testor condition and then verifying the typical condition (exclusion evaluation) [16,18,19], and those evaluating the exclusion first and then the testor condition [11,12,20]. We include in our comparative study the best algorithms of these two families: fast-CT_EXT and fast-BR, respectively. Their candidate evaluation process is illustrated in Fig. 1.

In Fig. 1, it can be seen that both algorithms evaluate first the contribution of a new attribute to the current candidate (*contrib*), by using Proposition 1.

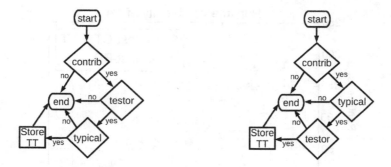

Fig. 1. Candidate evaluation flowchart for fast-CT_EXT (left) and fast-BR (right).

The time complexity of this procedure is $\Theta(m)$, where, m represents the number of rows in the basic matrix. In fast-CT_EXT, the testor condition (*testor*) of candidates with a contributing new attribute is evaluated by using Definition 1. The time complexity of this procedure is also $\Theta(m)$. Then, in fast-CT_EXT, the exclusion evaluation (*typical*) is performed only for the testors found by the algorithm, to determine whether they are or not typical testors. This final evaluation is supported by Proposition 4, and its time complexity is $\Theta(nm)$, where, n represents the number of columns in the basic matrix. In fast-BR, the order of *testor* and *typical* evaluations is inverted, as it can be seen by comparing the two flowcharts shown in Fig. 1. Thus, since the number of testors is usually a small fraction of the total evaluated candidates, we can state that most of the times fast-BR will perform more exclusion evaluations than fast-CT_EXT.

3.1 Comparison over Synthetic Basic Matrices

This experiment was designed to explore the impact of the basic matrix density, i.e., the number of 1's divided by the total number of cells of the matrix; on the relative performance of fast-CT_EXT and fast-BR. We conducted this experiment over 500 randomly generated basic matrices with 2000 rows and 30 columns. The size of these matrices was selected in order to keep reasonable runtime for both algorithms. Our 500 matrices were generated with densities of 1's uniformly distributed in the range (0.16–0.80). The algorithms for this study have been implemented in Java[1].

The 500 matrices were split into 15 groups by discretizing the range of densities, each group having approximately 33 basic matrices. Figure 2 shows the average runtime of all matrices in each group for the two algorithms, as a function of the density of 1's in the basic matrix. In this figure, the vertical bars show the standard deviation of each bin. A third degree polynomial least square fit to the algorithm's runtime was used to determine their intersection, which was in the density value of 0.31. From Fig. 2, it can be seen that fast-CT_EXT

---

[1] The source code as well as all the basic matrices and datasets used in our experiments can be downloaded from http://ccc.inaoep.mx/~ariel/CTBRES.

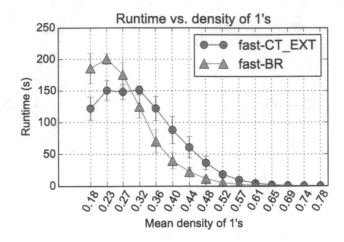

Fig. 2. Runtime vs. density for basic matrices with 2000 rows and 30 columns.

was faster for basic matrices with density under 0.31, while fast-BR was faster for matrices with density above 0.31.

To understand this behavior, we must look back to Fig. 1. The exclusion evaluation has the highest computational complexity in both algorithms $\Theta(nm)$. Exclusionary attributes are found when there is at least one column in the basic matrix, considering only those attributes in the current candidate, that can be removed without increasing the number of empty rows. This condition is more frequent in matrices with higher density, where overlapping of 1's is most likely. For matrices with a relative high density, the heavier candidate evaluation process of fast-BR pays off, because exclusionary attributes are avoided from subsequent evaluations. For matrices with lower density, the simpler approach of fast-CT_EXT results in a faster execution.

In order to provide clarity for the above intuitive explanation of the relation found, we will see two examples in detail. In Table 1, we present a basic matrix with 5 rows and 5 columns which has a density of 0.2. According to our previous result, the strategy followed by fast-CT_EXT should be the fastest for this basic matrix. In Table 2, we present a basic matrix with the same size, which has a density of 0.44. In this way, for this last basic matrix, the strategy followed by fast-BR should be the fastest. Since these are two really small cases, we will not pay attention to the fastest algorithm for each case but we are going to show the conditions that explain the result shown in Fig. 2.

Five candidates need to be evaluated in the lexicographical order to find the only typical testor of the basic matrix shown in Table 1. The candidates are $\{x_0\}$, $\{x_0, x_1\}$, $\{x_0, x_1, x_2\}$, $\{x_0, x_1, x_2, x_3\}$ and the typical testor $\{x_0, x_1, x_2, x_3, x_4\}$. For each candidate, fast-CT_EXT evaluates the contribution and the testor condition which are the procedures with the lowest time complexity ($\Theta(m)$). This algorithm evaluates the exclusion only for the last candidate which is the only testor among the candidates. The strategy followed by fast-BR, evaluates the

Table 1. Identity matrix (5 × 5).

x_0	x_1	x_2	x_3	x_4
1	0	0	0	0
0	1	0	0	0
0	0	1	0	0
0	0	0	1	0
0	0	0	0	1

Table 2. Higher density matrix (5 × 5).

x_0	x_1	x_2	x_3	x_4
1	1	0	0	0
0	0	0	1	1
0	1	1	0	0
1	0	1	1	0
0	0	1	0	1

exclusion for each candidate and the testor condition is evaluated only once for the typical testor. Since the exclusion evaluation has a higher time complexity $(\Theta(nm))$, this is a priori a worse strategy. The key element that makes fast-BR slower for this basic matrix with low density is, however, that no advantage can be taken from the exclusion evaluations since there are not superfluous attributes in this matrix. Let us take, for instance, the candidate $\{x_0, x_1, x_2\}$ which has the compatibility mask $cm_{\{x_0,x_1,x_2\}} = (1,1,1,0,0)$ according to Definition 4. Noting that $cm_{\{x_0,x_1,x_2\}} \wedge c_{x_0} = (1,0,0,0,0)$, $cm_{\{x_0,x_1,x_2\}} \wedge c_{x_1} = (0,1,0,0,0)$ and $cm_{\{x_0,x_1,x_2\}} \wedge c_{x_2} = (0,0,1,0,0)$, the supersets of the candidate $\{x_0, x_1, x_2\}$ cannot be excluded from the search by using Proposition 3. The same result is obtained by evaluating the other candidates.

Looking at the basic matrix shown in Table 2, we can understand the advantage of the strategy followed by fast-BR on matrices with higher density. Taking the candidate $\{x_0, x_1, x_2\}$ again, we compute $cm_{\{x_0,x_1,x_2\}} = (0,0,0,0,1)$ and it can be seen that the higher presence of 1's in the basic matrix leads to a higher presence of 0's in the compatibility mask, according to Definition 4. This time, $cm_{\{x_0,x_1,x_2\}} \wedge c_{x_0} = (0,0,0,0,0)$ and thus, the three supersets of the candidate $\{x_0, x_1, x_2\}$ can be excluded from subsequent evaluations, according to Proposition 3. In fast-CT_EXT, the supersets of this candidate need to be evaluated since it is not a testor and the exclusion is not evaluated for it. The most important aspect about this prune is that the number of excluded candidates in this way is exponentially related to the number of attributes in the dataset. Thus, this results in a great advantage for fast-BR specially in those matrices which have a high number of superfluous (exclusionary) attributes. We have associated this kind of basic matrices with densities above 0.31 in our previous experiment.

For basic matrices with a density value close to 0.31, any algorithm can be the fastest; but we may expect a small runtime difference. The overlapping of algorithms' runtime for this region can be seen in Fig. 2. Then, the rule obtained from synthetic matrices may be expressed as: fast-CT_EXT is faster than fast-BR for basic matrices which have a density of 1's under 0.31, otherwise, fast-BR is faster than fast-CT_EXT.

4   Evaluation on Standard Datasets

In order to evaluate the rule obtained from synthetic basic matrices, we selected 10 datasets from the UCI machine learning repository [5]. The first four datasets have basic matrix densities clearly under 0.31, the fifth dataset (mushroom) is close to this value and the last five datasets have densities clearly above 0.31. Numerical attributes were discretized using the Weka's equal width discretization filter with 10 bins, as described in [8].

Table 3. Fast-CT_EXT and fast-BR runtime for standard real datasets.

| Dataset | | | Basic Matrix | | TTs | Fast-CT_EXT | Fast-BR |
Name	Atts	Instances	Rows	Density		Runtime(ms)	Runtime(ms)
Cpu-act	22	8192	16	0.09	6	**20**	35
Spect-train	23	80	15	0.10	26	**22**	29
Vote	17	435	15	0.11	3	**2**	8
pbc	19	418	45	0.24	88	**28**	41
Mushroom	23	8124	28	0.33	264	**24**	30
Student-por	32	649	8158	0.41	851584	1874570	**161350**
Sponge	46	76	68	0.42	10992	630	**140**
Student-mat	32	395	6904	0.43	679121	1003870	**81820**
Lung-cancer	57	32	237	0.47	4183355	188200	**7340**
Cylinder-bands	40	512	1147	0.55	23534	5030	**530**

Table 3 shows the number of attributes and instances for each dataset, the number of rows and the density of the basic matrix, the number of typical testors and the runtime for fast-CT_EXT and fast-BR. Datasets in Table 3 are sorted by the density of the basic matrix.

We can see from these results that the rule obtained from synthetic matrices is fulfilled in these standard real datasets. Notice that fast-CT_EXT was faster than fast-BR for those basic matrices with a density of 1's under 0.31, while fast-BR was faster than fast-CT_EXT for those basic matrices which have a density of 1's over 0.31, except for the mushroom dataset. However, the density of the basic matrix for mushroom is 0.33, a value too close to the boundary value expressed in our rule of 0.31, and as it was expected, in this region any algorithm could be the fastest. This fact corroborates our predictions for the boundary region. Notice in Table 3 that the bigger differences in runtime are connected to higher basic matrix dimensions.

5   Conclusions

This paper presents an empirical study on the performance of two algorithms for computing all typical testors, regarding the density of 1's of the basic matrix. We selected for this study the best algorithms of two candidate evaluation strategies for external scale algorithms: fast-CT_EXT and fast-BR. A set of synthetic basic matrices were used to explore their performance. From our experimental results, we obtained a simple rule for selecting a priori the fastest algorithm for a given dataset. Then, we successfully confirm this rule over standard datasets from the UCI machine learning repository. We think, however, that the main contribution of this paper is the introduction of a new approach to assess the performance of algorithms for computing typical testors. This approach, which consists in evaluating the algorithms' performance as a function of the density of the basic matrix, provides more substantial information than the traditional evaluation over a random sample of datasets.

Future work can be directed to study the algorithms' relative performance regarding the number of rows and columns in the basic matrix, for different densities.

Acknowledgements. This work was partly supported by National Council of Science and Technology of Mexico under the scholarship grant 399547.

References

1. Águila, L., Ruíz-Shulcloper, J.: Algorithm CC for the elaboration of k-valued information on pattern recognition problems. Math. Sci. J. **5**(3) (1984). (in Spanish)
2. Alba-Cabrera, E., Ibarra-Fiallo, J., Godoy-Calderon, S.: A theoretical and practical framework for assessing the computational behavior of typical testor-finding algorithms. In: Ruiz-Shulcloper, J., Sanniti di Baja, G. (eds.) CIARP 2013. LNCS, vol. 8258, pp. 351–358. Springer, Heidelberg (2013). doi:10.1007/978-3-642-41822-8_44
3. Alba-Cabrera, E., Ibarra-Fiallo, J., Godoy-Calderon, S., Cervantes-Alonso, F.: YYC: a fast performance incremental algorithm for finding typical testors. In: Bayro-Corrochano, E., Hancock, E. (eds.) CIARP 2014. LNCS, vol. 8827, pp. 416–423. Springer, Cham (2014). doi:10.1007/978-3-319-12568-8_51
4. Ayaquica, I.O.: A new external scale algorithm for typical testor computation. In: Memories of the Second Ibero American Workshop on Pattern Recognition, Havana, pp. 141–148 (1997). (in Spanish)
5. Bache, K., Lichman, M.: UCI machine learning repository (2013)
6. Bravo-Martínez, A.: Algorithm CT for calculating the typical testors of k-valued matrix. Math. Sci. J. **4**(2), 123–144 (1983). (in Spanish)
7. Chikalov, I., Lozin, V.V., Lozina, I., Moshkov, M., Nguyen, H.S., Slowron, A., Zielosko, B.: Three Approaches to Data Analysis: Test Theory, Rough Sets and Logical Analysis of Data. Intelligent Systems Reference Library, vol. 41. Springer, Heidelberg (2013)
8. Flores, M.J., Gámez, J.A., Martínez, A.M., Puerta, J.M.: Analyzing the impact of the discretization method when comparing bayesian classifiers. In: García-Pedrajas, N., Herrera, F., Fyfe, C., Benítez, J.M., Ali, M. (eds.) IEA/AIE 2010. LNCS, vol. 6096, pp. 570–579. Springer, Heidelberg (2010). doi:10.1007/978-3-642-13022-9_57
9. González-Guevara, V.I., Godoy-Calderón, S., Alba-Cabrera, E., Ibarra-Fiallo, J.: A mixed learning strategy for finding typical testors in large datasets. In: Pardo, A., Kittler, J. (eds.) Progress in Pattern Recognition, Image Analysis, Computer Vision, and Applications. LNCS, vol. 9423. Springer, Cham (2015). doi:10.1007/978-3-319-25751-8_86

10. Lazo-Cortés, M., Ruíz-Shulcloper, J., Alba-Cabrera, E.: An overview of the evolution of the concept of testor. Pattern Recognit. **34**, 753–762 (2001)
11. Lias-Rodríguez, A., Pons-Porrata, A.: BR: a new method for computing all typical testors. In: Bayro-Corrochano, E., Eklundh, J.-O. (eds.) CIARP 2009. LNCS, vol. 5856, pp. 433–440. Springer, Heidelberg (2009). doi:10.1007/978-3-642-10268-4_50
12. Lias-Rodríguez, A., Sanchez-Díaz, G.: An algorithm for computing typical testors based on elimination of gaps and reduction of columns. Int. J. Pattern Recognit. Artif. Intell. **27**(08), 1350022 (2013)
13. Martínez-Trinidad, J.F., Guzmán-Arenas, A.: The logical combinatorial approach to pattern recognition an overview through selected works. Pattern Recognit. **34**, 741–751 (2001)
14. Rodríguez-Diez, V., Martínez-Trinidad, J.F., Carrasco-Ochoa, J.A., Lazo-Cortés, M., Feregrino-Uribe, C., Cumplido, R.: A fast hardware software platform for computing irreducible testors. Expert Syst. Appl. **42**(24), 9612–9619 (2015)
15. Ruíz-Shulcloper, J., Alba-Cabrera, E., Lazo-Cortés, M.: Introduction to typical testors theory. Green Series, No. 50. CINVESTAV-IPN, México (1995). (in Spanish)
16. Ruíz-Shulcloper, J., Aguila, L., Bravo, A.: BT and TB algorithms for computing all irreducible testors. Math. Sci. J. **2**, 11–18 (1982). (in Spanish)
17. Ruíz-Shulcloper, J.: Pattern recognition with mixed and incomplete data. Pattern Recognit. Image Anal. **18**(4), 563–576 (2008)
18. Sanchez-Díaz, G., Lazo-Cortés, M.: CT-EXT: an algorithm for computing typical testor set. In: Rueda, L., Mery, D., Kittler, J. (eds.) CIARP 2007. LNCS, vol. 4756, pp. 506–514. Springer, Heidelberg (2007). doi:10.1007/978-3-540-76725-1_53
19. Sanchez-Diaz, G., Piza-Davila, I., Lazo-Cortes, M., Mora-Gonzalez, M., Salinas-Luna, J.: A fast implementation of the CT_EXT algorithm for the testor property identification. In: Sidorov, G., Hernández Aguirre, A., Reyes García, C.A. (eds.) MICAI 2010. LNCS, vol. 6438, pp. 92–103. Springer, Heidelberg (2010). doi:10.1007/978-3-642-16773-7_8
20. Santiesteban, Y., Pons-Porrata, A.: LEX: a new algorithm for the calculus of typical testors. Math. Sci. J. **21**(1), 85–95 (2003). (in Spanish)

Assessing Deep Learning Architectures
for Visualizing Maya Hieroglyphs

Edgar Roman-Rangel[(✉)] and Stephane Marchand-Maillet

Department of Computer Science, University of Geneva, Geneva, Switzerland
{edgar.romanrangel,stephane.marchand-maillet}@unige.ch

Abstract. This work extends the use of the non-parametric dimensionality reduction method t-SNE [11] to unseen data. Specifically, we use retrieval experiments to assess quantitatively the performance of several existing methods that enable out-of-sample t-SNE. We also propose the use of deep learning to construct a multilayer network that approximates the t-SNE mapping function, such that once trained, it can be applied to unseen data. We conducted experiments on a set of images showing Maya hieroglyphs. This dataset is specially challenging as it contains multi-label weakly annotated instances. Our results show that deep learning is suitable for this task in comparison with previous methods.

Keywords: t-SNE · Deep learning · Visualization

1 Introduction

Visualizing high-dimensional data is a challenging problem commonly addressed by: (1) feature selection [5], where the data is represented using only a subset of the original feature space, which is defined by the most relevant dimensions of the high-dimensional data according to certain criteria; or (2) feature extraction [5], where a low-dimensional representation results from a mapping function able to retain properties of interest from the high-dimensional representation of the data. Often, 2 or 3 dimensions are kept for visualization purposes.

Several methods have been proposed merely with the purpose of visualizing a fixed dataset. Among them, t-distributed Stochastic Neighbor Embedding (t-SNE) [11] stands out due to the intuitive reasoning behind it and its relative easy implementation. Another characteristic that makes t-SNE popular is that it is non-parametric, which makes possible to use it for any type of data including instances with no labels [5]. However, it is unclear how to apply it to new instances that were not seen during the learning stage. To address this issue, only a few previous works have focused on developing out-of-sample extensions of t-SNE [4–6,10], obtaining promising results that have been evaluated mainly subjectively based on their resulting low-dimensional visualization.

Since the preservation of local structure is at the heart of t-SNE, we rely on retrieval experiments to evaluate the quality of the resulting low-dimensional representations. This is, we use a quantitative criterion. We also propose the use

© Springer International Publishing AG 2017
J.A. Carrasco-Ochoa et al. (Eds.): MCPR 2017, LNCS 10267, pp. 137–146, 2017.
DOI: 10.1007/978-3-319-59226-8_14

of deep learning methods for learning to replicate a t-SNE mapping function previously computed. By comparing their retrieval performance, we are able to assess the mapping quality of several deep learning architectures and functions. Since we consider no label information for training the deep networks, our approach remains fully unsupervised, and naturally follows t-SNE in this regard.

The experiments on this work were performed on a challenging dataset of binary images containing Maya hieroglyphs, which are very complex in visual terms. Specifically, they are compounds of individual signs whose arrangement varies arbitrarily. Often, this results in partial overlaps of classes, both at the visual and semantic spaces, i.e., an instance of a weakly annotated scenario.

Our results show that: (1) evaluating the retrieval performance is adequate for assessing dimensionality reduction techniques; and (2) deep learning can approximate t-SNE mapping function better that previous methods, such it can be applied to unseen data, while remaining fully unsupervised.

The remaining of this paper is organized as follows. Section 2 discusses related work on out-of-sample dimensionality reduction. Section 3 presents our assessment of methods for out-of-sample dimensionality reduction. Section 4 details the experimental protocol we followed for the assessment. Section 5 discusses the results. And Sect. 6 presents our conclusions.

2   Related Work

Only a few attempts have been made to enable t-SNE with out-of-sample extensions, mainly because it was designed for visualization purposes rather than learning purposes. This is, the interest is only on visualizing the underlying distribution of points from a fixed dataset. Good review papers, including other dimensionality reduction techniques besides t-SNE, are found in [1,11].

One of such works [10] consists in learning a parametric model constructed upon stacked autoencoders, with a latter fine tuning step that takes into account the divergence between the distribution of point in both the high- and the low-dimensional representations. This approach is in fact, designed to improve the use of traditional autoencoders for dimensionality reduction [6,12], where the optimization was only driven by sparsity constraints. Both of these methods are of easy implementation and have attained decent results in the MNIST dataset. However, they seem to fail when dealing with multi-instance images.

Specific works focusing on out-of-sample extension of t-SNE [4,5] rely on regression approaches, either linear [4] or kernel-based [2]. These methods are simple and can obtain competitive results when trained with supervision, e.g., supervised t-SNE [7]. However, we are interested in an unsupervised scenario.

Therefore, we took a step further and combined intuitions behind these two types of approaches. Specifically, we use regression function embedded in neural networks to enable out-of-sample extensions to unsupervised t-SNE. The details of our approach are presented in Sect. 3.

3 Our Approach

This section presents our methodology. First, it provides a brief description of the t-SNE computation, and then it presents our proposed deep learning approach.

T-SNE [11] estimates a mapping function that provides low-dimensional representations, such that, for a given point of interest x_i, the probability p_{ij} of its neighbors $(x_j, \forall j \neq i)$ in the high-dimensional space is the same as the probability q_{ij} of its neighbors in the low-dimensional space. This mapping is obtained by minimizing the Kullback-Leibler divergence $D_{KL}(p\|q)$ between the probabilities across both spaces.

Here, probabilities in the original high-dimensional space are estimated by,

$$p_{ij} = \frac{p_{i|j} + p_{j|i}}{2n}, \tag{1}$$

where, n is the number of points in the full set, and

$$p_{j|i} = \frac{\exp\left(-0.5\|x_i - x_j\|^2/\sigma_i^2\right)}{\sum_{k \neq i} \exp\left(-0.5\|x_i - x_k\|^2/\sigma_i^2\right)}, \tag{2}$$

where, x_i denotes the i-th high-dimensional vector, and σ_i^2 is the variance of a Gaussian centered at x_i.

Complementary, the probabilities in the low-dimensional space are estimated using the Student t-distribution,

$$q_{ij} = \frac{\left(1 + \|y_i - y_j\|^2\right)^{-1}}{\sum_{k \neq i} \left(1 + \|y_i - y_k\|^2\right)^{-1}}, \tag{3}$$

where, y_i is the expected low-dimensional representation of point x_i.

As previously mentioned, this approach provides low-dimensional representations distributed similarly to their high-dimensional counterparts, thus it is suitable for unlabeled data. However, it is not suitable for replicating such transformation for unseen data.

Deep Learning approaches are able to approximate arbitrary complex transformation functions with high degree of precision [8]. The reason for their success in such complex tasks results from their multilayer arrange that process multiple non-linear transformations, which properly designed and trained, might achieve high levels of abstraction.

Here, we propose to exploit this characteristic to replicate t-SNE mapping function for unseen data. Mathematically, we optimize a loss function $\mathcal{L}(f_\Omega(x), y)$, where the loss $\mathcal{L}$ could refer to the mean square error (mse), or any other error, between the high-dimensional input vector x and the low-dimensional expected vector y; and the function $f_\Omega(\cdot)$ corresponds to a deep neural network, such that,

$$f_\Omega : x \mapsto y, \tag{4}$$

and it is parametrized by a set of weights Ω.

In Sect. 4 we detail several neural networks evaluated as the function f_Ω.

4   Experiments

This section details the procedure followed to validate our approach. First, it introduces the dataset used for experiments, and then the experimental protocol.

4.1   Dataset

The data used for our experiments is a collection of binary images containing Maya hieroglyphs. Specifically, Maya glyph-blocks, which correspond to compounds of individual glyph-signs visually arranged together to express full sentences. These images were generated by our team by applying binarization to a subset of glyph-blocks from the collection of the Maaya project [3][1]. Currently, we are working towards the release of this collection for research purposes.

In particular, our dataset consists of 155 classes, which correspond to 155 different types of glyph-blocks. More precisely, each class is defined by the sequence of names of its individual glyph-signs. Here the Thompson catalog [9] is used as naming convention, e.g., T0759b-T0025-T0181 refers to a glyph-block with three individual signs: 759b, 25, and 181; and the prefix 'T', for Thompson, is always used. These annotations were defined by expert epigraphers following a reading convention for Maya hieroglyphs. Note that following this definition of a class, two classes might partially overlap, i.e., two glyph-blocks might contain the same individual signs one or several times, or they might have the same set of signs but on different arrangement.

The 155 selected classes correspond to those glyph-blocks with at least 20 instances in the original Maaya corpus, from which we randomly picked exactly 20. Later, we generated 20 more instances by applying erosion, and 20 more with dilation, both convolving with filter of size fixed to 9×9 pixels. Therefore, our datasets contains 9300 instances, thus 60 per class. Furthermore, the 40 synthetically generated instances are always used as *training set* in our experiments, and the original 20 instances as *test set*. Figure 1 shows some examples of glyph-blocks.

Fig. 1. Examples of Maya glyph-blocks. Each image exemplifies a specific class.

Representation. We used the VGGnet [8] as feature extractor, which has been previously trained on the ImageNet dataset. More precisely, we fed the images of the glyph-blocks to the VGGnet, and recorded the output of the last convolutional layer (conv5). This representations is a 4096-dimensional vector, which for our specific dataset of binary images, happens to be a very sparse

---
[1] www.idiap.ch/project/maaya/.

vector as only those units processing edge information are activated through the network. We use this representation as the input for all of our experiments.

Note that this corresponds to an instance of a transfer learning scenario. Or in other words, the pipeline for feature extraction is fully unsupervised since the VGGnet was trained on a totally different dataset.

4.2 Protocol

We used 6200 instances of glyph-blocks (40 per class) as training set, and the remaining 3100 instances (20 per class) for testing. Specifically, we computed an initial t-SNE dimensionality reduction on the training set, and then used the proposed approach (or the baseline methods) to estimate the corresponding mapping function that could replicate the initial t-SNE reduction. Finally, we evaluated its performance on unseen data, i.e., test set.

The Matlab implementation of t-SNE [11] that we used outputs 2-dimensional representations whose values are zero-centered and between -150 and 150. We scaled these representations to be between -1 and 1 before feeding them to the neural network. And we used hyperbolic tangent and rectified linear transfer functions as the units of the hidden layers in the network.

Table 1 shows the list of neural architectures and transfer functions that we evaluated for the replication of the t-SNE dimensionality reduction.

Table 1. Architectures and transfer functions evaluated for replicating t-SNE dimensionality reduction. Logistic and ReLu transfer functions where used with different combinations of hidden layers and units.

Network	Number of units in the hidden layers		
A	200 tanh	100 tanh	50 tanh
B	200 tanh	100 tanh	100 tanh
C	100 tanh	100 tanh	100 tanh
D	200 tanh	100 tanh	–
E	200 relu	100 relu	50 relu
F	200 relu	100 relu	100 relu
G	200 relu	100 relu	–
H	200 tanh	100 relu	100 tanh

We used gradient descent for all of our experiments and mean square error (mse) as loss function. Also, we used the following fixed hyperparameters: learning rate equals to 0.01; momentum rate of 0.9. We ran our experiments for 1000 epochs, and relied on early stopping if the optimization reached 10 iterations without improvement, or if the gradient reached a value of 10^{-10}.

We compared our results using retrieval experiments as evaluation of performance. We report the mean Average Precision (mAP) obtained for the first 10 and the first 20 retrieved elements, $mAP@10$ and $mAP@20$ respectively.

5   Results

Our first experiment, which later we use as baseline, is an initial comparison of the retrieval performance achieved by the output of the VGGnet before and after applying t-SNE in the traditional sense. Table 2 shows the mean Average Precision (mAP) obtained for the first 10 and the first 20 retrieved elements, $mAP@10$ and $mAP@20$ respectively.

Table 2. mAP obtained on the training set alone and the training set combined with the test set. For both cases using the 4096-D vector provided by the VGGnet, and a 2-D vector after applying t-SNE.

Set	Dimensionality	mAP	
		@10	@20
Training	4096-D	0.9591	0.9258
Training	(t-SNE) 2-D	0.9509	0.9311
Training & Test	4096-D	0.9578	0.9245
Training & Test	(t-SNE) 2-D	0.9455	0.9287

The first observation is that squeezing the output of 4096 dimensions of the VGGnet into only 2 dimensions, results in a slight decrease in the retrieval performance (i.e., discriminative potential) for the first 10 retrieved elements. This behavior was already expected since, intuitively, 2 dimensions might be insufficient for discriminative representations of the very complex shapes in our dataset. However, it is interesting to notice that this drop is almost negligible. Furthermore, the retrieval performance shows a slight increment for the first 20 retrieved elements, i.e., from 0.9258 to 0.9311 when using only the training set, and from 0.9245 to 0.9287 when the training and test sets are merged. Another observation is that the retrieval performance decreases by adding the test set in the experiments. This is a consequence of the addition of noisy instances.

Since a new mapping function is obtained every time t-SNE is computed, results obtained using only the training set are, with high probability, different from those obtained using both the training and test sets together. This is to say, the retrieval performance is barely impacted when the test set is added to the training set, as t-SNE effectively estimates a mapping function considering all the elements in its input. However, such mapping function is different from that estimated only for the test set. Figure 2 shows two different point-clouds of 2-D coordinates for the training set only and the training and test sets together. It is clear that the location of colored clusters, which indicate different classes, is different in the two images.

In Table 3, we show the retrieval performance obtained with the several architectures described in Table 1. Compared to the baseline presented in Table 2, all of these approaches result in certain amount of loss of the discriminative capabilities of the vectors. Concretely, the best performing approach (B in Table 3)

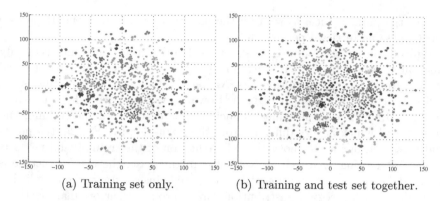

(a) Training set only.   (b) Training and test set together.

Fig. 2. Point-clouds showing different 2-D representations obtained after dimensionality reduction of the VGGnet outputs using t-SNE.

shows about 23% lower retrieval performance. This is the result of compressing 4096 dimensions into only 2 dimensions, which corresponds to a very *aggressive* transformation. Although we notice that using only 12 dimensions is still enough to achieve results above 90% in retrieval precision, in this work we are interested in generating representations of only 2 dimensions for visualization purposes.

Table 3. *mAP* obtained for different architectures of deep neural networks.

mAP	A	B	C	D	E	F	G	H
@10	0.674	**0.719**	0.552	0.718	0.692	0.702	0.326	0.706
@20	0.655	**0.706**	0.551	0.700	0.690	0.689	0.211	0.702

From Table 3, it is clear that approach B produces results whose performance is above all the others. In general terms, the use of the hyperbolic tangent function (*tanh*) attained better results in comparison to the rectified linear function (*relu*). Comparing approach B with the other approaches that also use the hyperbolic tangent function (i.e., A, C, D), one can notice that decreasing the number of units in the first layer, as done in approach C, has a larger impact on the retrieval performance than decreasing it in the last layer, as done in approach A. Surprisingly, removing one layer, as in approach D, has little effect as long as the number of units in the last layer remains the same. Intuitively, this happens because the number of parameters decreases, thus making the model easier to train although slightly less robust.

   This behavior is consistent with the results obtained using the rectified linear function, where approach F, which is similar to B in the number of layers and units, achieves higher performance with respect to other approaches using relu transfer functions (i.e., E, G). Finally, combining tanh and relu function has no effect in the performance, as seen by the fact that approach F and H provide almost the same retrieval performance.

Figure 3 shows the point-clouds obtained after using the dimensionality reduction approach B. To facilitate reading, Fig. 3a shows again the point-cloud resulting from applying traditional t-SNE on the training set (i.e., this is the same point-cloud as shown in Fig. 2a). However, different from Fig. 2b, which shows a full t-SNE dimensionality reduction applied on both the training and test sets, and which is very different from the point-cloud in Fig. 2a, the point-cloud presented in Fig. 3b shows the 2-D reduction of the training and test sets together after using approach B. Note that in this case, the color structure is conserved, i.e., the rough location of each class remains constant when applied to unseen data, with a regular distribution spread around their original locations. This is, the proposed method is able to generalize the use of t-SNE for unseen data up to certain degree.

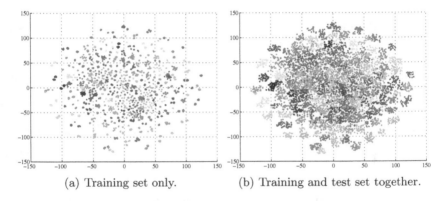

(a) Training set only.   (b) Training and test set together.

Fig. 3. Point-clouds showing the 2-D representations resulting from applying the proposed dimensionality reduction approach, i.e., approach B.

Specifically, the spreading in the location of the 2-dimensional representations, which results from using the proposed approach, and which is shown in Fig. 3b, is the reason for the drop in retrieval performance reported in Table 3.

For further validation of the proposed approach, we compared it against other methods previously reported in the literature. Table 4 shows the retrieval performance obtained from this comparison. In particular, the two autoencoder-based methods, autoencoder and parametric t-SNE, both obtained retrieval performance slightly below 70%. The ideas behind autoencoders and parametric t-SNE are very similar to each other, with the main difference that the fine tunning step of parametric t-SNE focuses on the optimization of the Kullback-Leibler divergence, whereas autoencoders are only constrained by a sparsity prior. However, this difference seems to be negligible, as suggested by the very similar performance obtained by these two methods.

From Table 4, one can also see that it is possible to achieve more than 50% retrieval precision by solving a simple linear regression formulation. Furthermore,

Table 4. mAP obtained for the best performing architecture according to Table 3, and other approaches for regression.

Method	$mAP@10$	$mAP@20$
[6] Autoencoder	0.668	0.695
[10] Parametric t-SNE	0.677	0.667
Linear regression	0.569	0.545
[4] MSI	0.615	0.609
[4] Kernel t-SNE	0.654	0.648
Our approach	**0.719**	**0.706**

the two methods presented in [4], which are kernel-based variations of linear regression, and that are referred to as MSI and kernel t-SNE, both obtained slightly improved retrieval precision with respect to the use of simple linear regression. However, all these three methods performs slightly worse than the autoencoder-based methods.

Finally, the proposed approach based upon the use of deep networks achieves about 72% of retrieval precision, which corresponds to a rough 23% of drop in performance when compared to traditional t-SNE. However, this high performance of t-SNE remains true as long as it is computed using the test set as part of the training set. In contrast, our approach enables the ability to generalize for unseen data within a range of acceptable loss in the discriminative potential of the resulting 2-dimensional representations.

6   Conclusions

We proposed the use of retrieval experiments for assessing the quality of unsupervised and non-parametric dimensionality reduction techniques. Specifically, for methods that replicate the mapping function of t-SNE for unseen data. We conducted experiments on a dataset of Maya hieroglyphs, whose main challenge lies on the visual complexity of its instances and the fact that they are weakly annotated. We also proposed an unsupervised method for estimating 2-D representations of very complex shapes, which can be readily applied to unseen data. More specifically, a deep network architecture able to approximate t-SNE mapping functions up to certain degree.

Our results show that using retrieval experiments is an adequate criteria for evaluation of non-parametric dimensionality reduction methods, and that the proposed approach performs better than previous methods. Furthermore, 2-dimensional representation is adequate for visualization purposes of complex shapes at the cost of lower retrieval performance.

For future work, more complex architectures and combinations of transfer functions, which could yield improved results should be evaluated. Also, it is expected that the evaluation of our approach on a supervised scenario will result in improved performance.

Acknowledgments. This work was supported by the Swiss-NSF MAAYA project (SNSF-144238).

References

1. Bengio, Y., Courville, A., Vincent, P.: Representation learning: a review and new perspectives. IEEE Trans. Pattern Anal. Mach. Intell. **35**(8), 1798–1828 (2013)
2. Bunte, K., Biehl, M., Hammer, B.: A general framework for dimensionality-reducing data visualization mapping. Neural Comput. **24**(3), 771–804 (2012)
3. Gatica-Perez, D., Gayol, C.P., Marchand-Maillet, S., Odobez, J.-M., Roman-Rangel, E., Krempel, G., Grube, N.: The MAAYA project: multimedia analysis and access for documentation and decipherment of maya epigraphy. In: Workshop DH (2014)
4. Gisbrecht, A., Lueks, W., Mokbel, B., Hammer, B.: Out-of-sample kernel extensions for nonparametric dimensionality reduction. In: Proceedings of the European Symposium on Artificial Neural Networks, Computational Intelligence and Machine Learning (2012)
5. Gisbrecht, A., Schulz, A., Hammer, B.: Parametric nonlinear dimensionality reduction using kernel t-SNE. Neurocomputing **147**, 71–82 (2015)
6. Hinton, G., Salakhutdinov, R.R.: Reducing the dimensionality of data with neural networks. Science **313**(5786), 504–507 (2006)
7. Kim, H., Choo, J., Reddy, C.K., Park, H.: Doubly supervised embedding based on class labels and intrinsic clusters for high-dimensional data visualization. Neurocomputing **150**, 570–582 (2015)
8. Simonyan, K., Zisserman, A.: Very deep convolutional networks for large-scale image recognition. CoRR abs/1409.1556 (2014)
9. Thompson, J.: A Catalog of Maya Hieroglyphs. University of Oklahoma, Norman (1962)
10. van der Maaten, L.: Learning a parametric embedding by preserving local structure. In: Proceedings of the International Conference on Artificial Intelligence and Statistics (2009)
11. Maaten, L., Hinton, G.: Visualizing data using t-SNE. J. Mach. Learn. Res. **9**, 2579–2605 (2008)
12. Wang, W., Huang, Y., Wang, Y., Wang, L.: Generalized autoencoder: a neural network framework for dimensionality reduction. In: Proceedings of IEEE CVPR (2014)

Image Processing and Analysis

Image Noise Filter Based on DCT and Fast Clustering

Miguel de Jesús Martínez Felipe, Edgardo M. Felipe Riveron,
Pablo Manrique Ramirez, and Oleksiy Pogrebnyak$^{(\boxtimes)}$

Centro de Investigacion en Computacion, Instituto Politecnico Nacional, Ave.
Miguel Othón De Mendizábal S/N, 07738 Mexico, D.F., Mexico
mjmf2402@hotmail.com,
{edgardo,pmanriq,olek}@cic.ipn.mx

Abstract. An algorithm for filtering images contaminated by additive white Gaussian noise in discrete cosine transform domain is proposed. The algorithm uses a clustering stage to obtain mean power spectrum of each cluster. The groups of clusters are found by the proposed fast algorithm based on 2D histograms and watershed transform. In addition to the mean spectrum of each cluster, the local groups of similar patches are found to obtain the local spectrum, and therefore, derive the local Wiener filter frequency response better and perform the collaborative filtering over the groups of patches. The obtained filtering results are compared to the state-of-the-art filters in terms of peak signal-to-noise ratio and structural similarity index. It is shown that the proposed algorithm is competitive in terms of signal-to-noise ratio and in almost all cases is superior to the state-of-the art filters in terms of structural similarity.

Keywords: Noise suppression · Collaborative filtering · Fast image clustering

1 Introduction

The main goal of this paper is noise suppression in images degraded by additive white Gaussian noise (AWGN). Noise is one of the main factors that degrades image quality [1, 2]. Various methods for image denoising are currently presented, but many researches are dedicated to design more efficient techniques [2–8]. The reason is that the quality of the restored images is still not completely satisfactory for the final users.

The image denoising techniques of state-of-the-art fall into two families [5]: (1) non-local filters [2] based on searching for similar patches and their joint processing, such as BM3D [3] and SA-DCT [4]; (2) those based on image clustering, kernel regression, singular value decomposition or principal component analysis for dictionary learning and sparse image representation minimization [5–8].

Among the family of non-local filters, nowadays the BM3D filter [3] has been shown to be the most efficient for processing the majority of grayscale test images [5, 9] and component-wise denoising of color test images [10] corrupted by AWGN. On the other hand, the filters based on sparse representation minimization show good results, which in some cases are superior to the BM3D technique [7], but these filters have a tremendous computational complexity, mainly due to very slow clustering stage

© Springer International Publishing AG 2017
J.A. Carrasco-Ochoa et al. (Eds.): MCPR 2017, LNCS 10267, pp. 149–158, 2017.
DOI: 10.1007/978-3-319-59226-8_15

and the following dictionary learning, localized histogram and other feature calculations. Besides, the iterative sparse minimization not only slows down the restoration but sometimes leads to worse denoising results.

Another aspect is denoising filter efficiency. Usually, a mean square criterion in the form of peak signal-to-noise ratio (PSNR) is used to estimate the filtered image quality. Unfortunately, such an approach not always means a better visual quality nor provides more appropriate image data for the posterior treatment and classification [12]. Furthermore, there are some relatively novel image quality criterions, such as based on the properties of human visual system (PSNR-HVSM) [13, 14], feature similarity index (FSIM) [15], structural similarity index (SSIM) [16] or multiscale structural similarity index (MSSIM) [17].

In this paper, we take advantages of both filter families. The presented filtering technique uses discrete cosine transform (DCT), the block matching algorithm in the transform domain of reduced computational complexity, Hadamard transform for patch group hard thresholding [11], and a fast clustering algorithm proposed to derive mean cluster spectra to form the local Wiener filter frequency responses. The obtained results with the proposed filter in terms of PSNR are similar to those obtained with BM3D, K-SVD [6] and NCSR [7], but are better in terms of SSIM.

The paper is organized as follows: in Sect. 2, image Wiener filter is considered and reduced to a hard thresholding filter in DCT domain. The proposed fast image clustering algorithm is described in Sect. 2. Section 3 explains the filtering technique, and numerical simulation results for the proposed filter in comparison to the best known filters of both mentioned above families are presented in Sect. 5. Finally, the conclusions follow.

2 Wiener Filtering and Thresholding in DCT Domain

Let us consider an additive image observation model

$$u(x,y) = s(x,y) + n(x,y) \tag{1}$$

where $u(x,y)$ is an observed noisy image, x, y are Cartesian coordinates, $s(x,y)$ denotes a noise-free image, and $n(x,y)$ is a white Gaussian noise not correlated with $s(x,y)$. The problem is to find an estimate of the noise-free image $\hat{s}(x,y)$ such that it minimizes mean square error (MSE) $E\left\{[s(x,y) - \hat{s}(x,y)]^2\right\}$, where $E\{\cdot\}$ denotes the expectation operator.

The optimal linear filter that minimizes the MSE is the well-known Wiener filter that in the spectral domain can be formulated as [9]:

$$H_W\left(\omega_x, \omega_y\right) = \frac{P_s\left(\omega_x, \omega_y\right)}{P_s\left(\omega_x, \omega_y\right) + P_n\left(\omega_x, \omega_y\right)}, \tag{2}$$

where $P_s(\omega_x, \omega_y), P_n(\omega_x, \omega_y)$ are power spectral densities of the signal and noise, respectively. When the noise is Gaussian, it is supposed $P_n(\omega_x, \omega_y) = \sigma_n^2$, where σ_n^2 is the variance of noise $n(x, y)$.

Unfortunately, the exact power spectrum density $P_s(\omega_x, \omega_y)$ is unavailable and often is substituted by its estimate $\hat{P}_s(\omega_x, \omega_y)$ that allows obtaining an estimate of the Wiener filter frequency response $\hat{H}_W(\omega_x, \omega_y)$. In the considered case of additive white Gaussian noise, the estimated Wiener filter response is approximated by

$$\hat{H}_W(\omega_x, \omega_y) = \frac{\hat{P}_s(\omega_x, \omega_y)}{\hat{P}_s(\omega_x, \omega_y) + \sigma_n^2}. \tag{3}$$

Usually, $P_u(\omega_x, \omega_y)$ is taken as an estimation of $P_s(\omega_x, \omega_y)$, but such a choice is not optimal. Here, we propose to estimate $P_s(\omega_x, \omega_y)$ over the set of DCT patches of 8×8 (this block size was selected for the comparison purposes) image blocks grouped in clusters; additionally, the estimate $\hat{P}_s(\omega_x, \omega_y)$ is adjusted using the mean spectrum of the found similar patches in a searched range of the current image block as in the technique BM3D [4, 11] but instead of searching in the space domain we used the search in DCT domain as in our previous work [18].

Moreover, the direct implementation of (3) in DCT domain can be substituted by the zeroing of DCT coefficients below some threshold [4]. This threshold was analytically derived to be of $6.5217615 \cdot \sigma_n^2$ [18]:

$$H_T(\omega_x, \omega_y) = \begin{cases} 1, & \text{if } |U(\omega_x, \omega_y)| \geq 2.55377397 \cdot \sigma_n \\ 0 & \text{otherwise} \end{cases}, \tag{4}$$

where $U(\omega_x, \omega_y)$ is the voltage DCT spectrum of the observed noisy image and σ_n is noise standard deviation. Also, the hard threshold (4) can adapt its characteristics to the local spectrum properties calculating to this end the Wiener filter frequency response [18]:

$$T(\omega_x, \omega_y) = \begin{cases} \beta_{\min} \cdot \sigma & \text{if } \hat{H}_{i,j}(\omega_x, \omega_y) > 0.87 \\ \beta_{\max} \cdot \sigma & f\hat{H}_{i,j}(\omega_x, \omega_y) < 0.3 \\ \beta \cdot \sigma & \text{otherwise} \end{cases}, \tag{5}$$

and $\beta_{\min} = 1, \beta_{\max} = 2.9; \beta = 2.55377397$ according to (4).

It was found by simulations that the image restoration quality highly depends on the quality of the signal spectral estimate $\hat{P}_s(\omega_x, \omega_y)$; so, the main problem is to estimate it better.

3 Local Spectra Estimation via Image Clustering

In this Section, we present a fast image clustering technique adequate for noise suppression in terms of local spectra $P_s(\omega_x, \omega_y)$ estimation to be used to form the local Wiener filter frequency response $\hat{H}_W(\omega_x, \omega_y)$ to process each 8×8 image block.

The proposed clustering algorithm starts with calculating 2D image histogram on image block mean and variance values in supposition that the image block data is described in terms of their first and second moments. To this end, the mean value range is assigned to be of 256 samples from 0 to 255; and the variance range is determined searching first for the maximum block variance value.

Then, the variance range is discretized; in this paper, we assume it is of 512 levels with respect to the maximum image block variance value, σ_{max}^2. The number 512 is considered to be sufficiently high to represent the full range of variance values and not too large for the following watershed transform. Using the mean and variance values of each 8×8 overlapping image block, 2D histogram of 256×512 bins is formed. Figure 1(a) shows an example of such a histogram of the standard test image "Lenna" having $\sigma_n^2 = 25$.

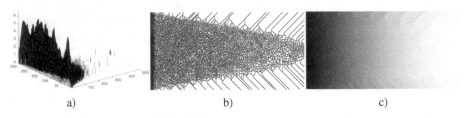

a)   b)   c)

Fig. 1. Transforming 2D image histogram: (a) 2d histogram of test image "Lenna"; (b) watershed transform; (c) the resulted cluster number map.

Next, the maximum of the histogram is found and the histogram is modified changing to the ratio of this maximum and the original histogram values. After this, the watershed transform [19] is performed over the modified 2D histogram (see Fig. 1(b)).

Finally, the watershed ridge points having zero value are substituted by there neighbor points searching the coordinates that have the maximal values in 2D histogram, as shown in Fig. 1(c).

As a result, the considered image is free of ridge lines and is clustered with an automatically determined number of groups that is produced by the watershed transform. Figure 2 shows the test image "Lenna" clustered using the described algorithm, where each pixel value is the cluster number of the corresponding 8×8 blocks of the original image (the blocks are overlapped).

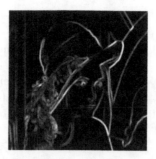

Fig. 2. Test image "Lenna" clustered with the proposed algorithm

4 Proposed Filtering Technique

The proposed advanced filtering technique is based on the BM3D [11] filtering strategy that assumes image clustering as described in Sect. 3: the search for similar blocks in the vicinity of the current image block, the formation of the lists of DCT coefficient patches and their processing using Hadamard transform, the hard thresholding and aggregation of the processed patches forming the filtered image.

The filtering technique starts with the pre-processing stage consisting of image clustering followed by calculation of the DCT patches for each image pixel in the range of $(M - m) \times (N - m)$ pixels, $M \times N$ is the image size, $m = 8$. Then, power spectrum estimations, $\hat{P}(c)$, of each cluster are calculated. Next, the search for similar image blocks using the DCT transformed data [18] is performed forming the list of patch coordinates, $\mathbf{l}_U(i,j) = \{l_1(i,j)\ldots l_{b_{\max}}(i,j)\}$, where $l_1(i,j)\ldots l_{b_{\max}}(i,j)$, $b_{\max}$ is the maximal number of similar patches, and the list of the corresponding distances $\mathbf{d}(i,j)$ for their usage in the filtering process.

After the pre-processing stage, the filtering is performed as described below.

At the proper filtering stage, first the spectral estimate of the current i,j-th image block data is calculated using the list of patches $\mathbf{l}_U(i,j)$ and the current cluster $c(i,j)$ spectral estimate, $\hat{P}(c(i,j))$:

$$\hat{P}_s(i,j) = \frac{\hat{P}(c(i,j)) + \sum_{k=1}^{b_{\max}/2} U_k(l_k(i,j)) \cdot \tilde{w}_k^P(i,j)}{2}, \tag{6}$$

where $b_{\max}$ is the maximal number of similar patches in the list $\mathbf{l}_U(i,j)$ for i,j-th image block, $\tilde{w}_k^P(i,j)$ is a normalized weighting coefficient calculated on the patch distances as

$$w_k^P(i,j) = \frac{\exp\{-d_k(i,j)/200\}}{\|\mathbf{w}^P(i,j)\|}, \tag{7}$$

and $\|\mathbf{w}^P(i,j)\|$ denotes the sum of the non-normalized coefficients $w_k^P(i,j)$,.

With the estimate $\hat{P}_s(i,j)$, the Wiener filter frequency response at i,j-th block position $\hat{H}_{i,j}(\omega_x, \omega_y)$ is formed according to (3). Next, Hadamard transform in the third

dimension is applied to the group of patches $\mathbf{U}(i,j) = \{\mathbf{U}_k(l_k(i,j)), \, k = \overline{1, b_{\max}}\}$:
$\mathbf{U}_{Hadamard} = Hadamard\{\mathbf{U}(i,j)\}$.

Then, we propose the following thresholding procedure:

$$\tilde{\mathbf{U}}_{Hadamard} = \begin{cases} \mathbf{U}_{Hadamard}(\omega_x, \omega_y) & if \; \omega_x = \omega_y = 0 \\ else \quad \mathbf{U}_{Hadamard}(\omega_x, \omega_y) & if \; |\mathbf{U}_{Hadamard}(\omega_x, \omega_y)| \geq T(\omega_x, \omega_y)|, \\ 0 & otherwise \end{cases} \quad (8)$$

where the thresholds $T(\omega_x, \omega_y)$ are formed according to (5). After the thresholding and the inverse Hadamard transform, $\tilde{\mathbf{U}}(i,j) = Hadamard^{-1}\{\tilde{\mathbf{U}}_{Hadamard}\}$, the group of filtered patches $\hat{\mathbf{U}}(i,j) = \{\hat{\mathbf{U}}_k(l_k(i,j)), \, k = \overline{1, b_{\max}}\}$ is obtained applying the inverse DCT transform to the patches. Finally, the collaborative filtering by the aggregation of the filtered blocks is performed. Note that, opposite to scanning window filtering, the filtered values are obtained simultaneously for all pixels of a given block and for all blocks in the group $\mathbf{l}(i,j)$. The averaged aggregation is performed as [18]

$$\hat{\mathbf{s}}_\Sigma(\mathbf{l}_U(i,j)) = \sum_{k=1}^{b_{\max}} (\hat{\mathbf{U}}_k(l_k(i,j))) \cdot \cdot w_k^A(i,j),$$

$$R_\Sigma(\mathbf{l}_U(i,j)) = \sum_{k=1}^{b_{\max}} w_k^A(i,j), \; w_k^A(i,j) = \exp\{-d_k(i,j)/500\}, \quad (9)$$

$$(\tilde{\mathbf{s}})_* = \frac{(\hat{\mathbf{s}}_\Sigma)_*}{(\mathbf{R}_\Sigma)_*}$$

where $\hat{\mathbf{s}}_\Sigma(\mathbf{l}(i,j))$ are the sum of the blocks defined by the list $\mathbf{l}(i,j)$, $R_\Sigma(\mathbf{l}(i,j))$ is the sum of weighting coefficients $w_k^A(i,j)$, $\tilde{\mathbf{s}}$ is the estimated image obtained from the pre-filtering stage, $()_*$ denotes element-wise operations.

5   Results

The numerical simulations using standard test images with plain regions "Lena", "F-16", "peppers" and textural images "aerial", "baboon", "bridge" are performed. For comparison purposes, three best state-of-the-art filters, K-SVD, BM3D (with the same search area defined by the parameter $shift = 21$ and overlapping step 1) and NCSR were used.

The filtering results are illustrated in Fig. 3, and the obtained values of PSNR and SSIM are presented in Tables 1 and 2. The visual quality of the images processed by the considered filters is very similar, although NCSR presents some over smoothing and the detail preservation is better with the proposed filter.

The simulation result analysis presented in Table 1 shows that in the majority of cases the very complex filter NCSR produced the best result, and the proposed filtering technique is competitive with the state-of-the-art filters in the PSNR sense. From the data presented in Table 2 it follows that in almost all cases the proposed filter is superior in terms of SSIM.

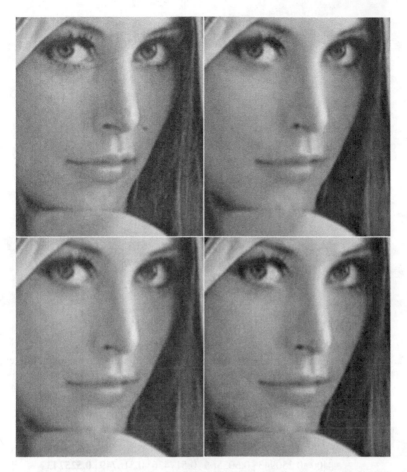

Fig. 3. Amplified fragments of image "Lena" distorted with $\sigma^{2|} = 225$ AWGN and processed by different filters (from left to right and top to bottom): original, NCSR, BM3D, proposed.

Table 1. Results of the filtering of the standard test images with different techniques in terms of PSNR. The best results are marked as bold.

Image	$\sigma^{2|}$	Noisy	K-SVD	BM3D	NCSR	Proposed
Lena	25	34.14	38.554	38.65	38.674	**38.8**
	100	28.132	35.428	35.722	35.866	**35.771**
	225	24.616	33.544	33.939	**34.175**	33.899
	400	22.137	32.21	32.621	**32.867**	32.479
F-16	25	34.14	39.067	39.125	39.232	**39.27**
	100	28.132	35.479	35.547	**35.849**	35.663
	225	24.627	33.4	33.479	**33.887**	33.537
	400	22.191	31.928	32.038	**32.443**	32.037

(continued)

Table 1. (*continued*)

Image	$\sigma^{2|}$	Noisy	K-SVD	BM3D	NCSR	Proposed
Peppers	25	34.146	37.657	37.568	**37.866**	37.789
	100	28.167	34.766	34.95	**35.081**	35.023
	225	24.681	33.239	33.512	**33.69**	33.488
	400	22.22	32.073	32.395	**32.613**	32.274
Aerial	25	34.145	36.673	36.809	**37.04**	36.921
	100	28.142	32.289	32.324	**32.789**	32.578
	225	24.644	29.879	29.912	**30.481**	30.234
	400	22.199	28.187	28.305	**28.812**	28.632
Baboon	25	34.141	35.177	35.145	35.264	**35.295**
	100	28.135	30.451	30.43	30.598	**30.668**
	225	24.618	27.960	27.957	28.232	**28.266**
	400	22.129	26.307	26.348	26.64	**26.698**
Bridge	25	34.159	35.578	35.545	**35.695**	35.687
	100	28.169	30.94	30.799	**31.143**	31.09
	225	24.682	28.544	28.413	**28.824**	28.762
	400	22.23	26.98	26.935	27.259	**27.294**

Table 2. Results of the filtering of the standard test images with different techniques in terms of SSIM. The best results are marked as bold.

Image	$\sigma^{2|}$	Noisy	KSVD	BM3D	NCSR	Proposed
Lena	25	0.650071	0.729343	0.717362	0.702052	**0.739799**
	100	0.434195	0.612688	0.625511	0.616173	**0.644212**
	225	0.321075	0.54803	0.565976	0.564288	**0.579691**
	400	0.250943	0.503566	0.517436	0.516749	**0.525743**
F-16	25	0.573694	0.677342	0.673943	0.664878	**0.697925**
	100	0.407647	0.571205	0.573248	0.574901	**0.593734**
	225	0.322104	0.516994	0.512244	0.520495	**0.527333**
	400	0.266426	0.477619	0.466926	0.470992	**0.477756**
Peppers	25	0.698161	0.742456	0.719933	0.736183	**0.745301**
	100	0.460156	0.588543	0.602402	0.588272	**0.623209**
	225	0.334244	0.526644	0.546989	0.527292	**0.563667**
	400	0.259053	0.490874	0.506874	0.487441	**0.518279**
Aerial	25	0.84257	0.899446	0.901742	0.900802	**0.90716**
	100	0.697669	0.81245	0.81954	0.826998	**0.830115**
	225	0.587101	0.740628	0.751832	**0.768144**	0.766316
	400	0.500772	0.671283	0.692477	0.707465	**0.709842**
Baboon	25	0.91839	0.929301	0.920775	0.924035	**0.929792**
	100	0.790511	0.824671	0.821622	0.81302	**0.840119**
	225	0.678611	0.733180	0.740913	0.734304	**0.767104**
	400	0.585645	0.654664	0.6706	0.65984	**0.704706**

(*continued*)

Table 2. (*continued*)

Image	$\sigma^{2|}$	Noisy	KSVD	BM3D	NCSR	Proposed
Bridge	25	0.913643	0.938448	0.935946	0.937837	**0.940305**
	100	0.780924	0.849257	0.844378	0.854613	**0.858391**
	225	0.658758	0.755813	0.755711	0.774548	**0.779514**
	400	0.556196	0.666819	0.677529	0.691471	**0.709657**

6 Conclusions

A filtering technique to process the images contaminated by additive white Gaussian noise has been presented. The algorithm uses discrete cosine transform and the groups of patches similar to the current image block, which are found using the proposed clustering algorithm and the search algorithm of the reduced complexity. The noisy components are rejected according to the Wiener filtering approach using Hadamard transform for thresholding and weighted aggregation. The obtained filtering results in comparison to the state-of-the-art filters, such as K-SVD, BM3D, NCSR show that the proposed algorithm is competitive in terms of signal-to-noise ratio and in almost all cases is superior in terms of structural similarity. Meanwhile, the considered filter results in terms of the visual quality are very similar, although the proposed filter preserves the image details better.

Acknowledgment. This work partially was supported by Instituto Politecnico Nacional as a part of research project SIP#20171559.

References

1. Pratt, W.K.: Digital Image Processing, 4th edn. Wiley-Interscience, New York (2007)
2. Buades, A., Coll, B., Morel, J.M.: A review of image denoising algorithms, with a new one. Multiscale Model. Simul. **4**(2), 490–530 (2005). doi:10.1137/040616024
3. Dabov, K., Foi, A., Katkovnik, V., Egiazarian, K.: Image denoising by sparse 3D transform-domain collaborative filtering. IEEE Trans. Image Process. **16**(8), 2080–2095 (2007). doi:10.1109/TIP.2007.901238
4. Foi, A., Katkovnik, V., Egiazarian, K.: Pointwise shape-adaptive DCT for high-quality denoising and deblocking of grayscale and color images. IEEE Trans. Image Process. **16**(5), 1395–1411 (2007). doi:10.1109/TIP.2007.891788
5. Chatterjee, P., Milanfar, P.: Is denoising dead? IEEE Trans. Image Process. **19**(4), 895–911 (2010). doi:10.1109/TIP.2009.2037087
6. Aharon, M., Elad, M., Bruckstein, A.M.: K-SVD: an algorithm for designing overcomplete dictionaries for sparse representation. IEEE Trans. Sig. Process. **54**(11), 4311–4322 (2006). doi:10.1109/TSP.2006.881199
7. Dong, W., Zhang, L., Shi, G., Li, X.: Nonlocally centralized sparse representation for image restoration. IEEE Trans. Image Process. **22**(4), 1620–1630 (2013). doi:10.1109/TIP.2012.2235847

8. He, N., Wang, J.B., Zhang, L.L., Xu, G.M., Lu, K.: Non-local sparse regularization model with application to image denoising. Multimedia Tools Appl. **75**(5), 2579–2594 (2016). doi:10.1007/s11042-015-2471-2

9. Pogrebnyak,O., Lukin., V.: Wiener discrete cosine transform-based image filtering. J. Electron. Imaging **21**(4), 043020-1–043020-15 (2012). doi:10.1117/1.JEI.21.4.043020

10. Fevralev, D., Lukin, V., Ponomarenko, N., Abramov, S., Egiazarian, K., Astola, J.: Efficiency analysis of color image filtering. EURASIP J. Adv. Sig. Process. **2011**(41), 1–19 (2011). doi:10.1186/1687-6180-2011-41

11. Leburn, M.: An analysis and implementation of the BM3D image denoising method. Image Process. Line **2**, 175–213 (2012). doi:10.5201/ipol.2012.l-bm3d

12. Lukin, V., Abramov, S., Krivenko, S., Kurekin, A., Pogrebnyak, O.: Analysis of classification accuracy for pre-filtered multichannel remote sensing data. Expert Syst. Appl. **40**(16), 6400–6411 (2013). doi:10.1016/j.eswa.2013.05.061

13. Egiazarian, K., Astola, J., Ponomarenko, N., Lukin, V., Battisti, F., Carli, M.: New full-reference quality metrics based on HVS. In: CD-ROM Proceedings of the Second International Workshop on Video Processing and Quality Metrics, Scottsdale, USA, 4 pages (2006)

14. Ponomarenko, N., Silvestri, F., Egiazarian, K., Carli, M., Lukin, V.: On between-coefficient contrast masking of DCT basis functions. In: CD-ROM Proceedings of Third International Workshop on Video Processing and Quality Metrics for Consumer Electronics, VPQM 2007, January, 4 pages (2007)

15. Zhang, L., Zhang, L., Mou, X., Zhang, D.: FSIM: a feature SIMilarity index for image quality assessment. IEEE Trans. Image Process. **20**(8), 2378–2386 (2011). doi:10.1109/TIP.2011.2109730

16. Wang, Z., Bovik, A.C., Sheikh, H.R., Simoncelli, E.P.: Image quality assessment: from error visibility to structural similarity. IEEE Trans. Image Process. **13**(4), 600–612 (2004). doi:10.1109/TIP.2003.819861

17. Wang, Z., Simoncelli, E.P., Bovik, A.C.: Multiscale structural similarity for image quality assessment. In: Conference Record of the Thirty-Seventh Asilomar Conference on Signals, Systems and Computers, vol. 2, pp. 1398–1402 (2003). doi:10.1109/ACSSC.2003.1292216

18. Callejas Ramos, A.I., Felipe-Riveron, E.M., Manrique Ramirez, P., Pogrebnyak, O.: Image filter based on block matching, discrete cosine transform and principal component analysis, Lecture Notes in Artificial Intelligence, Subseries of Lecture Notes in Computer Science. In: Advances in Artificial Intelligence, MICAI 2016 (To be published in 2017)

19. Beucher, S., Meyer, F.: The morphological approach to segmentation: the watershed transformation. In: Dougherty, E.R. (ed.) Mathematical Morphology in Image Processing, pp. 433–481 (1993)

Color-Texture Image Analysis for Automatic Failure Detection in Tiles

Miyuki-Teri Villalon-Hernandez, Dora-Luz Almanza-Ojeda[✉],
and Mario-Alberto Ibarra-Manzano

DICIS, Universidad de Guanajuato, Carr. Salamanca-Valle de Santiago
Km. 3.5+1.8, Comunidad Palo Blanco, 36885 Salamanca, GTO, Mexico
miyukivh@gmail.com, {dora.almanza,ibarram}@ugto.mx

Abstract. The defects in tiles are directly related with changes in the structure or color components producing spots or stains in the final product. Usually, a visual inspection is carried out in order to detect one of such common defects in tiles; however this process depends on the expertise and abilities of the operator on duty. In this paper, we present the automation of defect detection in tiles using vision algorithms and Artificial Neural Networks (ANN). Color and texture information extracted from real tile images are used as input to a classifier based on neural networks. Setting parameters for extracting the texture attributes are obtained performing detailed tests of different distances, orientations and window sizes. An initial architecture of the ANN is obtained using texture features extracted from Brodatz images. Next, the neural network parameters are computed using real images from the tile database. The experimental tests validate the global performance, accuracy and feasibility of our approach.

Keywords: Color-texture attributes · CIELab color space · Tile failure classification · Artificial Neural Networks (ANN)

1 Introduction

The production cost of the ceramic tile industry is expensive because of the cost of raw materials and energetics. In order to decrement the negative impact of this costs, industrial software for automatic vision inspection are proposed to classify the quality of the final product. There are many elements that can directly affect the quality of the final tile in the production line. That is, a failure could occurs in the beginning of the process, which is reflected in the structure of the tiles as irregular geometric composition. Middle and final stage of the process are related with failures that appears as small pinholes, surface cracks, presence of chromatic discrepancies in the tonality of surface and texture anomalies. For this reason, the last phase of manufacturing process requires that tiles be subject to a quality operation with the aim of identifying any defects. Usually, the quality inspection of the tiles in some industries is done by visual

J.A. Carrasco-Ochoa et al. (Eds.): MCPR 2017, LNCS 10267, pp. 159–168, 2017.
DOI: 10.1007/978-3-319-59226-8_16

inspection, that has a propensity to wrong classifications, thus, the experience, judgment, fatigue and visual capacity of the employee plays an essential role for a successful identification of fails. In other plants, a computer system support separates tiles that are in bad conditions from the other tiles. Such computer systems use different mathematical methods varying the precision in results and processing time [2].

The tiles fabrication process usually throws 4 different kinds of defects in the product related with the different RGB color components. The first class refers to spots which affect Green and Blue color components, the second class affects the Red and Blue components, the third one refers to Red and Green components, the fourth affects the three components. Finally, tiles without failures are considered as the fifth class.

1.1 Previous Works

In the 90's Boukavalas et al. [3–6], detected different kinds of failures (multi-classification) in the final tile, they used methods like color histograms, texture and color gradient. Other works in this area use different algorithms to calculate different features from tile images such as segmentation and Wavelet transformation. Common techniques used for classification are: Bayes functions [7], K-means [3,4], binary trees [6], K-NN [8] and neural networks [8–10].

In this work the CIELab color space is computed and complemented by common texture attributes obtained from the sum and difference images (SDH technique). Tile image attributes are tested for different combinations used as inputs to the ANN. The parameter configuration of the ANN is obtained after detailed tests.

Section 2 describes the color and texture feature computation. Section 3 explains the configuration and training process of the neural network. Then in the Sect. 4 it is presented the obtained results and it is made a comparison of the work. The conclusions are included at the end of this document.

2   Feature Extraction

The Fig. 1 illustrates the different steps for extracting color and texture features.

2.1 CIEL*ab Color Space

The RGB images are device-dependent and highly affected by illumination changes. To avoid this constraint, it is used the CIE L*ab color space which describes all the colors seen by the human eye.

As illustrated in the two first levels of Fig. 1, the RGB to $CIEL*ab$ transformation is performed through an intermediate space known as CIE-XYZ based on the tristimulus values. The three primary colors in the human eye are described

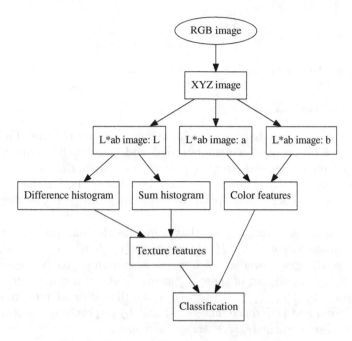

Fig. 1. Color and Texture features strategy

by the XYZ parameters of the Eq. 1. The numerical matrix in such equation represents the predefined parameters or factors to carry out the transformation.

$$\begin{bmatrix} X \\ Y \\ Z \end{bmatrix} = \begin{bmatrix} 0.4124 & 0.3575 & 0.1804 \\ 0.2126 & 0.7151 & 0.0721 \\ 0.0193 & 0.1191 & 0.9502 \end{bmatrix} \begin{bmatrix} R \\ G \\ B \end{bmatrix} \tag{1}$$

Then, the transformation CIE-XYZ to CIE L*ab is given by the following equations:

$$L* = 116 f\left(\frac{Y}{Y_w}\right) - 16 \tag{2}$$

$$a = 500 \left[f\left(\frac{X}{X_w}\right) - f\left(\frac{Y}{Y_w}\right) \right] \tag{3}$$

$$b = 200 \left[f\left(\frac{Y}{Y_w}\right) - f\left(\frac{Z}{Z_w}\right) \right] \tag{4}$$

where X_w, Y_w and Z_w are tristimulus of CIE-XYZ values with reference to the "white spot"; given by:

$$\begin{bmatrix} X_w \\ Y_w \\ Z_w \end{bmatrix} = \begin{bmatrix} 0.9504 \\ 1.0000 \\ 1.0887 \end{bmatrix} \tag{5}$$

and

$$f(t) = \begin{cases} t^{\frac{1}{3}} & if \left(\frac{6}{29}\right)^3 \\ \frac{1}{3}\left(\frac{29}{6}\right)^2 t + \frac{4}{29} & otherwise \end{cases} \tag{6}$$

where $f(t)$ is a time function.

2.2   Texture Features

The surface of an object is typically characterized by its texture. The texture is defined as a property in a neighborhood and is generally computed using gray scale or binary images. In this project, it is used the Sum and Difference Histograms (SDHs) algorithm proposed by Unser [1], for computing texture attributes. A general description of the SDH strategy is described in the following.

Be $K \times L$ a rectangular grid that contains the analyzed discrete texture of an image denoted by $I(k, l)$ where $(k \in [0, K-1]; l \in [0, L-1])$. Suppose that the gray level at each pixel is quantified to N_g levels, so let $G \in [0, ..., N_g - 1]$ be the set of these N_g levels. Next, for a given pixel (k, l), let $(\delta_k, \delta_l) = \{(\delta_{k_1}, \delta_{l_1}), (\delta_{k_2}, \delta_{l_2}), ..., (\delta_{k_M}, \delta_{l_M})\}$ be the set of M relative displacements. The Sum and Difference images, I_S and I_D respectively, associated with each relative displacement (δ_k, δ_l), are defined as:

$$\begin{aligned} I_S(k, l) &= I(k, l) + I(k + \delta_k, l + \delta_l) \\ I_D(k, l) &= I(k, l) - I(k + \delta_k, l + \delta_l) \end{aligned} \tag{7}$$

Thus, the range of the I_S image is $[0, 2(N_g - 1)]$, and for the I_D image is $[-N_g + 1, N_g - 1]$. From this, let define i and j as two any gray levels in the I_S and I_D image range respectively. Then, let D be a subset of indexes which specifies a region to be analyzed, so, the SDHs with parameters (δ_k, δ_l) over the domain $(k, l) \in D$ are, respectively, defined as:

$$\begin{aligned} h_S(i; \delta_k, \delta_l) &= h_S(i) = \#\{(k, l) \in D, I_S(k, l) = i\} \\ h_D(j; \delta_k, \delta_l) &= h_D(j) = \#\{(k, l) \in D, I_D(k, l) = j\} \end{aligned} \tag{8}$$

where, the total number of counts is

$$N = \#\{D\} = K \times L = \sum_i h_S(i) = \sum_j h_D(j) \tag{9}$$

The normalized SDHs is given by:

$$\widehat{P_S}(i) = \frac{h_S(i)}{N} \qquad \widehat{P_D}(j) = \frac{h_D(j)}{N} \tag{10}$$

Unser [1] has proposed a variety of features for extracting only useful texture information from the SDHs. The features most frequently used are: mean, variance, correlation, contrast, homogeneity, cluster shade and cluster prominence. In particular, all of these features are computed in this approach.

3   Failure Detection Strategy

The design of the classifier proposed is divided in two stages: (1) the configuration of the artificial neural network (ANN), and, (2) testing the texture tiles with the classifier obtained in the stage 1. The classifier is designed considering that there is no distance and optimal orientation for all texture features computed. Therefore, exhaustive tests are performed in order to find the best parameters to configure the ANN. Figure 2 illustrates the block diagram where each specific task performed to obtain the final classifier.

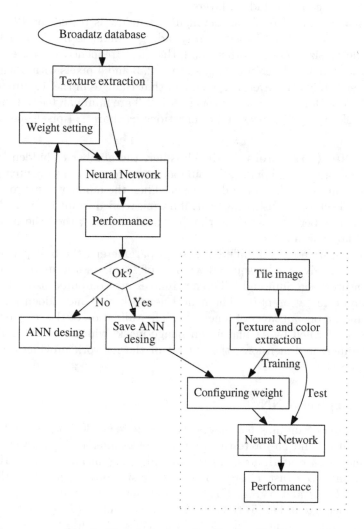

Fig. 2. Global strategy for tile classification

3.1 Design of the Artificial Neural Network (ANN)

The database for the development of this phase considers 5 different kinds of Brodatz images with the purpose of training the ANN to classify 5 classes as mentioned in Sect. 1. This number defines also the neurons at the output layer of the neural network, representing one neuron for each class. Furthermore, for each kind of image, 200 points are chosen randomly. Additionally, the parameters that describe the system are: (1) the size of the rectangular window to compute texture attributes, (2) the distance and orientation to compute sum and difference images, (3) the number of hidden layers in the neural network, and (4) the number of neurons in the hidden layers.

In this work, the size of the rectangular window is tested for 12 different values: 3, 5, 7, 9, ..., 25 and their corresponding combinations yielding 144 different window's sizes. On the other hand, the most frequently distances used to compute the sum and difference images are 1, 2, 3 and 4 pixels with orientations of $0°$, $45°$, $90°$, $135°$. However, it is assumed that changes in the texture features for $0°$ and $45°$ or for small distances such as $1, 2, 3$ are minimal due to their proximity. Therefore, the distances and orientations used are $1, 4$ pixels and $0°, 90°$, respectively.

Setting ANN Parameters. In the classifier, the number of hidden layers is limited to one or two, and finally, the number of neurons per layer contemplate 2, 3, 4, ..., 40 neurons in the first hidden layer. Then the number of neurons in the second layer is chosen considering that: if an unsatisfied result is reached (using a particular number of neurons in the first hidden layer), then the neurons in the second layer are increased.

After exhaustive tests of different system parameters, the best performance and accuracy is given for a squared window of 13 pixels size used to compute texture features, the Sum and Difference images are computed using 1 pixel of distance and orientation of $0°$. The neural network uses one hidden layer of 28 neurons, and an output layer of 5 neurons. The validation of the selected neural network is performed by evaluating an image-mask, which contains 5 Brodatz images in different dimensions, the output of the network provides an overall accuracy of 84.5%.

3.2 Training the ANN

The tile database contains 99 images used as follows: 30 images for training, 14 for validation and 55 for testing. Thus, twelve different inputs were tested: (1) the nine texture attributes, (2) the five most significant texture attributes (mean, variance, contrast, cluster shade and cluster prominence), (3) the color components $L*$, a and b from $CIE - L * ab$ space, (4) the test (2) and the color components a and b, (5) the test (2) and four difference colors, (6) the test (5) and four difference hue and, finally tests (7) to (12) use the same order as the six earlier experiments, but with the equalized images.

The output of the neural network delivers a class label for every pixel. A gray level value is assigned to each class in order to generate the resulted image.

A tile without defects differs from a tile with defects quantitatively. Therefore, it is possible to analyze the resulted image by calculating the rate among pixels detected without fail and total pixels on the image. This rate varies depending on the test. However, this value define if the tile have or not defects. In case of the detection of failures in tiles, a class of fail is assigned by analyzing the axis length of the spot. Finally, if different classes of failures are found in the same tile, a second test analyses the larger one, in order to assign a final unique class. A particular case occurs when the tile image is labeled with faults, but without spots, in this case, the tile is directly assigned to class 2, corresponding to defects in the color components due to bad illumination conditions.

The Table 1 shows the confusion matrix for the experimental test number 5 (the five most significant texture attributes and four color differences). Note that, tile images without faults (class 5) are correctly classified (100%) for all cases. On the contrary the worst performance is obtained for the class 1 (labeled as class 3), and for the class 2 (labeled as class 1). By analyzing such errors, it is found that it is possible to locate irregular spots on the tile due to an illumination problem which affect more the class 1 and 2 than the others. Despite of these errors in the classification, the diagonal of the matrix conserves the highest classification values, as it is expected.

Table 1. Confusion matrix of the images used in the experimental test 5.

Desired\Obtained	1	2	3	4	5	
1	6	2	3	0	0	54.54%
	6.06%	2.02%	3.03%	0.0%	0.0%	45.46%
2	5	6	0	0	0	54.54%
	5.05%	6.06%	0.0%	0.0%	0.0%	45.46%
3	2	2	29	0	0	87.87%
	2.02%	2.02%	29.29%	0.0%	0.0%	12.13%
4	2	4	1	26	0	78.78%
	2.02%	4.04%	1.01%	26.26%	0.0%	21.22%
5	0	0	0	0	11	100.0%
	0.0%	0.0%	0.0%	0.0%	11.11%	0.0%
	40.0%	42.85%	87.87%	100.0%	91.66%	78.78%
	60.0%	57.15%	12.13%	0.0%	8.34%	21.22%

4   Experimental Results

The experimental tests are performed using tile database provided by the company Daltile at Monterrey in Mexico, consisting of 99 color digital images of tiles taken at the end of the glazing process. The Figs. 3 and 4 illustrates the classification results, detected failures are highlighted in squares. In particular, the fail in the Fig. 3 (second row) is even more difficult to see by the human eye. The dark color pixels in the resulted images make reference to different tons and no homogeneous zones of color in the image tile. However, such pixels do not belong to one specific class of the ANN, that is, they represent an unknown failure.

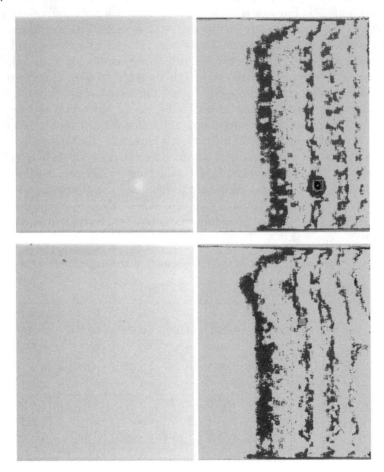

Fig. 3. Left column: tiles belonged to class 1 and 4, respectively. Right column: resulted output images, the True Positives are highlighted.

4.1   Performance Evaluation and Comparative Analysis

Results in Table 2 points out that the classes 3, 4 and 5 obtain the highest scores for all metrics. The low score in precision and sensitivity for classes 1 and 2 is mainly due to a few number of images belonged to this class, one-third in comparison to the rest of the classes.

This approach can be compared with other similar projects that use different neural network architectures. For instance in [9,11], the authors perform tile classification but only for two classes. Another similar work is presented by

Table 2. Performance evaluation of the classifier

Measures\Classes	Class 1	Class 2	Class 3	Class 4	Class 5
Precision	25.0%	16.6%	92.3%	100.0%	100.0%
Sensitivity	33.3%	33.3%	96.0%	76.19%	100.0%
Specificity	94.23%	90.38%	93.33%	100.0%	100.0%
Accuracy	90.90%	87.27%	94.54%	90.90%	100.0%

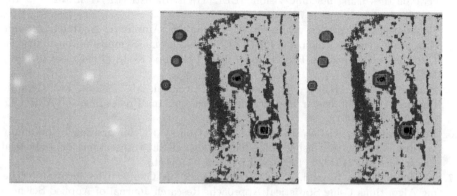

Fig. 4. Left: tile belonged to class 3 with several fails (TP). Middle: resulted image classifies one of the spots as class 1. Right: several spots classified as class 3.

Kukonen et al. [8] in which a multi-classification work is proposed, also 5 classes. Such approach uses a K-NN and a neural network SOM with spectral characteristics from the tile, obtaining a 30.0% of error in the classification using a 1-NN and 2-D SOM. Second tests are presented obtaining 20.0% of error using a 7-NN and 1-D SOM, in the system proposed has an error rate is 20.0% also.

5   Conclusions

A neural network is used as a classifier of color and texture features from tile images. A database of tiles with and without different kind of fails are used for experimental tests, validating the accuracy and feasibility of the proposed approach in comparison with other approaches. This project validates that texture features like contrast, variance, homogeneity complements the color information provided by the CIE-Lab color space. The classifier achieves high performance for detecting classes 3 to 5. However, some of the tile images present changing light conditions that is hard to classify. Finally, even if the approach is a first proposal strategy, good classification results are provided for tackling this typical industrial problem. Future researches include an increment of the number of images per class and to have a balanced number of cases among classes.

References

1. Unser, M.: Sum and difference histograms for texture classification. IEEE Trans. Pattern Anal. Mach. Intell. **8**, 118–125 (1986). doi:10.1109/TPAMI.1986.4767760
2. Hocenski, Z., Keser, T.: Failure detection and isolation in ceramic tile edges based on contour descriptor analysis. In: Mediterranean Conference on Control Automation, MED 2007, pp. 1–6 (2007). doi:10.1109/MED.2007.4433713.
3. Boukouvalas, C., Kittler, J., Marik, R., Petrou, M.: Automatic color grading of ceramic tiles using machine vision. IEEE Trans. Industr. Electron. **44**, 132–135 (1997). doi:10.1109/41.557508
4. Boukouvalas, C., Kittler, J., Marik, R., Petrou, M.: Automatic grading of ceramic tiles using machine vision. In: 1994 IEEE International Symposium on Industrial Electronics, Symposium Proceedings, ISIE 1994, pp. 13–18 (1994). doi:10.1109/ISIE.1994.333123.
5. Boukouvalas, C., Kittler, J., Marik, R., Mirmehdi, M., Petrou, M.: Ceramic tile inspection for colour and structural defects. In: Proceedings of AMPT95, pp. 390–399 (1995)
6. Boukouvalas, C., Kittler, J., Marik, R., Petrou, M.: Color grading of randomly textured ceramic tiles using color histograms. IEEE Transactions on Industrial Electronics **46**, 219–226 (1999). doi:10.1109/41.744415
7. Aborisade, D.O., Ibiyemi, T.S.: Ceramic Wall Tile Quality Classification Training Algorithms Using Statistical Approach. Research Journal of Applied Sciences **2**, 1255–1260 (2007). http://medwelljournals.com/abstract/?doi=rjasci.2007.1255.1260
8. Kukkonen, S., Kälviäinen, H., Parkkinen, J.: Color features for quality control in ceramic tile industry. Opt. Eng. **40**, 170–177 (2001). doi:10.1117/1.1339877
9. Andrade, R., Eduardo, C.: Methodology for automatic process of the fired ceramic tile's internal defect using IR images and artificial neural network. J. Braz. Soc. Mech. Sci. Eng. **33**, 67–73 (2011). doi:10.1590/S1678-58782011000100010
10. Smith, M.L., Stamp, R.J.: Automated inspection of textured ceramic tiles. Comput. Ind. **43**, 73–82 (2000). doi:10.1016/S0166-3615(00)00052-X, Elsevier
11. Rimac-drlje, S., Keller, A., Nyarko, K.E.: Self-learning system for surface failure detection. In: 13th European Signal Processing Conference, pp. 1–4 (2005)

ROIs Segmentation in Facial Images Based on Morphology and Density Concepts

Jesús García-Ramírez[✉], J. Arturo Olvera-López, Ivan Olmos-Pineda, and Manuel Martín-Ortíz

Faculty of Computer Science, Benemérita Universidad Autónoma de Puebla, Puebla, Mexico
gr_jesus@outlook.com, {aolvera,iolmos,mmartin}@cs.buap.mx

Abstract. In computer vision, facial images have several applications such as Facial Expression Recognition and Face Recognition. The segmentation of Regions Of Interest (ROIs) in face images are relevant, because those provide information about facial expressions. In this paper a method to segment mouth and eyebrows in face images based on edge detection and pixel density is proposed. According to the experimental results, our approach extracts the ROIs in face images taken from different public datasets.

Keywords: Face images · Face ROIs segmentation · Image processing · Expression recognition

1 Introduction

Nowdays, computer vision is an interesting area for computing researchers, where image processing is the baseline for that area. Some examples of applications in computer vision are: facial expression recognition and face recognition. Different works related to these applications have been reported in the literature [1,2].

The human face is a part of the body with a great scientific interest, because of many expression such as angry, happiness, fear, among others, are reflected on this region. Regions Of Interest (ROIs) in the human face are the eyes, nose, eyebrows, and mouth, these regions describe features from human expressions. Finding these ROIs in digital images is not an easy task because of the low contrast between skin color and those ROIs.

For the last reason, the image processing filters such as thresholding and border extraction do not work well at all in face images, those filters need to be improved in order to get a better performance in this kind of images.

Border extraction in ROIs is a very important task in order to get descriptive data from the face; this information can be used for either face recognition or to find facial expressions. It is difficult to apply a global analysis to face images, because of the ROIs have different features about illumination and density, then a regional analysis is a better way for detecting borders in ROIs for face images.

© Springer International Publishing AG 2017
J.A. Carrasco-Ochoa et al. (Eds.): MCPR 2017, LNCS 10267, pp. 169–178, 2017.
DOI: 10.1007/978-3-319-59226-8_17

For mouth detection, different approaches have been proposed, for example, chromatic information and the Expectation-Maximization algorithm are used to segment the mouth in face images [3]. A different method analyzes image histogram, where ROIs are detected from color energy [4], using R and G channels from the RGB color model. Other approaches use information such as blood concentration [5], where regions such as lips are detected. The disadvantage of this strategy is the fact that it is necessary to capture the input images with a monochrome camera equipped via an acousto-optic device that captures blood concentration. A different strategy for detecting lips and mouth is based on active shape models and active contour models [6], which is robust to different conditions of illumination.

On the other hand, for detecting eyebrows, different authors propose a binarization strategy, using different color space brands (L and b from CIELAB color space), then the Otsu algorithm is applied to find the eyebrow region [7]. In [8] a method based on local active shape model is proposed, in this approach different angles and distances between the eyes and the eyebrows are used to find the shape of a ROI.

In this paper a method for segmenting mouth and eyebrows is proposed, where the aim is to minimize the error (noise pixels detected as ROI, and ROI pixels detected as noise). It is important to mention that eyebrows and mouth contribute with relevant information about facial expressions. The method consist of applying different pixel operations, such as edge detectors and contrast modification, in order to obtain a binary image. Finally, a method based on density and morphological techniques is proposed for getting descriptive information from ROIs including a stage for eliminating noise regions.

This paper is structured as follows: in Sect. 2 the proposed methodology is presented; in Sect. 3 the experimental results obtained by applying our approach over three public datasets are introduced. Finally, conclusions and future work are discussed in Sect. 4.

2   Methodology for Face ROIs Segmentation

In this section a couple of algorithms to find the ROIs in facial images (mouth and eyebrows) are introduced. The proposed methodology can be seen in Fig. 1 and it is described in the following paragraphs. We assume that input digital images are represented according to the RGB color model. However, in our process, channels R and G for the gray scale conversion are used, because B channel does not provide information about edges [3]. The face is located by the Viola & Jones algorithm, the eyebrows and mouth regions are segmented, finally the obtained regions are denoised. Details about our proposed algorithms are described in the following lines.

Digital image processing is a computational expensive process, because all the pixels in the image are taken into account to apply a filter, this process depends on the image size. For that reason, segmentation is an important task to process fewer pixels in the image. In order to reduce the ammount of pixels to process, the face region is located with the Viola & Jones algorithm which is based on intensities of the pixels related to mouth and eyes regions, which are darker than another around them [9].

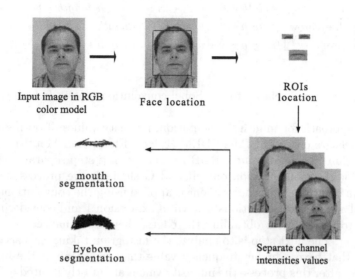

Fig. 1. Proposed Methodology which consists of the stages: Face location, ROIs Location, ROIs segmentation. Input images were taken from the MMI database [12].

In order to find mouth and eyebrows regions in the image, a template with values related to initial and final points of a rectangular area are proposed and they are shown in Table 1, where width and height are values from the output image by the Viola & Jones algorithm. Before getting the values of the Table 1, other values were tested, the better values are presented in the Table 1, these values are proposed for the used databases according to the geometrical features; these values can be modified depending on the dataset.

After the template is used to locate the ROIs, the next step is to apply edge detectors and filters in order to find edge information for mouth and eyebrow regions. These regions have different features, the mouth is a region where edge detectors are applied to get descriptive information about that, on the other hand eyebrow region is not a dense region because shadows between it and eye region are visible in a face image; due these facts different approaches to get descriptive information from each region need to be applied.

In face images the transitions between the ROIs such as mouth and the skin color are not visible, for this reason, the regional filters need to be modified, to take into account a higher area for applying the convolution process, in order to have a better performance in edge detection.

For detecting edge information in face images, we propose an extended convolution matrix, considering dimensions $(2u + 1)$ by $(2u + 1)$, where u is the value that determine the matrix dimension. In Fig. 2 it is shown an example

Table 1. Points to process the template to find the eyebrow and mouth regions in face images.

	Initial point of the region	Final point of the region
Mouth	$(heigth * .6, width * .2)$	$(heigth, heigth * .8)$
Rigth eyebrow	$(heigth * .2 + width * .1)$	$(heigth * .4, heigth * .5)$
Left eyebrow	$(heigth * .2 + width * .5)$	$(heigth * .4, heigth * .9)$

with $u = 1$ and $u = 2$ related to the Sobel convolution matrix (which is used in our experiments).

In our approach for mouth segmentation, intensity values of both green and blue channels are processed. The EDEM (Edge DEtection in Mouth) algorithm (see Algorithm 1) has as input a RGB image, the first step separates the three channels, for the mouth region only R and G channels are processed. To find the edges in mouth regions the gradient is applied using the convolutional masks (horizontal and vertical directions) shown in expression 1 and considering $u = 2$ according to Fig. 2. After obtaining the edges, they are enhanced by applying the sine filter. The next step is to analyze the histogram taking into account the intensities that have a higher frequency value than the mean of the histogram (lines 8–15), after this process, the intensity values are mostly located either near to zero (low regions) or near to 255 (high intensities), then a binarization process ($x = 0$ if $x \leq threshold$, $x = 255$ otherwise) is applied with a $threshold = 127$, finally the algorithm returns an image with the edges of the mouth and some noise, this noise will be removed in the next step.

-1	-2	-1
0	0	0
1	2	1

a)

-1	0	-2	0	-1
0	0	0	0	0
0	0	0	0	0
0	0	0	0	0
1	0	2	0	1

b)

Fig. 2. Modification of the convolution matrices used in our approach, (a) Region with $u = 1$, (b) Region with $u = 2$.

$$\begin{bmatrix} -1 & 0 & 1 \\ -2 & 0 & 2 \\ -1 & 0 & 1 \end{bmatrix} , \begin{bmatrix} -1 & -2 & -1 \\ 0 & 0 & 0 \\ 1 & 2 & 1 \end{bmatrix} \tag{1}$$

Now the eyebrows region will be segmented, for this region other process is applied because the features like the density of the region or the lightning are different among them. In this process only the R channel is used to find the region.

Algorithm 1. Edge DEtection in Mouth (EDEM) algorithm.

Require: Image in RGB color model I
Ensure: Binary image with the mouth edges I_s
  1: Separate the intensities values of channels $I = I_R, I_G, I_B$
  2: **for** each value in I_R and I_B **do**
  3:    G_V =Apply the vertical gradient (Equation 1)
  4:    G_H =Apply the horizontal gradient (Equation 1)
  5:    value=$\sqrt{(G_V)^2 + (G_H)^2}$
  6:    The new pixel value will be: $255 * Sin(\frac{\pi * value}{2 * 255})$
  7: **end for**
  8: Compute the histogram (H_R, H_G) of each new images
  9: Compute the mean of I related to the intensities values μ: $\frac{width * heigth}{255}$
10: **for** each value in I_R and I_B **do**
11:   **if** The frequency of the histogram value is minor than μ **then**
12:     The new value of the pixel in I_s will be 255
13:   **end if**
14: **end for**
15: Binarization with $threshold = 127$
16: Return I_s

For segmenting the eyebrows, ERED (Eyebrow REgion Detection) algorithm is proposed and it is shown in Algorithm 2. This algorithm takes as input a RGB image but only the R channel is processed, to increase the contrast the hyperbolic tangent filter is applied followed by the thresholding function in Eq. 2 where f' is the sine filter, this process is applied to remove the shadow between the eye and the eyebrow region.

$$f(x,y) = \begin{cases} f(x,y) \; if \; f(x,y) < threshold \\ \\ f'(x,y) \; if \; f(x,y) > threshold \end{cases} \quad (2)$$

The eyebrow is not a dense region, for that reason a morphological closing operation is applied to smooth the contour and eliminate thin holes in the image. The structure element considered in our approach can be seen in Eq. 3, this is an element commonly used in morphological operations. Then Otsu algorithm is applied to binarize the image in order to find eyebrow information [10]. Finally, the image with some noise and the eyebrow region is returned.

$$Structure \; Element = \begin{bmatrix} 0 & 1 & 0 \\ 1 & 1 & 1 \\ 0 & 1 & 0 \end{bmatrix} \quad (3)$$

EDEM and ERED algorithms have as output a binary image with the ROI and some noise of other parts in the face. To denoise the image an algorithm based on the clustering algorithm DBscan [11] is proposed. This process can be seen in the Algorithm 3 (DEnse Regions in Binary Image, DERBI) which has as input a binary image, in this case the black pixels are the edges information of the ROIs, the main objective of this algorithm is to minimize the noise and obtain the ROIs in a binary image. This algorithm has as output a list with the clusters in the image. The algorithm analyzes the black pixels and their

Algorithm 2. Eyebrow REgion Detection (ERED) algorithm.

Require: Image in RGB color model I
Ensure: Binary image with the eyebrow region I_s
1: Separate the channel of I $\{I_R, I_G, I_B\}$
2: Apply the hyperbolic tangent filter to I_R.
3: **for** each value in I_R **do**
4:    Apply the threshold filter of the Equation 2
5: **end for**
6: Apply the closing morphological filter with the structure element of equation 3 I_M
7: Compute the Histogram H of I_M and apply the Otsu algorithm to get a threshold th
8: **for** each value x in I_R **do**
9:    **if** x is minor than th **then**
10:       the new pixel value will be 255
11:    **else**
12:       the new pixel value will be 0
13:    **end if**
14: **end for**
15: Return the new image I_s

---

neighbours, taking as density those pixels surrounding with the same color (in this case black), all reachable pixels (with a distance equal to 1) are added to a list and then all the pixels in the list are analyzed in a similar way. If there is not more black pixels a new list is created and the process is repeated with other black pixel. The process finishes when all the black pixels are analyzed.

As output of DERBI algorithm a list of clusters with the coordinates of black pixels is returned, those clusters contain the dense regions in a binary image. After the list of clusters is found the next step is to apply some metrics to determine the clusters corresponding to the ROI. The used metrics are the density with respect to the rectangular area of the cluster, the region in the image related to this rectangular area and the proximity to the center of the image. A range to each metric is established to get the ROI information, if the cluster fulfils these ranges it will be depicted in the binary image.

Algorithm 3. DEnse Regions in Binary Image (DERBI) algorithm.

Require: Binary image I
Ensure: List with the clusters in I (*Clusters*)
1: A list with the points coordinates of the black pixels in I (*Points*)
2: **for** each point x in *Points* **do**
3:    **if** x is not visited **then**
4:       Create a cluster and add to *Clusters*
5:       Add the coordinates to the current cluster with their neighbours in *Points*
6:       Set the pixels coordinates as visited
7:       **for** each value in the current cluster **do**
8:          Add the neighbours coordinates to the current cluster of the not visited neighbours and then set as visited
9:       **end for**
10:   **end if**
11: **end for**
12: Return *Clusters*

---

3   Experimental Results

In this section the results of applying the methods described in Sect. 2 are reported. In the experiments the MMI database was used, this database consist of 474 images, the images of the database are captured from five subjects [12]. In addition Jaffe and VidTIMIT databases are used to compare the result of our approach to segment mouth region. Jaffe Database contains 213 images from 60 Japanese subjects of 7 expressions [13], VidTIMIT contains video and audio recording from 43 persons [14].

To determine the accuracy of the approaches, a test set of the database was selected, the set consist in five images of each subject, the ROIs in the images was manually segmented to compare the segmented images with our approach. To compare the images a polynomial is found with the divided differences method, then the polynomial coefficients from the control images and the segmented images are compared. For mouth region two quadratic polynomials are computed (one for the upper region of the mouth and other for the lower one), and for the eyebrow region only one cubic polynomial is computed. An example for these two polynomials is shown in Fig. 3, the mathematical expression to get points is presented in the top of the images.

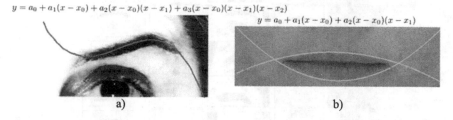

Fig. 3. Examples of interpolated functions using polynomial expressions: (a) eyebrow region and (b) mouth region.

The points of each ROI are translated to the origin, so the first point is located in $(0, 0)$, for that reason the first coefficient of all the polynomials will be 0 and this is not taken into account in the comparison, for mouth region three points have been used for the interpolation (the initial, the final and the center points), in eyebrow region four points are found to get the cubic polynomial that describes the region.

The comparison of the coefficients with the mean and the standard deviation of each coefficient in the polynomial is shown in Tables 2 and 3. The values of mean and standard deviation are near to zero, because of that control images and output images are similar.

In Fig. 4 results of our approach with some MMI database images are shown, the first column depicts to the input images. The second column are the output of EDEM algorithm and it can be seen that the noise in the images are due the illumination and in some cases the beard. In the ERED algorithm the noise is

Table 2. Mean of the error in the comparison of the polynomials of mouth and eyebrow regions.

	Coefficient 1	Coefficient 2	Coefficient 3
Lower mouth	0.101	.00070	X
Upper mouth	0.160	0.00104	X
Eyebrows	0.190	0.0016	0.000019

Table 3. Standard deviation of the error in the comparison of the polynomials if mouth and eyebrow regions.

	Coefficient 1	Coefficient 2	Coefficient 3
Lower mouth	0.105	.00071	X
Upper mouth	0.1518	0.001005	X
Eyebrows	0.177	0.0018	0.000035

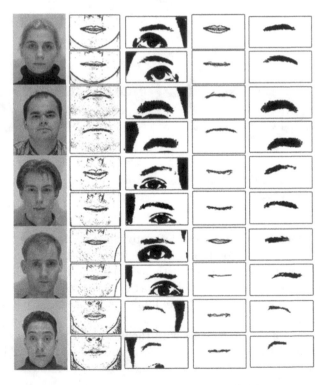

Fig. 4. Results of the MMI database: first column shows the input images; second column depicts the results from the EDEM algorithm in G and R channels; column 3 shows EDEM algorithm results; the last two columns show the DERBI algorithm segmentation.

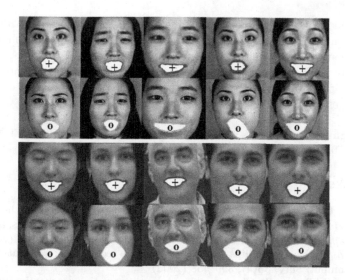

Fig. 5. Comparison of segmentation results, our results are marked by "o", results obtained by [6] are marked by "+".

related to eye region and the borders of the hair, the output images are depicted in the third column. Finally the images from the DERBI algorithm are shown in the last two columns. It can be seen that the found regions can describe the segmented ROI, and they could be used to find numeric features that define the region.

Additionally, we report a comparison with the results reported in [6] with jaffe and VidTIMIT databases. For this databases image equalization over mouth region it is applied, since these images have less quality than the MMI database and jaffe database was taken in gray scale. The mouth segmented region are indicated by the white area and the contour of the found region. Our results are depicted in Fig. 5 marked by "o", results reported in [6] are marked with "+". It can be seen that our approach is able to segment the mouth ROI in most of the cases. Particularly, the wrong segmentation results are due the illumination region in nose, this is because there is shadow between the nose and mouth, other reason is the result of the edge detector.

4 Conclusions and Future Work

In this paper a method to segment mouth and eyebrows is presented, the proposed method is based on transformations of the traditional regional and point filters to find the borders in mouth region and the eyebrows in face images.

Additionally, a method to find the ROIs in a binary image is proposed, this method is based on morphological operations and in DBScan algorithm to detect clusters in a binary image, then some metrics were used to detect the ROI in the image.

According to the experiments, our approach is able to segment mouth and eyebrows in most of the cases for the three datasets and the performance of our method is competetive when comparing with that proposed in [6].

As future work we are going to extract features from the segmented regions in order to either train or combine supervised learners for predicting expressions in face images.

Aknowledgments. This work was partially supported by the CONACyT Mastering Scholarship 701191, and the project OLLJ-ING17-I, VIEP-BUAP.

References

1. Corneau, C., Oliu, M., Cohn, J., Escalera, S.: Survey on RGB, 3D, thermal, and multimodal approaches for facial expression recognition: history, trends, and affect-related applicants. IEEE Trans. Pattern Anal. Mach. Intell. **99**, 2–20 (2015)
2. Wenyi, Z., Rama, C., Phillips, P., Azriel, R.: Face recognition: a literature survey. ACM Comput. Surv. **35**(4), 399–458 (2003)
3. Lucey, S., Sridharan, S., Chandran, V.: Adaptive mouth segmentation using chromatic features. Pattern Recogn. Lett. **23**(11), 1293–1302 (2002)
4. Panning, A., Niese, R., Al-Hamadi, A., Michaelis, B.: A new adaptive approach for histogram based mouth segmentation. Int. J. Electr. Comput. Energ. Electron. Commun. Eng. **3**(8), 1564–1569 (2009)
5. Danielis, A., Giorgi, D., Larsson, M., Strömberg, T., Salvetti, O.: Lip segmentation based on Lambertian shadings and morphological operators for hyper-spectral images. Pattern Recogn. **63**(1), 355–370 (2017)
6. Le, H., Savvides, M.: A novel Shape Constrained Feature-based Active Contour model for lips/mouth segmentation in the wild. Pattern Recogn. **54**(1), 23–33 (2016)
7. Martins, P., César, F., Nardênio, A.: A real-time eyebrow segmentation and tracking technique to support an electric wheelchair interface. In: Proceedings of International Conference on Computer as a Tool (EUROCON), pp. 1–6 (2015)
8. Hoang, L., Prabhu, U., Savvides, M.: A novel eyebrow segmentation and eyebrow shape-based identification. In: Proceedings of IEEE International Joint Conference on Biometrics, pp. 1–8 (2014)
9. Viola, P., Jones, P.: Rapid object detection using a boosted cascade of simple features. In: Proceedings of Computer Vision and Pattern Recognition, pp. 511–518 (2001)
10. Otsu, N.: A threshold selection method from gray-level histogram. IEEE Trans. Syst. Man Cibernetics **9**(1), 62–66 (1979)
11. Ester, M., Kriegel, H.P., Sander, J., Xu, X.: A density-based algorithm for discovering clusters in large spatial databases with noise. KDD **96**(34), 226–231 (1996)
12. Pantic, M., Valstar, M., Rademarker, R., Maat, L.: Web-based database for facial expression Analysis. In: Proceedings of International Conference on Multimedia and Expo, pp. 5–10 (2005)
13. Michael, J., Shigeru, A., Miyuki, K., Jiro, G.: Coding facial expressions with gabor wavelets. In: Proceedings of International Conference on Automatic Face and Gesture Recognition, pp. 200–205 (1998)
14. Conrad, S., Brian, L.: Multi-region probabilistic histograms for robust and scalable identity inference. In: Tistarelli, M., Nixon, M.S. (eds.) ICB 2009. LNCS, vol. 5558, pp. 199–208. Springer, Heidelberg (2009). doi:10.1007/978-3-642-01793-3_21

A Pathline-Based Background Subtraction Algorithm

Reinier Oves García[1]([✉]), Luis Valentin[1], Carlos Pérez Risquet[2],
and L. Enrique Sucar[1]

[1] Computer Science Department, Instituto Nacional de Astrofísica Óptica y
Electrónica, Sta. María Tonantzintla, 72840 Puebla, Mexico
{ovesreinier,luismvc,esucar}@inaoep.mx
[2] Universidad Central "Marta Abreu" de Las Villas, Santa Clara, Cuba
cperez@uclv.edu.cu

Abstract. Background subtraction is an important task in video
processing and many algorithms are developed for solving this task. The
vast majority uses the static behavior of the scene or texture information for separating foreground and background. In this paper we present
a novel approach based on the integration of the unsteady vector field
embedded in the video. Our method does not learn from the background
and neither uses static behavior or texture for detecting the background.
This solution is based on motion extraction from the scene by plane-curve
intersection. The set of blobs generated by the algorithm are equipped
with local motion information which can be used for further image analysis tasks. The proposed approach has been evaluated with a standard
benchmark with competitive results against state of the art methods.

Keywords: Background subtraction · Motion detection · Optical flow ·
Vector field integration

1 Introduction

Video analysis has become an active research topic in Computer Science with
application in robotics, video surveillance, pose estimation, human computer
interaction, etc. One of the first step in all these video applications is the back-
ground/foreground subtraction of the scene. Nevertheless, performing this task,
in an automatic way, remains an important and difficult challenge.

A common way to perform background subtraction is to train a model with
images while their are appearing, and then use the last one for testing if it
is adjusted to the model. Another simpler strategy is to compute the difference
between the current frame and the last one. Modern change detection algorithms
are generally split into three parts [16]: first, a background model of the scene is
created and periodically updated by analyzing frames from the video sequence.
Then, preliminary foreground/background segmentation labels *(or probabilities)*
are assigned to all pixels of every new frame based on their similarity to the

© Springer International Publishing AG 2017
J.A. Carrasco-Ochoa et al. (Eds.): MCPR 2017, LNCS 10267, pp. 179–188, 2017.
DOI: 10.1007/978-3-319-59226-8_18

model. Finally, regularization is used to combine information from neighboring pixels and to make sure uniform regions are assigned homogeneous labels. Because of the wide range of possible scenarios and the parameters that control model sensitivity, the foreground/background segmentation can be very difficult to obtain in some cases, especially when illumination variations, dynamic background elements and camouflaged objects are all present in a scene at the same time.

In this paper we present a novel background subtraction algorithm based on motion change detection through optical flow integration. Our algorithm computes a set of trajectories from the velocity field described by the optical flow. The resulting set of trajectories are used to extract the background/foreground of the video stream. Additionally, they provide a model of the foreground objects characterized by their punctual *velocity, curvature and torsion* which could represent useful features for several applications, such as video surveillance.

2   Vector Fields

Vector fields are classified by its time dependency in two main groups, steady and unsteady vector fields. Steady vector fields represent time-independent flows *(e.g. laminar flows)* while the unsteady or time-dependent *(e.g. turbulent flows)* case are the most complex and represent the changes of the flow over the space-time domain.

An unsteady vector field in $\mathbb{R}^n$ is mathematically represented by a continuous vector-valued function $V_{time}(X, t)$ on a manifold M where the spatial component is represented by $X \in \mathbb{R}^{n-1}$ and temporal dimension is given by $t \in \mathbb{R}$. For such a reason V_{time} is modeled as a system of Ordinary Differential Equations *(ODE)* $\frac{dX}{dt} = V_{time}(X, t)$. In other words V_{time} is a map $\varsigma : \mathbb{R}^{n-1} \times \mathbb{R}$ with initial conditions $X(t_0)$ where its solutions are called characteristic curves, tangent curves or orbits [18,24].

On the other hand a video can be seen like a map $V : \mathbb{R}^2 \times \mathbb{R}$ where all frames $F_i \in \mathbb{R}^2$ evolve in time $T \in \mathbb{R}$. The color information represented by each frame F_i simulates an apparent movement over the temporal domain from F_i to F_{i+1}. The pattern of motion at each pixel on the scene is computed through the partial derivatives of F [1,4]. The optical flow is known as a 2-dimensional steady vector field densely sampled over the frame space.

Massless particles trajectories are computed integrating V_{time} in space as well as in the time-space dimension. These trajectories are classified in four different types, depending on the integration space; all of them differ for the unsteady case, while for the steady case all trajectories coincide. For our purpose we only have to consider *pathlines* which represent the movement of massless particles over the space–time domain. The arc-length parametrization of a pathline P starting at point $p(x(t_0))$ is defined as:

$$p(x(t_0)) = x_0 + \int_{t_0}^{t_n} V(x(t), t)dt \tag{1}$$

As trajectories model the motion of the flow, other local characteristics are defined over V_{time}. Curvature, torsion and instantaneous velocity are intrinsic magnitudes of the vector field which are defined at each non critical point. These properties can be assigned to each curve in every visited point [18, 23].

3   Related Work

Background Subtraction (BS) is the first step applied to detect regions of interest in a video stream. It consists on creating a background model, so that it is possible to discriminate between the static elements and the moving objects in a video sequence. The simplest way to do this is to subtract the current image from a reference image. However, this method is susceptible to illumination changes. Finding a good reference image for BS is complicated due to the dynamic nature of real-world scenes.

Instead of using a fixed reference image for BS, some of the first adaptive methods used pixel intensity average and filtering to create a parametric background model. This kind of approach is robust to noise and can slowly adapt to global illumination variations, but is generally inadequate against shadows. Gaussian mixture models [5, 17] were introduced to solve the latter problem, this approach can handle a dynamic background by using a mixture of Gaussian probability density functions over the color intensities of the image. This approach remains to this day a very popular solution.

BS based on non-parametric models have also been proposed [3]. Unlike parametric models, these rely directly on the local intensity of observations to estimate background probability density functions at individual pixel locations.

Another approach that deals with a multimodal background model is the so-called codebook [8, 25]. This method assigns to each background pixel a series of key color values (codewords) stored in a codebook. These codewords will take over particular color in a certain period of time. BS has also been achieved by other methodologies. In [15] the authors improve the subtraction by superpixels and Markov Random Fields; [7] proposes an approach based on region-level, frame-level [12, 20] or hybrid frame/region-level [11, 19] comparisons to explicitly model the spatial dependencies of neighboring pixels.

The use of methods based on artificial neural network have achieved good results [9, 10] on different scenarios without prior knowledge. However, this kind of approach requires a very large training period. Hybrid approaches have also been proposed. For instance, in [22] the authors combine flux tensor-based motion detection and classification results from a Split Gaussian Mixture Model, and use object-level processing to differentiate static foreground objects from ghost artifacts.

In contrast with previous work, the proposed approach focuses on detecting the foreground based on motion detection through optical flow; once the foreground is detected, the rest of the image is the background. This does not require a training stage, and as a byproduct gives motion information on the moving blobs, which could be used for further analysis.

4 A Pathline-Based Approach for Background Subtraction

The core idea of this method is based on the computation of pathlines behind the unsteady vector field embedded in the video. The flow field construction works over the optical flow extracted in a few continuous sequence of frames. The methodology used in this work is shown in Fig. 1 and described below.

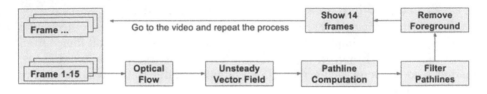

Fig. 1. Methodology of the pathline-based background subtraction strategy over online and offline videos.

Optical Flow: In order to compute the optical flow [4] we take a segment of 15 frames from the video. We select only 15 frames for reducing computational time and drifting [21], but the algorithm is able to work with as many frames as required. From the optical flow we compute the steady vector field *(non-time-dependent vector field)* over each pair of consecutive frames [4] obtaining a list of N steady vector fields.

Unsteady Vector Field: Over the list of steady vector field is generated in the previous step we compute a 2-dimensional unsteady vector field. This field is transformed into a 3-dimensional steady vector field with time implicitly in the *z-direction*. To accomplish this task we generate a grid in R^3 and put in each position $p(x, y, z)$ of the grid the vector $v(u, v, 1)$ [24]. This dimensionality augmentation in Eq. 1 facilitates the integration of the vector field and increases the accuracy and speed of the numerical integration schema. The new dimension incorporated to each vector ($z = 1$) integrates over t which represents the temporal component of the data set.

Pathline Computation: The integration of the vector field with initial conditions produces a set of trajectories called pathlines. These pathlines describe the motion of the massless particle over the space-time domain. Here we sample the grid in a regular way for producing a set of curves parametrized by time. This set of curves represents the motion of color information *(pixels)* over the selected frames (See Subsect. 5.1).

Pathline Filtering: At the same time that pathlines are being generated, a filtering process is taking place. The orientation of this kind of trajectories is important for its later discrimination. Background pixels move more in *z-direction* than in the other two. For that reason we seed a pathline anywhere in our sampled grid but we restrict its travel over the *space-time* taking into account its orientation (See Subsect. 5.2).

Removing Foreground: The filtered pathline bundle contains only trajectories associated with the motion of the scene represented by the set of frames. These trajectories were generated over a continuous space and hence they store inter-pixel information as well inter-frame information. A discretization process of the curve takes place in this step as well as the plane-curve intersection for detecting motion in each frame. Each curve intersects at least one frame in only one point. These intersection points are taken as foreground points (See Sect. 6 for more details).

In each frame, the coordinates of moving blobs (foreground) is given by the intersection points of the pathline bundle with each frame. Once a set of (15) frames (chunk) is processed, the next video segment is analyzed. It is important to note that N frames produce only $N - 1$ steady vector fields. For this reason we have to start the next set taking one value before the previous set ends. This consideration guaranties that motion never ends between each pair of video segments. It is possible that the last set does not fulfill the size of our sample and we recommend for that cases to join the residual set with the previous one so that no motion information is lost.

5   Tracking Motion over an Unsteady Vector Field

Considering pixels as massless particles moving over the flow described by the embedded unsteady vector field we can track their information by means of integrating the ODE system V_{time} with initial conditions over the sampled grid. For our purpose we select Runge-Kutta [2] of 4th order to integrate the ODE system. This method is applied for CFD by the *Computational Fluids Dynamics* community for its accuracy and speed. Our image grid is equipped with a trillinear interpolation schema [14]. This interpolation algorithm allow us to reach a good approximation of trajectories in points where the field is not explicitly defined.

5.1   Dense Pathline Computation

For the sake of capturing the motion in more detail a dense set of pathlines is computed [13]. The dense characteristic of the set is given by the seeding strategy used by the integration algorithm. The grid that contains V_{time} is sampled in the x and y directions by a step factor that separates one spatial initial condition from another. For the temporal domain, we sample the grid one-to-one, so that no temporal information is lost.

Once the initial condition is set, the stop condition must be defined. Given a pathline P starting at $p_0(x_0, y_0, t_0)$ we advect the flow until the next computed point fulfills one of these three conditions: p_n goes out of the domain, p_n reaches a critical point or the vector from p_n to p_{n-1} is almost normal to the XY plane. For instance, suppose we have a video of 340×240 and a duration of 10 frames. A video like this produces a *3-dimensional* steady vector field V_{time} over the domain $240 \times 480 \times 10$. (It is evident that beyond that limits V_{time} it is not defined as well as in critical points ($||V_{time}(p)|| = 0$) [6].

5.2   Filtering Pathlines by Their Orientation

In the previous section it is omitted the third stop rule of the integration algorithm because of its importance. The first two rules only guaranty that every evaluated point in V_{time} is defined and are mandatory to use them while in vector field integration. For our context we add a third rule to isolate automatically background and foreground.

As we mention earlier, background information can be approximated by a straight line parallel to the z-axis. For such a reason our third rule states that given two consecutive points, p_i and p_{i+1}, of the same curve, the distances are defined as: $x_{distance} = |x_i - x_{i+1}|$, $y_{distance} = |y_i - y_{i+1}|$ and $z_{distance} = |z_i - z_{i+1}|$. The orientation condition is reached when the $z_{distance}$ is larger than a threshold (three in our experiments).

Identifying this kind of points while integrating reduces dramatically the execution time of the algorithm and avoids regions where the texture of the objects in the scene coincides with the background texture.

6   Background Subtraction

At this point all the pathlines are filtered and the resulting bundle only represents the motion of the objects over a continuous space. A good representation of the motion at each frame is achieved by means of a discretization of the curve space in the z-direction. This process is given by moving a plane over the frame space and computing the intersections of the curves with the plane for each movement.

Given the plane equation $F = (1, 0, t) + \lambda(0, 1, t) + \beta(0, 0, t)$ and a pathline $P(s)$ parametrized by its z-component we can compute the intersection point $I(x, y, t)$ where P crosses F, $t \in [1..15]$. The foreground in frame F at time t is defined by the set of points I generated by the intersection of the pathline bundle with the image plane.

The point set generated by the intersection of pathlines with the list of frames conforms the motion of the scene but it is a sparse representation of the motion. To eliminate "holes" in the frame a morphological operation –dilatation– is applied. This operation is accomplished by means of drawing a square at each foreground point in each frame with a size of $GridSpaceX/2$. This strategy guaranties the connection between all points in each moving blob producing a continuous region (see Fig. 2(c)).

It is evident that blobs generated by this algorithm are larger than the original moving objects. For reducing this information we compute the frame difference between F_t and F_{t-1}, in order to generate a mask that contains the original moving objects but also noise. Using a bitwise AND operation between the resulting image from our approach and the image produced by frame difference it is possible to eliminate noise and outliers, resulting in a more robust blob *(See Fig. 3 last column)*.

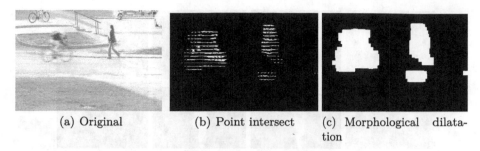

(a) Original   (b) Point intersect   (c) Morphological dilatation

Fig. 2. Visual representation of the plane-curve intersection in the *Change Detection Dataset* (frame 475): (a) Original image. (b) Points where the pathline bundle intersects the frame. (c) Morphological dilatation applied over the point set

7   Experimental Results

To evaluate our approach, we use the benchmark introduced for the 2012 CVPR Workshop on Change Detection. This data set offers different types of realistic condition scenarios, along with an accurate ground truth data. From this benchmark, we are using four categories: highway, office, pedestrians and PETS2006; evaluating our method with 6049 images. This data set has been used to test several state-of-the-art background subtraction methods. Figure 3 illustrates an example from each category, with the results of the proposed background subtraction method as well as another state-of-the-art method.

For evaluation we used two sets of outdoor and two sets of indoor image sequences. Considering that our method is based on motion detection, the outdoor sets present a significant challenge due to the moving background elements. Experimental results are presented using precision and recall *(see Fig. 4)*. Besides, we are comparing these results with another two state of the art methods [16,17], and also with the simplest background subtraction, frame difference (FD). All our experiments were done on an Intel i7 CPU at 3.3 GHz, using the OpenCV library.

Figure 4(a) depicts a plot with the precision metrics for each category of the data set. We observe that our method obtains *high* values for every category. Although it does not have the highest precision value, it is about three times faster than the one that gets the best results. In Fig. 4(c) we show the time needed to process each one of the data set categories.

On the other hand, we don't get the best results in the recall metric, as it is shown in Fig. 4(b). This is because our approach is based on motion detection, and when an object stops is not possible to detect it. However, this background subtraction strategy not only separates moving objects from the background, but also provides relevant information about the moving objects, for instance, velocity, curvature and torsion. These local properties are presented intrinsically in the unsteady vector field embedded in the video and are computed during integration. Our resulting image is modeled as a *3-channel* matrix where the

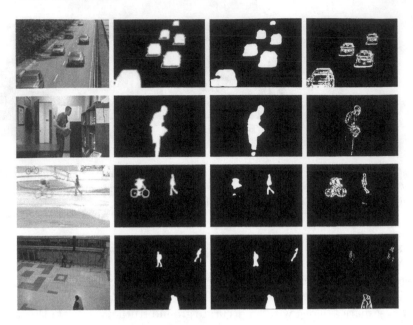

Fig. 3. Background subtraction results. The first column shows a set of images from the data set. In the second column we can see the ground truth. The third column depicts the results of SuBSENSE [16], and in the last column we can see the results of our method.

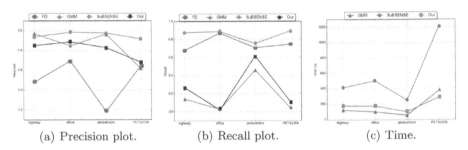

(a) Precision plot.   (b) Recall plot.   (c) Time.

Fig. 4. Precision (a) and Recall (b) plots for each category in the data set. Blue (Hexagon), red (Triangle) and magenta (Circle) lines represents the Frame Difference, Gaussian Mixture Model and SuBSENSE methods, respectively; while the black (Square) line is for our method. We also show the time (c) required by every method to process each category. (Color figure online)

first channel stores the instantaneous velocity, the second channel stores the curvature and the last one stores the torsion. In that way, our blob is more informative than classical blobs.

8   Conclusions

In this paper we presented a novel approach for background subtraction based on the integration of the unsteady vector field embedded in the video. Our method is not based on the steadiness of the scene and neither on texture information; but on motion detection based on optical flow. For such a reason this proposal is particularly useful for scenarios where the background change constantly. Experimental results in a benchmark data set that includes dynamic scenarios, demonstrate that our method is efficient and competitive with other state-of-the-art techniques; in particular it is able to capture all moving blobs no matter how small they are. Besides the blob identification, our method is able to compute local scalar quantities *(velocity, curvature, torsion)* that increase the blob information, providing useful features for further processing, such as object tracking and classification.

Acknowledgments. This work was supported in part by FONCICYT (CONAYT and European Union) Project SmartSDK - No. 272727. Reinier Oves García is supported by a CONACYT Scholarship No.789638

References

1. Beauchemin, S.S., Barron, J.L.: The computation of optical flow. ACM Comput. Surv. (CSUR) **27**(3), 433–466 (1995)
2. Butcher, J.: Runge-Kutta method. Scholarpedia **2**(9), 3147 (2007)
3. Elgammal, A., Harwood, D., Davis, L.: Non-parametric model for background subtraction. In: Vernon, D. (ed.) ECCV 2000. LNCS, vol. 1843, pp. 751–767. Springer, Heidelberg (2000). doi:10.1007/3-540-45053-X_48
4. Farnebäck, G.: Two-frame motion estimation based on polynomial expansion. In: Bigun, J., Gustavsson, T. (eds.) SCIA 2003. LNCS, vol. 2749, pp. 363–370. Springer, Heidelberg (2003). doi:10.1007/3-540-45103-X_50
5. Friedman, N., Russell, S.: Image segmentation in video sequences: a probabilistic approach. In: Proceedings of the Thirteenth Conference on Uncertainty in Artificial Intelligence, UAI 1997, San Francisco, CA, USA, pp. 175–181. Morgan Kaufmann Publishers Inc. (1997)
6. Helman, J.L., Hesselink, L.: Visualizing vector field topology in fluid flows. IEEE Comput. Graph. Appl. **11**(3), 36–46 (1991)
7. Iodoin, J.-P., Bilodeau, G.-A., Saunier, N.: Background subtraction based on local shape. CoRR, abs/1204.6326 (2012)
8. Kim, K., Chalidabhongse, T.H., Harwood, D., Davis, L.: Real-time foreground-background segmentation using codebook model. Real-Time Imaging **11**(3), 172–185 (2005)
9. Maddalena, L., Petrosino, A.: A self-organizing approach to background subtraction for visual surveillance applications. IEEE Trans. Image Process. **17**(7), 1168–1177 (2008)
10. Maddalena, L., Petrosino, A.: The SOBS algorithm: what are the limits? In: CVPR Workshops, pp. 21–26. IEEE Computer Society (2012)
11. Nonaka, Y., Shimada, A., Nagahara, H., Taniguchi, R.: Evaluation report of integrated background modeling based on spatio-temporal features. In: CVPR Workshops, pp. 9–14. IEEE Computer Society (2012)

12. Oliver, N.M., Rosario, B., Pentland, A.P.: A Bayesian computer vision system for modeling human interactions. IEEE Trans. Pattern Anal. Mach. Intell. **22**(8), 831–843 (2000)
13. Peng, Z., Laramee, R.S.: Higher dimensional vector field visualization: a survey. In: TPCG, pp. 149–163 (2009)
14. Rajon, D.A., Bolch, W.E.: Marching cube algorithm: review and trilinear interpolation adaptation for image-based dosimetric models. Comput. Med. Imaging Graph. **27**(5), 411–435 (2003)
15. Schick, A., Bäuml, M., Stiefelhagen, R.: Improving foreground segmentations with probabilistic superpixel Markov random fields. In: 2012 IEEE Computer Society Conference on Computer Vision and Pattern Recognition Workshops, pp. 27–31. IEEE, June 2012
16. St-Charles, P.-L., Bilodeau, G.-A., Bergevin, R.: Subsense: a universal change detection method with local adaptive sensitivity. IEEE Trans. Image Process. **24**(1), 359–373 (2015)
17. Stauffer, C., Grimson, W.E.L.: Adaptive background mixture models for real-time tracking. In: CVPR, pp. 2246–2252. IEEE Computer Society (1999)
18. Theisel, H., Weinkauf, T., Hege, H.-P., Seidel, H.-P.: Topological methods for 2d time-dependent vector fields based on stream lines and path lines. IEEE Trans. Vis. Comput. Graph. **11**(4), 383–394 (2005)
19. Toyama, K., Krumm, J., Brumitt, B., Meyers, B.: Wallflower: principles and practice of background maintenance. In: ICCV, pp. 255–261 (1999)
20. Tsai, D.-M., Lai, S.-C.: Independent component analysis-based background subtraction for indoor surveillance. IEEE Trans. Image Process. **18**(1), 158–167 (2009)
21. Wang, H., Schmid, C.: Action recognition with improved trajectories. In: Proceedings of the IEEE International Conference on Computer Vision, pp. 3551–3558 (2013)
22. Wang, R., Bunyak, F., Seetharaman, G., Palaniappan, K.: Static and moving object detection using flux tensor with split Gaussian models. In: Proceedings of IEEE CVPR Workshop on Change Detection (2014)
23. Weinkauf, T., Theisel, H.: Curvature measures of 3d vector fields and their applications (2002)
24. Weinkauf, T., Theisel, H.: Streak lines as tangent curves of a derived vector field. IEEE Trans. Vis. Comput. Graph. **16**(6), 1225–1234 (2010)
25. Mingjun, W., Peng, X.: Spatio-temporal context for codebook-based dynamic background subtraction. AEU Int. J. Electron. Commun. **64**(8), 739–747 (2010)

Robotics and Remote Sensing

Perspective Reconstruction by Determining Vanishing Points for Autonomous Mobile Robot Visual Localization on Supermarkets

Oscar Alonso-Ramirez$^{(\boxtimes)}$, Maria Dolores Lopez-Correa,
Antonio Marin-Hernandez, and Homero V. Rios-Figueroa

Artificial Intelligence Research Center, Universidad Veracruzana,
Sebastían Camacho No. 5, Centro, 91000 Xalapa, VER, Mexico
oscalra_820@hotmail.com, mbsoft.xal@gmail.com, {anmarin,hrios}@uv.mx

Abstract. Mobile robots are more and more used on diverse environments to provide useful services. One of these environments are supermarkets, where a robot can help to find and carry products, maintain the account of them and to mark out from a list, the products already in the shopping car (maybe the same robot). However, a common problem on these environments is the autonomous localization, due to the fact that supermarkets are a set of aisles, and most of them look the same for laser range finders; sensors commonly used for localization. On this paper, we present an approach to localize autonomous mobile robots on supermarket by using a perspective reconstruction of the shelves and then an statistical comparison of the products present in them. In order to detect the shelves, the vanishing points are estimated to provide a fast and efficient way to segment products on them. To avoid multiple vanishing points on this kind of environments, result of the variety of products present, a variation of a RANSAC approach is proposed. Once a vanishing point has been determined, an homography process is applied to the shelves in order to rectify images. And finally, by horizontal histograms the robot is able to segment individual products to be compared to the data base. Then the robot will be able to detect by a probability function the correct aisle where it is.

Keywords: Mobile service robots · SLAM · RANSAC methods

1 Introduction

Recently, supermarkets (and other similar environments), have been interested in service robots that can provide efficient services to their costumers. However, in order to provide an effective service on these environments, mobile robots face a new localization challenge. A regular supermarket consist of a set of long parallel aisles with shelves on both sides. Normally this distribution is common in most supermarkets and they do not change the distribution of the aisles very often, making it a very static environment (Fig. 1).

© Springer International Publishing AG 2017
J.A. Carrasco-Ochoa et al. (Eds.): MCPR 2017, LNCS 10267, pp. 191–200, 2017.
DOI: 10.1007/978-3-319-59226-8_19

Fig. 1. Aisles of a supermarket

Due to the repeated structure of the supermarket aisles the classic localization techniques commonly based on laser range finders are not enough, due to the fact that many aisles of this environment, viewed with its laser sensor, look exactly the same, making the localization a very difficult task. Moreover, coupled with the problem known as closed loop, i.e. recognize when the robot is in the same place that it has visited previously; these problems make the robot easily lost.

The main idea is to try to use the same information as humans. Normally humans use the products on the shelves to identify the aisle where they are, so if a robot is equipped with a camera and a pan/tilt unit, they can try to segment and identify the products to improve its localization.

Even when the aisles are statics the distribution of the products on the shelves is not. Due to the marketing strategies some changes can occur, for example:

- Products relocations for sales or promotions
- A new presentation of an existing product
- Presentation of a new brand or new products
- Display of seasonal products

The changes with the distribution of the products makes impossible to obtain a static visual map for the robot with such information, so the robot will have to identify the objects at the same time as he goes through the aisles and compute the probability of being on a particular aisle.

To identify the different products on the shelves of the supermarkets is not a trivial task for the robot. On the shelves, we can find different products with different colors, sizes and shapes; and normally they are placed one next to each other, which makes difficult for the robot to segment the different objects. Also, it is expected when the robot follows a client, he will not be looking at the products directly, because the client normally moves in a straight line (or as most as possible) through the aisle. Therefore the robot needs to be able to recognize the products with a considerable distortion from the perspective of the image.

In order to improve the autonomous mobile robot localization, we propose a technique to get additional information from the aisles by detecting at first, the vanishing points from the shelf lines and then segmenting products from the shelves, in order to estimate the correct aisle in function of the quantity and type of products present.

This paper is organized as follow, in Sect. 2, we present some significant related works. On Sect. 3 are explained the proposal approach, beginning with the problem of perspective reconstruction and then by segmenting the products in shelves. Section 4 refers to results and evaluations and finally on Sect. 5 are given the conclusions and future work.

2   Related Work

The incorporation of visual information to the maps, has been already tackled by some research groups; for example, Mariottini and Roumeliotis [7] present a strategy for active vision-based localization. They localize the robot in a large visual-memory map using a series of images. To reduce localization ambiguity that comes from very similar images when the robot navigates in the environment, it is used a Bayesian approach. However, this approach needs to have previous information of the places where he would navigate and changes in the shelves would make that the robot does not recognize the place.

Other works have focussed on an hybrid localization approach, for example, Li et al. [5] and the robot HERB [9] use visual landmarks to improve the robot localization. Both of them present good results, but the visual landmarks are not always suitable for real environments, since they can be blocked, damaged or removed by the humans.

In [2], Lucas, an autonomous mobile robot developed as a library assistant is presented. His localization is based on the fusion of odometry, monocular vision and sonar data, validated with an extended Kalman Filter. The robot uses a simple a priori map and does not use pre-recorded images to aid localization.

The robot TOOMAS, presented in [4], is a shopping guide for home improvement stores. The robot uses a Map-Match-SLAM algorithm [8]. To create the map the robot navigate trough the store and later, based on the store management system, it label the locations of all the articles in the store.

Robocar [3] is another service robot designed as a shopping assistant, in this case for helping blind people to do their shopping independently. In this case, robot localization is done by using RFID tags along the aisles of the supermarket. Even though, to equip the stores with RFID tags, indoor GPS or other active or passive components, could be a practical solution to improve the localization, most store owners prefer not to install markers [4]. They prefer a simple plug & play solution using only the robot's on-board sensors and advanced navigation techniques. The work presented in [10] proves the importance of identify and localize relevant objects and later incorporate them to the robot map.

In this work, it is proposed to solve navigation ambiguity using geometry of the aisles, in order to be able to segment products on the shelves and then by a probabilistic method determine the most probable position of the robot on the supermarket in real-time and avoiding to not pay attention to their primary user, the client.

3   Perspective Reconstruction

The proposed approach is divided on two main issues: (a) a correct detection of the shelves based on the scene vanishing points, and (b) products segmentation from the shelves.

3.1   Detecting Vanishing Points

As vanishing points are invariant to translation and changes in orientation, they may be used to determine the orientation of the robot. In [2] is presented a robot in a library environment (which has many similarities with a supermarket). On this work the robot's onboard camera is fixed to the robot axis, then the vanishing point of the image will tell the orientation of the robot with respect to the orientation of the bookshelves. Using the vanishing point detection, the relationship between 2D line segments in the image and the corresponding 3D orientation in the object plane can be established.

Behan [2] states "all the dominant oblique lines will share a common vanishing point due to the parallelepiped nature of the environment", so the same characteristic can be expected for the supermarket environment. A popular approach for detecting the vanishing points is the one presented by Barnard [1] where based on a Gaussian-sphere he determines the vanishing points to find the orientation of parallel lines and planes. In [6] Magee and Aggarwal present another computationally inexpensive approach for detecting the vanishing points.

However, when the camera of the robot is not fixed, i.e. it's placed over a pan/tilt unit (PTU), determining the vanishing points on the scene viewed by the robot do not correspond with the robot's orientation. The robot used (Fig. 2) along this work, has its cameras over a PTU. With this cameras, robot should follow user to help him when necessary and simultaneously it needs to locate itself.

In order to detect vanishing points, the lines corresponding to the shelves are identified. However, the presence of the products in the shelves create a very noisy images, so, in order to have a good line detection, a smoothing filter is applied. A bilateral filter was chosen because it present a better conservation of the edges compared with the most common filters. Then to find the lines in the image the Hough line transform was used. From the set of lines detected by the algorithm (denoted by L), all their angles are calculated in order to reject vertical lines as well as some lines not corresponding to the possible shelves orientations. Then the remaining lines R are grouped in pairs to calculate their intersection points I. But a common problem is that depending on the complexity of the

Fig. 2. The robot UVerto, the cameras are placed over a PTU.

Fig. 3. Examples of the lines and intersection points detected

scene, the number and distribution of intersection points I can vary a lot, as seen on Fig. 3.

Thus to face this problem a methodology similar to RANSAC has been proposed to detect vanishing points in the scene. With the lines R and the intersection points I, the vanishing points can be calculated by following the next steps:

- First the standard deviation from the set of points I is calculated.
- If the standard deviation of I is smaller than the threshold Th then the mean of I is considered as a vanishing point and the algorithm ends, if not, continue.
- From I a random subset C is selected.
- The standard deviation from C is calculated.
- If the standard deviation of C is smaller than the threshold Th then the mean of C is considered as a vanishing point and the algorithm ends, if not, return to the third step.

After the vanishing point is detected in the image, the robot can segment the shelves into different levels, which are called "slices". Since the robot will be hardly moving in the center of the aisle, the vanishing points is used to determinate which side of the image will be processed, choosing the side closer to the robot, as it has a bigger part of the visual field.

The slices are formed by following the next steps: from all the lines L in the image, those who have a smaller distance to the vanishing point than a threshold $(Th2)$ are selected and denoted by LF. All the elements in LF are sorted in function of the axis and the image plane. If the distance between two consecutive lines is smaller than a threshold $(Th3)$ they are considered as part of the same shelve and then not processed as a slice.

3.2   Recovering Front Covers of Products

In the image sections we refer as "slices"' we can find the products displayed on the shelves however a few considerations must be taking into account. First, the image resolution is not the same for all the products, the ones who are closer to the robot have a higher resolution but can be incomplete since the lines sometimes hit the upper or the lower edge of the image. Second, the slices that give more information are generally the ones closer to the center of the image, i.e. shelves in the center of the image.

Fig. 4. Characteristics of the products that make difficult the product segmentation

One intuitive idea for segmenting the products would be to find the vertical lines on the slices but due to perspective deformation of the image and the characteristics of the products (the different shape and size of the boxes, the way they are placed in the shelves, etc., as it is shown in Fig. 4) it is necessary to improve this detection.

Fig. 5. Examples of the vertical lines detected

From the set of edges previously detected in the whole image, we perform a search to detect vertical lines within the slice. We select a pair of edges that intersect the two lines of a slice and are closer to the edge of the image as shown in Fig. 5. Having these four intersections the perspective deformation can be corrected.

In Fig. 6, it is shown the result of extracting and transforming the products on the shelve. The red lines mark the shelve detected with the previously explained approach, the blue lines are the vertical lines selected to perform the perspective correction and in green the intersection points used to perform an homography transformation.

Once it has been rectified this part of the image, the "sub-image" of the slice containing the products is processed to find the edges to separate individual products. Tests performed with the Canny and Sobel algorithms, were not good enough to this end.

So, to improve this detection, the image is transformed to a different space color and it is proceed to analyze one of the channels to extract the edges. For the experiment with the HSV space, the channel V was analyzed and for the CIE L*a*b the channel L was analyze with good results.

The Figs. 7 and 8 show the results for the analysis with the CIE L*a*b space color and an edge detection with the Scharr operator. Finally, histogram of edge points are performed, as can be observed on the images of previous figure. The next step is to classify the objects segmented from the shelves. Once the robot is able to recognize the products he can infer by probability the aisle in which he is located. However a more detail explanation of the object recognition and its implementation is out of the scope of this paper. At this stage of the work, only these histograms are used to compare products on the shelves to the ones captured on a previously build database with the supermarket products.

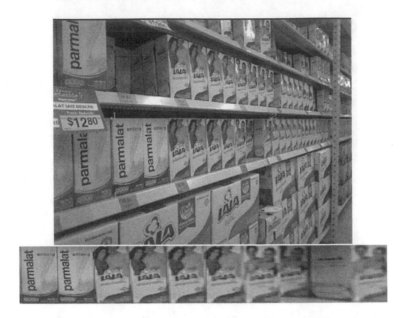

Fig. 6. Examples of the products transformed (Color figure online)

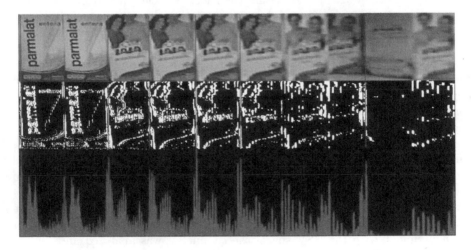

Fig. 7. Example of the products segmentation and analysis

The use of SIFT or similar features are not encourage since they are applied to static environments.

4   Experiments and Results

For the experiments around 3000 images were taken in a real environment at a local supermarket, using a Kinect like sensor mounted on the robot. The images

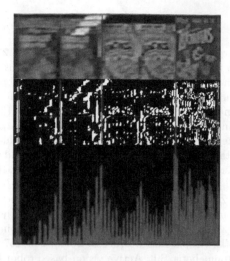

Fig. 8. Example of the products segmentation and analysis

were capture while the robot was moving through the aisles. No customers were present during the data acquisition, however it was tried to emulate the way a robot would navigate in the environment while helping a customer by avoiding to move in a straight line, as he should avoid obstacles, and without facing directly to the shelves.

A few images were selected from the total of images captured in the supermarket for testing our algorithm, the images present small distortion due to movement and belong to cases were the vanishing point is evident for the human and case were it is not.

5   Conclusion

In this paper it has been presented a method for autonomous service robot localization on the aisles of a supermarket. The method uses the products on shelves, in order to compute the probability of been on a certain aisle of the supermarket by counting the number and type of them. Products are segmented form the shelves based on a simple but efficient RANSAC kind algorithm to determine the vanishing points on the images. This work was tested on real images on a semi-structured environment in the presence of a lot of noise, working between 12 to 15 Hz. We were able to combine existing algorithms to the one we propose for the detection of the vanishing points, in order to identify the main lines on the shelves. Having this "slices" reduces the search space and simplifies the object segmentation. The time consumed by our method makes it possible for the robot to perform it on real time. As future work a further classification of the segmented products is proposed. A research for the existing techniques must be performed to select an algorithm that adapts better to our particular problem, giving the images characteristics.

References

1. Barnard, S.T.: Interpreting perspective images. Artif. Intell. **21**(4), 435–462 (1983)
2. Behan, J., OKeeffe, D.T.: The development of an autonomous service robot. Implementation: Lucas the library assistant robot. Intell. Serv. Robot. **1**(1), 73–89 (2008)
3. Gharpure, C.P., Kulyukin, V.A.: Robot-assisted shopping for the blind: issues in spatial cognition and product selection. Intell. Serv. Rob. **1**(3), 237–251 (2008)
4. Gross, H.M., Boehme, H., Schroeter, C., Müller, S., König, A., Einhorn, E., Martin, C., Merten, M., Bley, A.: Toomas: interactive shopping guide robots in everyday use-final implementation and experiences from long-term field trials. In: IEEE/RSJ International Conference on Intelligent Robots and Systems, IROS 2009, pp. 2005–2012. IEEE (2009)
5. Li, X., Zhang, X., Zhu, B., Dai, X.: A visual navigation method of mobile robot using a sketched semantic map. Int. J. Adv. Rob. Syst. **9**(4), 138 (2012)
6. Magee, M.J., Aggarwal, J.K.: Determining vanishing points from perspective images. Comput. Vis. Graph. Image Process. **26**(2), 256–267 (1984)
7. Mariottini, G.L., Roumeliotis, S.I.: Active vision-based robot localization and navigation in a visual memory. In: 2011 IEEE International Conference on Robotics and Automation (ICRA), pp. 6192–6198. IEEE (2011)
8. Schröter, C., Böhme, H.J., Gross, H.M.: Memory-efficient gridmaps in Rao-Blackwellized particle filters for slam using sonar range sensors. In: EMCR (2007)
9. Srinivasa, S.S., Ferguson, D., Helfrich, C.J., Berenson, D., Collet, A., Diankov, R., Gallagher, G., Hollinger, G., Kuffner, J., Weghe, M.V.: Herb: a home exploring robotic butler. Auton. Rob. **28**(1), 5–20 (2010)
10. Stückler, J., Biresev, N., Behnke, S.: Semantic mapping using object-class segmentation of RGB-D images. In: 2012 IEEE/RSJ International Conference on Intelligent Robots and Systems (IROS), pp. 3005–3010. IEEE (2012)

On the Detectability of Buried Remains with Hyperspectral Measurements

José Luis Silván-Cárdenas[✉], Nirani Corona-Romero,
José Manuel Madrigal-Gómez, Aristides Saavedra-Guerrero,
Tania Cortés-Villafranco, and Erick Coronado-Juárez

Centro de Investigación en Geografía y Geomática "Ing. Jorge L. Tamayo" A.C.,
Contoy 137, Lomas de Padierna, Tlalpan, 14240 Mexico D.F., Mexico
jlsilvan@centrogeo.org.mx
http://www.centrogeo.org.mx

Abstract. In this study we tested some methods for detecting clandestine graves using hyperspectral remote sensing technology. Specifically, we addressed three research questions: What is the true potential of hyperspectral images for detecting buried remains? What is the useful information in hyperspectral images for detecting buried remains? When they should be acquired following a burial? For this matter, we simulated seven graves with varying number of carcasses of domestic pigs and monitored the spectral reflectance of the surface during a period of six months. A total of twelve hyperspectral images were formed and analyzed using standard pattern recognition methods. Results indicated that hyperspectral data can indeed have a true potential for detecting buried remains, but the detection can succeed only after three months from burial, and the useful wavelength intervals are mainly distributed along the spectral range of 700–1800 nm and with several narrow intervals that could not have been discovered using multispectral sensors.

Keywords: Remote sensing · Clandestine graves · Hyperspectral images · Partial least squares

1 Introduction

In order to assist the search of clandestine graves several methods have been developed, of which non-intrusive methods are preferred because they reduce terrain disturbance and search times. One of the most used methods is the ground-penetration radar, which is a near-surface geophysical technique that identifies underground anomalies through variations of dielectric permittivity using the backscatter of a microwave beam [13]. Other studies have taken advantage of the chemical properties of the soil in the presence of human (or animal) remains, which shows significant differences in the concentrations of several elements such as N and M, among others, besides the presence of several volatile organic compounds (toluene, benzene and hexane) that reduce or modify the growth pattern of vegetation, which has also been exploited to identify massive

© Springer International Publishing AG 2017
J.A. Carrasco-Ochoa et al. (Eds.): MCPR 2017, LNCS 10267, pp. 201–212, 2017.
DOI: 10.1007/978-3-319-59226-8_20

graves [2, 12, 13]. Although all these approaches can be very effective in detecting common graves, they are often impractical, costly and risky, since they need to be implemented in situ, which seriously limits the area of exploration and exposes the search team to a high risk.

Remote sensing and pattern recognition methods offer a low risk alternative that allows increasing the area of exploration and reduce the associated costs and search times. In this way, vegetation patterns can be identified, soil mineral concentration measured, or some of the physical features mentioned above detected [6, 9]. For instance, aerial photographs have been used to detect abnormal changes of graves vegetation, as well as abnormal landmarks on the graves [4]. Ultraviolet photographs have also been used to map the maturity of vegetation, since graves present younger vegetation than other areas due to a change in soil pH [13]. Obviously the detection of clandestine graves through aerial photographs presents limitations in terms of spectral sensitivity since they detect the reflected light signal in three spectral bands, generally of the visible spectrum. More recent investigations have evaluated the effectiveness of multispectral and hyper-spectral images in the detection of human remains [8–10]. For instance, [9] used in situ data acquired with the field spectrometer ASD Fieldspec FR (in the spectral range of 350–2500 nm) and the HyMap II sensor (125 spectral bands in the range of 450–2500 nm at 4.7–5.2 m spatial resolution) for detecting simulated animal mass graves. They found that, on both scales (in situ and airborne), the graves with bodies did have a signature that distinguishes it from graves that did not contain bodies. They also observed that regeneration of vegetation was strongly inhibited by the presence of residues. A more recent study [10] had demonstrated the feasibility of hyperspectral sensors (CASI and SASI covering the spectral ranges of 408–905 nm and 883–2524 nm respectively) for the detection of single graves, but also acknowledged that additional research was needed to consider other environments and areas with different types of vegetation [10].

Although it is recognized that hyperspectral images constitute a rich source of radiometric information, it is also recognized that in most cases such information can be redundant or useless so that one must question the actual utility of the data for each type of problem. For the case of the detection of clandestine graves, not only we what to know What is the true potential of the hyperspectral images to detect buried remains? But also, What is the useful information in the hyperspectral images? And What is the time window for a better detection? These questions were addressed through simulated graves with domestic pig carcasses, for which the evolution of soil reflectance over a period of six months was measured and spectral signatures were analyzed using standard pattern recognition techniques. In the rest of the paper we describe the experimental setting used (Sect. 2), the methods used for grave detection (Sect. 3), and the results obtained (Sect. 4).

2   Experimental Setting

2.1   Graves Simulation

In February of 2016, we simulated seven graves near Yautepec, Morelos. The location of graves fulfils a number of required characteristics such as accessibility, security, representativeness and climate. Temperature and humidity regimes in this site ensures a relatively fast decomposition of buried remains, so that its effect on the upper surface, if any, could be observed within a monitoring period of around six months. The land was bounded by a solid wall which prevented removal of remains by scavengers and/or human intrusions. Pits of 2 by 2 m wide by 1.2–1.5 m depth were created with a backhoe in an area of 400 m² (Fig. 1). Then, ten carcasses of domestic pigs (pietrain breed) were distributed into four of the seven pits, as indicated Fig. 1, and Table 1 and pits were covered back with the same soil. The other three empty pits were also covered, and where meant to simulate soil disturbances caused by other non-burial process such as plow or constructions. The graves were marked to avoid walking over them during the whole monitoring period and other areas apart from graves were also designated to kept undisturbed.

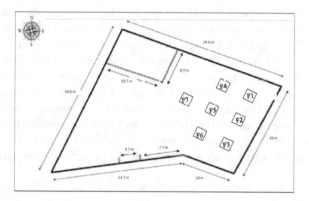

Fig. 1. Location and distribution of simulated graves. Graves are labeled as F1 through F7 and their characteristics are summarized in Table 1. Main entrance (South) and interior walls (North) are also indicated.

2.2   Hyperspectral Measurements

After burial, the evolution of soil reflectance was measured every other week using a full-range field spectroradiometer (Field Spec 4 Std Res by ASD Inc. @ 350–2500 nm). In order to ensure that the same locations were measured every time, a squared threads mesh was fixed at 1.5 m above aground, and with the threads separated by one meter in each direction. Note that the graves were distributed systematically so that 2-by-2 vertices of the mesh fell in the center of each grave, thus ensuring the maximum number of samples per grave. The

Table 1. Characteristics of simulated graves.

Grave ID	Depth [m]	♯ Carcasses	Weight [kg]
F1	1.2	0	0
F2	1.2	3	255
F3	1.5	0	0
F4	1.0	2	170
F5	1.1	0	0
F6	1.2	1	85
F7	1.1	4	340

reflectance measurements were made on each vertex of the mesh by pointing the fiber optic vertically towards the soil surface and using a standard calibration procedure based on a labertian surface of known reflectance (Spectralon). Since measurements were taken using the natural illumination, the time of data acquisition was approximately within 11:00 am–13:00 pm, and whenever the sky conditions was clear or mostly clear, the winds blew slowly (0–5 kmh) and there was no rainfall.

A total of twelve hyperspectral images of 19-by-19 pixels and 2151 spectral bands were formed from the measurements of every sampling date, from February 12 to July 29 (Fig. 2). Although the ground sampling distance was 1 m, the instantaneous field of view of sensor was 25° giving an effective image resolution of 0.67 m. By July, the vegetation had already over passed the mesh height so that some measurements were carried out under the canopy of vegetation. This was actually the main reason to stop monitoring as measurements under the canopy are not compatible with airborne or satellite measurements. Images were not georeferenced as this was not necessary to the project purpose, yet they can be easily aligned to North with a counter-clockwise rotation by an angle of 19° (See Fig. 1).

2.3 Data Preparation

A number of correction and calibration procedures were applied to the images in order to perform a consistent analysis. We first corrected the spectral misalignments between individual sensors of the instrument (VNIR: 350–1000 nm, SWIR1:1001–1800 nm and SWIR2:1800–2500 nm). This was observed only in a couple of images and was caused by an insufficient warm-up of the instrument. The problem was corrected by applying an offset to the reflectances captured by the SWIR1 and SWIR2 sensors. We also eliminated spectral bands of low atmospheric transmissivity in which reflectance values fluctuated erratically (1350–1480 nm, 1780–2032 nm and 2450–2500 nm). Some images also presented the so-called banding artifacts in which the pixels of an entire line appear lighter or darker than those of neighbour lines. This was caused by a poor calibration of sensor and was corrected by multiplying the target line by a coefficient that was estimated using neighbour lines. Finally, all the measurements were normalized as if they were taken under same illumination conditions. The radiometric variability was caused, among other things, by the seasonal variation of the incidence angle of direct sun rays. Fortunately, one of the points of the mesh fell on a piece of concrete which was regarded as spectrally invariant in the spectral range 400–500 nm (blue band). Using as a reference the first acquired image, the other images were re-calibrated through the empirical line method, so that following calibration all images presented similar reflectances in the blue band for the spectrally-invariant point.

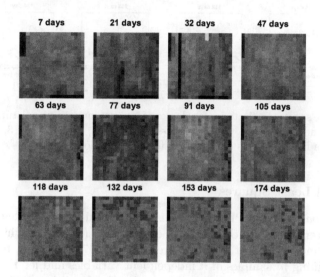

Fig. 2. Color infrared composite of corrected and normalized images for each acquisition date. The number of days from burial is shown on top of each image and the burial date was February 5, 2016. (Color figure online)

The color-infrared composite of corrected and normalized images are shown in Fig. 2, where black pixels in the upper-left corner correspond to missing data that were not measured, whereas those of the vertical line in the 32-days image (March 7) were eliminated due to measuring errors. Figure 3 shows plots of average spectral reflectance for classes of interests including disturbed and undisturbed ground and graves with n carcasses denoted as n-grave, for $n = 1, 2$, and 4.

3   Methods for Detecting Buried Remains

The detection of buried remains with hyperspectral measurements was formulated in terms of an estimate for number of buried carcasses, while the estimation was based on the so-called Partial Least Square method, where the independent variables corresponded to reflectance at each wavelength. This method was chosen for its relative simplicity and because it is appropriate for the case when the number of variables is larger than the number of samples, as in our case.

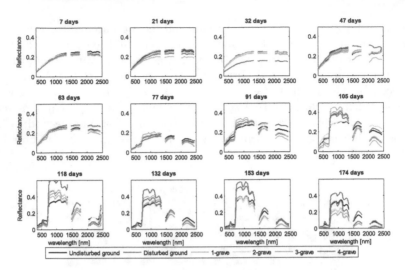

Fig. 3. Average spectral reflectance for classes of interest including undisturbed ground, disturbed ground and graves with n carcasses (n-grave), for $n = 1, 2, 3$, and 4. The number of days from burial is indicated on top of each plot.

3.1 Partial Least Squares

The Partial Least Square (PLS) method is a statistical technique that consists in estimating latent variables that serve as components (scores) to which both the independent and dependent variables are projected [7,14]. Let X be an $n \times m$ matrix containing m samples of n independent variables and let Y be an $n \times p$ matrix containing the corresponding n samples of p dependent variables (in our case $p = 1$). The PLS model relates X and Y through the system of equations:

$$X = TP^T + E \tag{1}$$
$$Y = TQ^T + F \tag{2}$$

where T is an $n \times l$ matrix of l scores, P and Q are orthogonal loading matrices E and F are independent and identically distributed random normal variables. There are several algorithms to estimate the scores and loading matrices in the PLS model above, which also provide estimates of the dependent variable in the form of linear regression $Y = XB + B_0$, where B and B_0 are regression coefficients.

In this study we used the PLSREGRRESS function implemented in MAT-LAB software (The Math Works Inc. 2010), which is based on mean-subtracted variables X_0 and Y_0, so that $P = X_0^T T$, $Q = Y_0^T T$ and $T = X_0 W$ for a weighting matrix W of order $m \times l$ so that T is orthonormal. This implementation also allows to determine the optimal number of components l through a k-fold cross-validation approach. With this method, the dataset is divided into k subsets, and the holdout method is repeated k times. Each time, one of the k subsets is used as the test set and the other $k-1$ subsets are put together to form a training set. Then the average error across all k trials is computed. The process is repeated for $1, 2, \ldots, l_{max}$ components, and the number of optimal components is selected as the one with lowest prediction error.

3.2   Variable Importance of Projection

One advantage of using PLS method is that there are many ways to reduce the number of variables in the final model [11]. One of such method is based on the so-called variable importance of projection (VIP), which measures the contribution of each variable according to its relative importance given by its loading and the relative variance explained by each PLS component. More specifically, the VIP of the jth variable is computed as:

$$v_j = \sqrt{\frac{m \sum_{i=1}^{l} q_i^2 t_i^T t_i (w_{ij}/\|w_j\|)^2}{\sum_{i=1}^{l} q_i^2 t_i^T t_i}} \qquad (3)$$

where $w_{i,j}$ is the loading weight of the j variable in the i component, and t_i, q_i and w_i are the ith column vectors of matrices T, Q and W, respectively. In our case, since T is orthonormal, then $t_i^T t_i = 1$ for all i.

Theoretically, variables with VIP values greater or equal than 1 must be included in the final model. In practice, however, this variable selection strategy may leave out some important variables or include unimportant variables by chance because the computation of the VIP is based on a relatively small sample. A more robust alternative named the Bootstrap-VIP technique [5] was used here. This method is based on repeated computations of the VIP for subsamples of around 40% from the entire sample. Then, variables are selected if their average VIP plus one standard deviation exceed the unit.

4   Results

4.1   Temporal Window of Optimal Detection

We computed the average spectral separability of graves and disturbed ground with respect to undisturbed ground pixels and of graves with respect to disturbed ground. The spectral separability was computed as 1 minus the cosine of the spectral angle between a pair of reflectance signatures. This measure becomes zero when the reflectance curves have similar shapes and differ only by a constant factor. On the other hand, it tends to unit for two reflectance curves with totally distinct shapes (orthogonal). We hypothesize that the spectral signatures of graves with remains have shapes that are distinct from those of other sites, so that the higher the separability the more likely the detection of graves will be.

An increase in spectral separability with respect to undisturbed pixels was observed for both graves (Fig. 4-left) and disturbed ground (center) within the first three months. This increase in spectral separability seems to be caused by

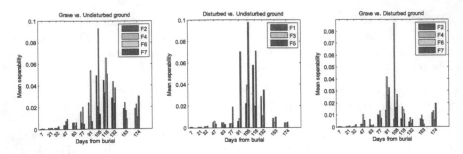

Fig. 4. Average separability of grave pixels with respect to undisturbed pixels (top) and average separability of grave with remains with respect to disturbed ground.

a faster vegetation growth on grave and disturbed ground than on undisturbed pixels. Nevertheless, the grave pixels exhibited also increasing separability with respect to disturbed ground (right). The latter shows the greatest separability within the period between 90 to 120 days after the burial. This seemed to be caused also by a relative greater growth rate of vegetation for grave pixels. The second mode, which occurred towards the end of the monitoring period (174 days), was due to shadows cast by vegetation that had already overpassed the measuring height. A third and lowest mode occurred around the 7 weeks from burial, which coincides with the period of advanced decay of pig carcasses [1], but also with the early appearance of vegetation over the graves with the remains.

4.2   Useful Spectral Intervals

In order to determine the most useful wavelengths for detecting buried remains, we applied the Bootstrap-VIP method as follows. A training sample was first selected consisting of around 50% of image pixels that were randomly selected from three images acquired within the temporal windows of higher spectral separability (90–120 days). Due to the lower representativity of graves, all of available pixels on graves were added to the sample. This dataset was used with a 30-fold cross-validation strategy, which yielded an optimal number of PLS components of 21. With that number of components, the Bootstrap-VIP method was then applied and the most important variables were selected. Finally, the PLS regression coefficients were determined for selected variables.

Figure 5-top shows results from the Bootstrap-VIP method, where the upper envelop of the gray area indicates the mean VIP plus one standard deviation, which was used to select the model variables (values above the red line). Figure 5-bottom shows a plot of the regression coefficients B for selected variables. As a reference, the typical spectral bands used in remote sensing are highlighted. It should be noted that selected wavelength intervals are distributed across the full spectral range, with the major contributions within the NIR (700–1000 nm) and the SWIR1 (1001–1800-nm) spectral regions. The blue, yellow and red bands had little or no use in this model. Some spectral regions that are not measured from space at all turned out to be important, whereas some intervals are too

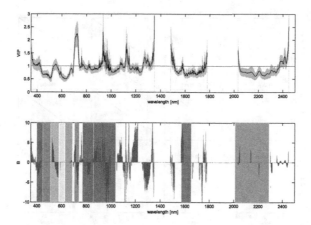

Fig. 5. The top plot shows the mean VIP (black line) and its uncertainty (shaded area) estimated with Bootstrap-VIP method. The red line indicates the variable selection threshold. The bottom plot shows the regression coefficients B for selected variables (reflectance). Typical spectral bands of space-borne remote sensing are highlighted in colors. From left to right they are: Coastal, Blue, Green, Yellow, Red, Red Edge, NIR-1, NIR-2, SWIR-1, SWIR-2 (Color figure online)

narrow ($\ll 100$ nm) for any multi-spectral sensor, thus stressing the need for hyperspectral sensing.

4.3 Detection Accuracy

Constructed model was applied to all images and estimates were compared with actual number of buried carcasses. Figure 6 shows the scatter plots between estimates and actual values. In most cases, the model underestimated the number of bodies and did not present a clear positive trend, except for one date (118 days). This suggests that the model presents a very narrow time window for accurately estimating the number of buried carcasses (black line). Fortunately, the correct detection of graves requires estimating at least one carcase (red line), for which the temporal window can be wider. This can be shown best by computing some measures of the estimation errors and of the detection accuracy. In order to compensate the unbalanced proportion of pixels from graves with respect to those from other sites, we computed the average per-class RMSE (Classwise RMSE) and the kappa coefficient of agreement [3]. In addition, and just for comparison, we computed the root mean squared error over all pixels (Overall RMSE) and the percent of correctly detected pixels (Overall Accuracy).

If one only considers the Overall RMSE, the estimations have virtually the same low error throughout the monitoring period except for the last date (Fig. 7, left, blue bars). This is because the vast majority of pixels are associated with undisturbed ground. However when considering the Classwise RMSE, the pattern reflect more clearly what is observed in the scatterplots of Fig. 6. The Classwise RMSE demonstrates that only in two dates the estimation error

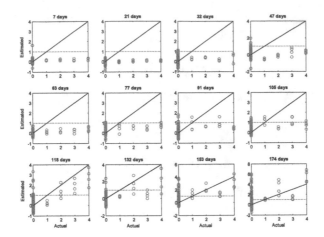

Fig. 6. Scatterplots between actual and estimated number of carcasses. The black line indicates the ideal case of perfect estimation and the red line indicates the detection threshold applied. The number of days from burial is indicated on top of each plot. (Color figure online)

was below one. Likewise, when using the overall accuracy one observes that more than 90% of the pixels are classified correctly, in all but the last sampling date (Fig. 7, right, blue bars). This apparent high detection accuracy is due the much higher number of pixels for one class for which one should expect a very high accuracy by chance. The kappa coefficient compensates the amount of accuracy by chance and thus it is a more reliable measure. The kappa value remains near zero for the first two and a half months afterwards it starts to increase steadily until day 118, and then it decreases again but in a slower and irregular pattern. This irregularity could be associated to the fact that on these dates some measures were already made below the vegetation canopy, where shadows produced a greater variation in reflectance.

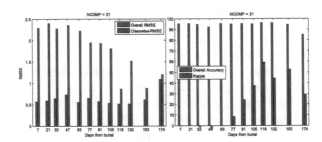

Fig. 7. Root mean square error of estimated carcasses with PLS (left) and detection accuracy after thresholding estimations. (Color figure online)

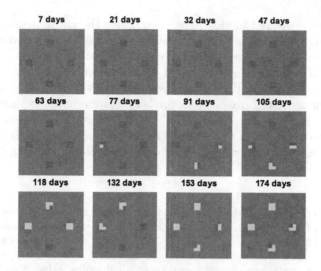

Fig. 8. Distribution of errors/successes detection for each sampling date, where date is specified in terms of the number of days from burial and the colors represent: incorrect true (red), incorrect false (blue), correct true (light gray) and correct false (dark gray). (Color figure online)

The spatial distribution of the errors can be seen in Fig. 8 for each sampling date. The correct detection of the graves (light gray tone) was made possible only on images acquired after 100 days from burial. It is also interesting to note that the date of optimum detection (118 days) did not correspond to the date of greatest spectral separability (Fig. 4), and that the grave F6 was not detected on this date, the grave F4 was partially detected (3 out of 4 pixels), while the graves F5 and F7 were fully detected, thus suggesting that the amount of buried mass was a factor for successful detection on that date. Yet, that the four graves were fully detected only on the last date (072916) suggested that the detection key on that date was the shadowing of taller vegetation that developed on those sites, and which also influenced neighbouring pixels causing false detections.

5   Conclusions

We have presented preliminary results from a study aiming at assessing the potential of hyperspectral imagery for detecting clandestine graves in Mexico. Through a controlled experiment, we demonstrated that the timing for image acquisition is very important as the effect of decay of buried remains can become apparent only after some time period of around three months. Through a least square analysis of monitored surface reflectance, we demonstrated that the NIR and SWIR1 spectral regions are critical for estimating the concentration of buried remains. Although some spectral intervals that were relevant to the detection are wide enough to be measured by multispectral sensors, there were several narrow intervals that stressed the need for hyperspectral sensing.

The results presented in this paper may guide the construction of spectral indices that can be used for the delimitation of search areas by forensic anthropologists. Given the costs involved in hyperspectral image acquisition, a further examination of simulated multispectral data needs to be conducted in order to determine to what extend the multispectral images, which are much cheaper, can be used for detecting graves. Also, further research should explore non-linear models, since the PLS model may have limited power for representing the complexity of the data used in this study. One of such alternatives is the support vector machine with nonlinear kernel, which is being investigated in this project.

References

1. Carter, D.O., Tibbett, M.: Cadaver decomposition and soil: processes. In: Soil Analysis in Forensic Taphonomy, pp. 29–52. CRC Press, Boca Raton (2008)
2. Carter, D.O., Yellowlees, D., Tibbett, M.: Cadaver decomposition in terrestrial ecosystems. Naturwissenschaften **94**(1), 12–24 (2007)
3. Cohen, J.: A coefficient of agreement for nominal scales. Educ. Psychol. Measur. **20**(1), 37–46 (1960)
4. France, D.L., Griffin, T.J., Swanburg, J.G., Lindemann, J.W., Davenport, G.C., Trammell, V., Armbrust, C.T., Kondratieff, B., Nelson, A., Castellano, K., et al.: A multidisciplinary approach to the detection of clandestine graves. J. Forensic Sci. **37**(6), 1445–1458 (1992)
5. Gosselin, R., Rodrigue, D., Duchesne, C.: A bootstrap-vip approach for selecting wavelength intervals in spectral imaging applications. Chemom. Intell. Lab. Syst. **100**(1), 12–21 (2010)
6. Jahn, B., Upadhyaya, S.: Determination of soil nitrate and organic matter content using portable, filter-based mid-infrared spectroscopy. In: Rossel, R.A.V., McBratney, A.B., Minasny, B. (eds.) Proximal Soil Sensing, pp. 143–152. Springer, New York (2010)
7. de Jong, S.: Simpls: an alternative approach to partial least squares regression. Chemom. Intell. Lab. Syst. **18**, 251–263 (1993)
8. Kalacska, M., Bell, L.: Remote sensing as a tool for the detection of clandestine mass graves. Can. Soc. Forensic Sci. J. **39**(1), 1–13 (2006)
9. Kalacska, M.E., Bell, L.S., Sanchez-Azofeifa, G.A., Caelli, T.: The application of remote sensing for detecting mass graves: an experimental animal case study from Costa Rica. J. Forensic Sci. **54**(1), 159–166 (2009)
10. Leblanc, G., Kalacska, M., Soffer, R.: Detection of single graves by airborne hyperspectral imaging. Forensic Sci. Int. **245**, 17–23 (2014)
11. Mehmood, T., Liland, K.H., Snipen, L., Sæbø, S.: A review of variable selection methods in partial least squares regression. Chemom. Intell. Lab. Syst. **118**, 62–69 (2012)
12. Molina, C.M., Pringle, J.K., Saumett, M., Evans, G.T.: Geophysical and botanical monitoring of simulated graves in a tropical rainforest, Colombia, South America. J. Appl. Geophys. **135**, 232–242 (2016)
13. Pringle, J., Ruffell, A., Jervis, J., Donnelly, L., McKinley, J., Hansen, J., Morgan, R., Pirrie, D., Harrison, M.: The use of geoscience methods for terrestrial forensic searches. Earth-Sci. Rev. **114**(1), 108–123 (2012)
14. Rosipal, R., Krämer, N.: Overview and recent advances in partial least squares. In: Saunders, C., Grobelnik, M., Gunn, S., Shawe-Taylor, J. (eds.) SLSFS 2005. LNCS, vol. 3940, pp. 34–51. Springer, Heidelberg (2006). doi:10.1007/11752790_2

An Airborne Agent

Daniel Soto-Guerrero$^{(\boxtimes)}$ and José Gabriel Ramírez-Torres$^{(\boxtimes)}$

Cinvestav Tamaulipas, Ciudad Victoria, Mexico
{dsoto,grtorres}@tamps.cinvestav.mx

Abstract. In order to change the control approach that commercially available Unmanned Aerial Vehicles (UAVs) use to execute a flight plan, which is based on the Global Positioning System (GPS) and assuming an obstacle-free environment, we propose a hierarchical multi-layered control system that permits to a UAV to define the flight plan during flight and locate itself by other means than GPS. The work presented on this article aims to set the foundation towards an autonomous airborne agent, capable of locating itself with the aid of computer vision, model its environment and plan and execute a three dimensional trajectory. On the current stage of development we locate the vehicle using a board of artificial markers, the flight plan to execute was defined as either a cubic spline or a Lemniscate. As results, we present the resultant flight data when the proposed control architecture drives the vehicle autonomously.

1 Introduction

An autonomous agent acts on its environment by its own means, relying from little to nothing on other parties. An agent has to model its environment and locate itself in it, then it can create a plan and execute it to achieve a certain goal.

This translates to a set of subproblems to make out of a UAV an agent: *(a)* model its environment and estimate its current status; *(b)* given a certain goal, generate a feasible plan for the UAV to execute; *(c)* control the vehicle so that it can execute the plan. In the following sections we describe the multilayered hierarchical approach we propose to give a solution to all this subproblems; more importantly, the use of computer vision to locate the vehicle and the possibility to operate different kinds of aerial vehicles. The results show how this approach allowed for the operation of heterogeneous vehicles while describing different three dimensional trajectories.

2 Related Work

A robot consists of a series of highly heterogenous systems that are complex in nature and require an orchestrated integration to function properly. For a robot with certain mechanical features, depending on the problem it is intended to solve, there are many approaches to control it, but the archetypes are only three [5]: hierarchical, behavioral and hybrid. The hierarchical approach follows

© Springer International Publishing AG 2017
J.A. Carrasco-Ochoa et al. (Eds.): MCPR 2017, LNCS 10267, pp. 213–222, 2017.
DOI: 10.1007/978-3-319-59226-8_21

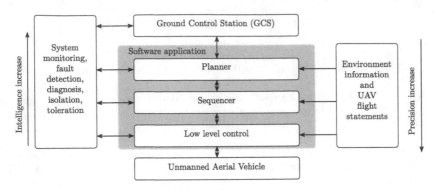

Fig. 1. The three-layer software architecture for an autonomous robot.

a sense-plan-act scheme, prioritizing deliberative control above all else, making it not very flexible. The behavioral approaches follow a bottom-up priority scheme, they use *behaviors* to react to the environment in real-time; because its reactive nature, achieving complex objectives is often difficult. Hybrid architectures try to make the best out of the hierarchical and behavioral approaches, combining the deliberative skills from the first and the flexibility of the latter.

Some examples of high-level and computationally demanding features are: path-planning, human interaction and multi-robot coordination; examples for low-level routines are: sensor reading, actuator control and localization. A hybrid multi-layered architecture has proven successful on the field of mobile robotics because it allows to properly interface, upgrade and coordinate high-level and low-level routines [14], interacting with humans [6] and a group of identical robots [10,18]. This kind of architecture, like the one shown in Fig. 1 [3], consists of three layers: *(1)* the low-level control layer allows to directly manage and access all hardware peripherals in real-time, it also implement some reactive behaviors; *(2)* the planner represents a set of high-level processes that given the current status of the robot and its environment, create a plan for the robot to achieve a certain goal. *(3)* the sequencer is the intermediate layer between the low level control and the planner that will execute the steps of the plan in sequence. In case an error occurs it will update the planner with the current status and ask for a new course of action.

On the field of UAVs, control schemes have been tested following a reactive approach, *i.e.* they act proportionally to an error metric, usually defined by tracking and triangulating salient features with computer vision [19,20]. There are two schemes on how to profit from the payload of a UAV; in the first scheme, all generated information during the flight is stored in a non-volatile storage system for analysis after landing [9,13], in the second scheme, all gathered information is sent to the Ground Control Station (GCS) for further analysis and decision making [1,2]. Whatever the scheme, the UAV acts as a teleoperated entity with little autonomy to react to its flying conditions whether adverse or not. Even so, these schemes are popular and good enough for most civilian and military applications.

Nowadays, there's a growing research community working on a third scheme, on which the flight plan of the UAV is not only dictated by a set of georeferenced waypoints or remotely piloted. Instead, the UAV process the information collected from the onboard sensors to further understand its environment and react to or interact with it. This problem is known as *Simultaneous Localization and Mapping* (SLAM) [11,17]. Solving the SLAM problem means that a robot is able to navigate autonomously through its environment while it creates a virtual representation of its surroundings: the map [4,12]. The work presented on this article is related to the third scheme. On this article, we describe how we plan to go one step further from the reactive approach by introducing a three layer architecture for the control of UAVs and the first steps we have taken.

3   Hardware Description

This work was successfully tested with two different UAVs, the first one we tested was the Solo from 3D Robotics and the second vehicle we tested was the AR-Drone v2, manufactured by Parrot (see Fig. 2). Both vehicles are ready-to-fly UAVs and feature an onboard monoscopic camera. To communicate with these vehicles we only had to adapt the low-level control routines, for the AR-Drone we used the ROS package created to communicate with it. To gain access to the Solo we used Gstreamer[1] to receive the video feed and Dronekit (the software library to interface with UAVs compatible with the MAVlink protocol [15]). As a physical interface to operate the vehicle we used a hardware remote control, we gave a bigger priority to pilot commands over autonomous control; in case of unforeseen situations, the pilot can bypass the autonomous control immediately by operating the hardware controller. The software development was based on the Linux operating system and the Robotic Operating System (ROS) [16].

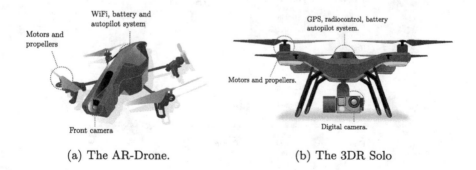

(a) The AR-Drone.   (b) The 3DR Solo

Fig. 2. The two drones tested.

---

[1] Webpage: http://gstreamer.freedesktop.org/.

4   Proposed Approach and Methodology

The test scenario is shown on Fig. 4, the UAV follows the desired trajectory marked in blue while, overflying artificial markers fixed on the ground, the downward looking camera captures the aerial view and transmits the video feed to the GCS. Figure 4 also shows the reference frames attached to the monoscopic camera C, the world reference frame W, the center of gravity (CoG) of the vehicle B and the NED frame (X: North, Y: East, Z: Down).

At the top level, the proposed hierarchical multi-layer architecture defines the desired trajectory $\mathbf{r}_d(t)$ with respect to W as a flight plan; and at the bottom level, it makes noisy estimations of the position of B with respect to W using computer vision. Noisy estimations are filtered using a Kalman Filter and then compared with the desired position, resulting in an error to be minimized by driving the vehicle close to the desired position.

Figure 3 shows the structure of the architecture proposed in this paper. At the top, we show the *high-level* planner node, in charge of computing the desired flight plan for the UAV. At the bottom, the low-level nodes including: the hardware interface to the vehicle and the camera, the computer vision localization nodes and the controller. The trajectory generator node, defines the desired position for the UAV according to the parameters defined by the planner. For now, the trajectory tracker incorporates the ability of generating a lemniscate or an spline trajectory; the sequencer is in charge of switching between the two, depending on the flight plan.

To deal with spatial relationships from frame C to frame W, we used rigid body transformations in homogeneous coordinates $^W_C\mathbf{T}$ denoted as:

$$^W_C\mathbf{T} = \begin{pmatrix} ^W_C\mathbf{R} & ^W\mathbf{t}_C \\ \mathbf{0} & 1 \end{pmatrix}$$

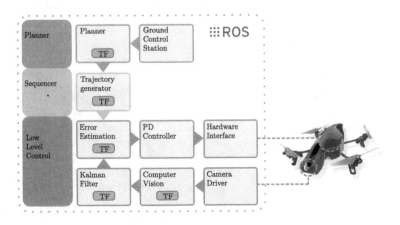

Fig. 3. The three-layer architecture for the UAV

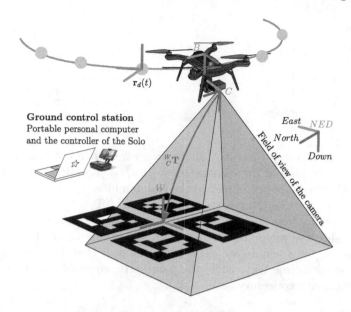

Fig. 4. The use case scenario for the UAV. For every reference frame, the color convention is: X axis red, Y axis green and Z axis blue. (Color figure online)

where ${}^{W}_{C}\mathbf{R} \in SO(3)$ and ${}^{W}\mathbf{t}_C$ are the rotation and translation components, respectively. Within the multi-layered architecture, we used the work from *Foote* [7] to manage all rigid body transformations.

Note that if we solve ${}^{W}_{C}\mathbf{T}$ with computer vision, we can locate B with respect to W because the camera is rigidly mounted on the UAV (the rigid body transformation from the camera to the center of mass of the vehicle ${}^{B}_{C}\mathbf{T}$ is known beforehand). To estimate ${}^{W}_{C}\mathbf{T}$ we used the technique developed by Garrido *et al.* [8]; which consists on segmenting from the images the artificial markers and estimate the pose of the camera from all detected corners. Then, the location of B with respect to W can be computed with ${}^{B}_{W}\mathbf{T} = {}^{B}_{C}\mathbf{T}\,{}^{W}_{C}\mathbf{T}^{-1}$.

As discussed earlier, we used a Kalman Filter over ${}^{B}_{W}\mathbf{T}$ to improve its accuracy and reduce the effects of errors when computing camera parameters, corner detection, image rectification and pose estimation. The state vector for the Kalman filter is $\mathbf{x}_W = [x, y, z, \psi, \dot{x}, \dot{y}, \dot{z}, \dot{\psi}]$, it defines the position, orientation and velocities of B with respect to W. From the onboard inertial measurement unit (IMU), we receive the horizontal velocity components with respect to B, flight's altitude and heading $\mathbf{z}_B = [v_x, v_y, h, \Psi]$; the *a priori* estimate of the Kalman filter was updated using only $\mathbf{z}_W = {}^{W}_{B}\mathbf{T}\,\mathbf{z}_B$, *i.e.* the inertial measurements with respect to W. The used state transition model, with k defining the time instant, is:

$$
\begin{bmatrix} x \\ y \end{bmatrix}_{k+1} = \begin{bmatrix} x \\ y \end{bmatrix}_k + \Delta_t\, {}^W_B\mathbf{T} \begin{bmatrix} v_x \\ v_y \end{bmatrix}_k \qquad \begin{bmatrix} \dot{x} \\ \dot{y} \end{bmatrix}_{k+1} = {}^W_B\mathbf{T} \begin{bmatrix} v_x \\ v_y \end{bmatrix}_k
$$

$$
\dot{z}_{k+1} = \frac{h_k - z_k}{\Delta_t} \qquad\qquad \dot{\psi}_{k+1} = \frac{\Psi_k - \psi_k}{\Delta_t}
$$

$$
z_{k+1} = h_k \qquad\qquad\qquad \psi_{k+1} = \Psi_k
$$

The *a posteriori* step runs at 24 Hz, a slower rate than the *a priori*, using as measurement the pose of the camera $\mathbf{z}_C = [x, y, z, \psi]_k$, estimated by computer vision [8]. After the innovation step in the Kalman filter, state vector $\mathbf{x}_W$ defines the latest estimation for the pose of the UAV, *i.e.* ${}^B_W\mathbf{T}$.

The desired trajectory $\mathbf{r}_d(t)$ to be described by the vehicle is dynamically computed using the waypoints delivered by the planner, joined together by a cubic spline or by a lemniscate. The spline is such that $\dot{\mathbf{r}}_d(t)$ is continuous, creating a smooth trajectory, while the parametric equation for the lemniscate defined the smooth trajectory as:

$$
\mathbf{r}_d(t) = \begin{bmatrix} x_d(t) \\ y_d(t) \\ z_d(t) \\ \psi_d(t) \end{bmatrix} = \begin{bmatrix} a\sin(\frac{t}{\epsilon}) \\ b\sin(\frac{2t}{\epsilon}) \\ c\sin(\frac{5t}{\epsilon}) \\ 0 \end{bmatrix} \tag{1}
$$

Figure 5 shows the resultant directed graph of the spatial relationships, using nodes as reference frames and labels on edges as the modules on the architecture that update the spatial relationship between two reference frames. The direction of every edge represents the origin and target frames of the homogeneous transform. In turn, the error measurement with respect to $\mathbf{B}$ is given by the rigid body transformation defined by:

$$
{}^{r_d}_B\mathbf{T} = \begin{bmatrix} \mathbf{R}_e & {}^e\mathbf{t}_B \\ \mathbf{0} & 1 \end{bmatrix} = {}^{r_d}_W\mathbf{T}\, {}^B_W\mathbf{T}^{-1}
$$

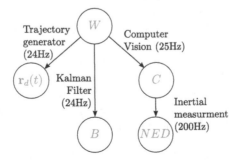

Fig. 5. The graph representing the rigid body transformation between frames.

After decomposing $\mathbf{R}_e$ on its three Euler angles $(\theta, \phi, \psi)_e$, we can compute a control command using a Proportional-Derivative controller:

$$\mathbf{u} = \mathbf{K}_p \begin{bmatrix} \mathbf{r}_d(t) - \mathbf{x} \\ \psi_e \end{bmatrix} + \mathbf{K}_d(\dot{\mathbf{r}}_d - \dot{\mathbf{x}})$$

where $\mathbf{x} = [x, y, z, \psi]$ and $\dot{\mathbf{x}} = [\dot{x}, \dot{y}, \dot{z}, \dot{\psi}]$ are estimated by the Kalman filter described before and $\mathbf{K}_P$ and $\mathbf{K}_D$ are diagonal matrices $\mathbb{R}^{4 \times 4}$

5   Results

The proposed approach was tested with the AR-Drone 2.0 and the 3DR Solo. We made the front camera of the AR-Drone to point downwards, so we could get a higher quality image from above. The Solo had a gimbal installed, as a result, we had to update $_C^B\mathbf{T}$ using the navigational data we received from the UAV. The camera settings for the GoPro are very versatile, for this exercise, we used a *narrow* field of view with a resolution of 1028×720 pixels. Furthermore, to better estimate the pose of the camera, the video feed was rectified using the camera intrinsic parameters. The computer vision algorithm was set to track a board of artificial markers, for the Solo the board measured 1.4×2.4 m (2×5 artificial markers), for the AR-Drone the board measured 4×4 m (20×21 markers, see Fig. 6a).

To observe the current status of the vehicle and its environment, we used Rviz to show the virtual representation of the world. What is shown on Fig. 7 is an screenshot of Rviz displaying: the location of the vehicle, the trajectory being followed and the detected board. The tests were done with a computer based on the Ubuntu Linux, i5 processor, 8 GB RAM.

On Figs. 8 and 9, we display the results as measured by the computer vision system while executing the spline and lemniscate maneuvers in x and y coordinates with respect to W. The $\mathbf{r}_d$ plot is the desired trajectory, corresponding to the spline generated from the waypoints. For completeness, we also display the

(a) Artifical markers and the UAV.   (b) The lemniscate trajectory.

Fig. 6. The software architecture working, creating a virtual representation of the real world and locating the drone with resect to the center of the board.

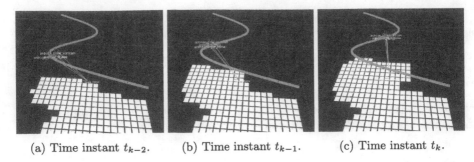

(a) Time instant t_{k-2}.   (b) Time instant t_{k-1}.   (c) Time instant t_k.

Fig. 7. Flight path along the spline. The markers are displayed as white squares on the ground ($z = 0$) at the moment they are detected by the computer vision system. The desired position and the estimated location of the drone are displayed as two coordinate frames, the two are almost always overlapping because of the adequate track of the trajectory.

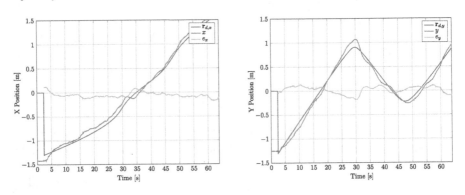

Fig. 8. Plot of the desired position vs. the estimated position of the vehicle while describing the spline trajectory.

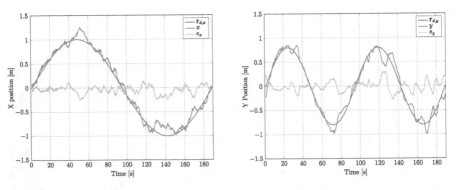

Fig. 9. Plot of the desired position vs. the estimated position of the vehicle while describing the Lemniscate trajectory.

error plot. The maximum measured error was 30 cm for the lemniscate trajectory and 22 cm for the spline trajectory. The waypoints used for the spline were a set of coordinates in 3D space: $p_1 = [-1.3, -1.3, 1.0]$, $p_2 = [-0.9, 0.0, 1.0]$, $p_3 = [-0.3, 0.9, 1.0]$, $p_4 = [0.45, -0.1, 0.0]$, $p_5 = [1.7, 1.0, 1.0]$. Parameters for the lemniscate trajectory with the AR-Drone were: $a = 1.0$, $b = 0.8$, $c = 0.2$, $\epsilon = 30.0$, with a height offset of $z = 1.2$.

6   Conclusions and Future Work

We have discussed a three layer architecture intended for the control of UAVs, that successfully guided the vehicle to describe the lemniscate and spline trajectories. Because the framework we used for this development runs on multiple platforms, including ARM on embedded computers, it is plausible to execute it onboard the UAV. Further development on the Sequencer and Planner layers would make the UAV and autonomous agent and leads the way towards a swarm of UAVs.

This document shows the results from the first step on our development and implementation roadmap. The next step is to execute it onboard the UAV. We are currently looking forward to extending the computer vision system with an visual odometry approach.

References

1. Ax, M., Thamke, S., Kuhnert, L., Schlemper, J., Kuhnert, K.-D.: Optical position stabilization of an UAV for autonomous landing. In: Proceedings of ROBOTIK; 7th German Conference on Robotics, pp. 1–6 (2012)
2. Blösch, M., Weiss, S., Scaramuzza, D., Siegwart, R.: Vision based MAV navigation in unknown and unstructured environments. In: IEEE International Conference on Robotics and Automation (ICRA), pp. 21–28 (2010)
3. Chen, H., Wang, X.M., Li, Y.: A survey of autonomous control for UAV. In: International Conference on Artificial Intelligence and Computational Intelligence (AICI 2009), vol. 2, pp. 267–271, November 2009
4. Correa, D.S.O., Sciotti, D.F., Prado, M.G., Sales, D.O., Wolf, D.F., Osorio, F.S.: Mobile robots navigation in indoor environments using kinect sensor. In: Second Brazilian Conference on Critical Embedded Systems (CBSEC), pp. 36–41 (2012)
5. Coste-Maniere, E., Simmons, R.: Architecture, the backbone of robotic systems. In: Proceedings ICRA. Millennium Conference. IEEE International Conference on Robotics and Automation. Symposia Proceedings (Cat. No.00CH37065), vol. 1, pp. 67–72 (2000)
6. Dumonteil, G., Manfredi, G., Devy, M., Confetti, A., Sidobre, D.: Reactive planning on a collaborative robot for industrial applications. In: 2015 12th International Conference on Informatics in Control, Automation and Robotics (ICINCO), vol. 02, pp. 450–457, July 2015
7. Foote, T.: tf: The transform library. In: 2013 IEEE International Conference on Technologies for Practical Robot Applications (TePRA), Open-Source Software Workshop, pp. 1–6, April 2013

8. Garrido-Jurado, S., Muñoz Salinas, R., Madrid-Cuevas, F.J., Marín-Jiménez, M.J.: Automatic generation and detection of highly reliable fiducial markers under occlusion. Pattern Recognit. **47**(6), 2280–2292 (2014)

9. Gong, J., Zheng, C., Tian, J., Wu, D.: An image-sequence compressing algorithm based on homography transformation for unmanned aerial vehicle. In: International Symposium on Intelligence Information Processing and Trusted Computing (IPTC), pp. 37–40 (2010)

10. Goryca, J., Hill, R.C.: Formal synthesis of supervisory control software for multiple robot systems. In: 2013 American Control Conference, pp. 125–131, June 2013

11. Grzonka, S., Grisetti, G., Burgard, W.: A fully autonomous indoor quadrotor. IEEE Trans. Robot. **PP**(99), 1–11 (2011)

12. Kamarudin, K., Mamduh, S.M., Md Shakaff, A.Y., Saad, S.M., Zakaria, A., Abdullah, A.H., Kamarudin, L.M.: Method to convert kinect's 3D depth data to a 2D map for indoor slam. In: IEEE 9th International Colloquium on Signal Processing and its Applications (CSPA), pp. 247–251 (2013)

13. Lou, L., Zhang, F.-M., Xu, C., Li, F., Xue, M.-G.: Automatic registration of aerial image series using geometric invariance. In: IEEE International Conference on Automation and Logistics, ICAL, pp. 1198–1203 (2008)

14. Luna-Gallegos, K.L., Palacios-Hernandez, E.R., Hernandez-Mendez, S., Marin-Hernandez, A.: A proposed software architecture for controlling a service robot. In: 2015 IEEE International Autumn Meeting on Power, Electronics and Computing (ROPEC), pp. 1–6, November 2015

15. Meier, L., Honegger, D., Pollefeys, M.: PX4: a node-based multithreaded open source robotics framework for deeply embedded platforms. In: 2015 IEEE International Conference on Robotics and Automation (ICRA), May 2015

16. Quigley, M., Conley, K., Gerkey, B.P., Faust, J., Foote, T., Leibs, J., Wheeler, R., Ng, A.Y.: ROS: an open-source robot operating system. In: ICRA Workshop on Open Source Software (2009)

17. Saska, M., Krajnik, T., Pfeucil, L.: Cooperative UAV-UGV autonomous indoor surveillance. In: 9th International Multi-conference on Systems, Signals and Devices (SSD), pp. 1–6 (2012)

18. Schöpfer, M., Schmidt, F., Pardowitz, M., Ritter, H.: Open source real-time control software for the Kuka light weight robot. In: 2010 8th World Congress on Intelligent Control and Automation, pp. 444–449, July 2010

19. Vanegas, F., Gonzalez, F.: Uncertainty based online planning for UAV target finding in cluttered and GPS-denied environments. In: 2016 IEEE Aerospace Conference, pp. 1–9, March 2016

20. Yang, L., Xiao, B., Zhou, Y., He, Y., Zhang, H., Han, J.: A robust real-time vision based GPS-denied navigation system of UAV. In: 2016 IEEE International Conference on Cyber Technology in Automation, Control, and Intelligent Systems (CYBER), pp. 321–326, June 2016

Natural Language Processing and Recognition

An Approach Based in LSA for Evaluation of Ontological Relations on Domain Corpora

Mireya Tovar[1]([✉]), David Pinto[1], Azucena Montes[2], and Gabriel González[3]

[1] Faculty of Computer Science,
Benemérita Universidad Autónoma de Puebla, Puebla, Mexico
{mtovar,dpinto}@cs.buap.mx
[2] TecNM, Instituto Tecnológico de Tlalpan, Mexico City, Mexico
amr@cenidet.edu.mx
[3] Centro Nacional de Investigación y Desarrollo Tecnológico (CENIDET),
Cuernavaca, Mexico
gabriel@cenidet.edu.mx

Abstract. In this paper we present an approach for the automatic evaluation of relations in ontologies of restricted domain. We use the evidence found in a corpus associated to the same domain of the ontology for determining the validity of the ontological relations. Our approach employs Latent Semantic Analysis, a technique based on the principle that the words in a same context tend to have semantic relationships. The approach uses two variants for evaluating the semantic relations and concepts of the target ontologies. The performance obtained was about 70% for class-inclusion relations and 78% for non-taxonomic relations.

Keywords: Ontology evaluation · Latent semantic analysis · Natural language processing

1 Introduction

The continuous increase in the number of documents produced on the Web makes it more complex and costly to analyze, categorize and retrieve documents without considering the semantics of each document. One way to represent knowledge of documents is through ontologies.

An ontology, from the computer science perspective, is "an explicit specification of a conceptualization" [1].

Ontologies can be divided into four main categories, according to their generalization levels: generic ontologies, representation ontologies, domain ontologies, and application ontologies. Domain ontologies, or ontologies of restricted domain, specify the knowledge for a particular type of domain, for example: medical, tourism, finance, artificial intelligence, etc. An ontology typically includes

This work is partially supported by the Sectoral Research Fund for Education with the CONACyT project 257357, by PRODEP-SEP ID 00570 (EXB-792) DSA/103.5/15/10854, and VIEP-BUAP project 00478.

J.A. Carrasco-Ochoa et al. (Eds.): MCPR 2017, LNCS 10267, pp. 225–233, 2017.
DOI: 10.1007/978-3-319-59226-8_22

the following components: classes, instances, attributes, relations, constraints, rules, events and axioms.

The ontologies are resources that allow to capture the explicit knowledge in the data, through concepts and relationships. In this paper we are interested in the process of discovering and evaluating ontological relations, thus, we focus our attention on the following two types: taxonomic relations and/or non-taxonomic relations. The first type of relations are normally referred as relations of the type "is-a" (hypernym/hyponymy or subsumption) or class-inclusion.

In order to evaluate concepts and semantic relations of three domain ontologies using Latent Semantic Analysis, in this research work we present two variants, the first one based on the cosine similarity, and second one based on clustering by committee.

The experiments carried out and the obtained results are discussed through the remaining of this paper, which is organized as follows: in Sect. 2 we present the related work, in Sect. 3 we present the concept of latent semantic analysis, whereas in Sect. 4 we describe the concept of clustering by committee, both employed in this research work. The method proposed is presented in Sect. 5. The experimental results are shown and discussed in Sect. 6. Finally, in Sect. 7 the conclusions of the work are given.

2   Related Work

Different approaches employing LSA for task related with ontologies can be found in literature. For example, in [2] it is presented an automatic method for ontology construction using latent semantic, clustering and Wordnet over a collection of documents.

In [3] they show methods for improving both, the recall and the precision of automatic methods for extraction of hyponymy (IS-A) relations from raw text. By applying latent semantic analysis (LSA) to filter extracted hyponymy relations, they reduce the error rate of their initial pattern-based hyponymy extraction by 30%, achieving precision of 58%. By applying a graph-based model of noun-noun similarity learned automatically from coordination patterns to previously extracted correct hyponymy relations, they achieve roughly a five-fold increase in the number of correct hyponymy relations extracted.

In [4], the authors describe an approach that extracts hypernym and meronym relations between proper nouns in sentences of a given text. Their approach is based on the analysis of the paths between noun pairs in the dependency parse trees of the sentences.

In [5] techniques of machine learning and statistical natural language processing are used to attempt to construct a domain concept taxonomy. They employ different evaluation measures such as: Precision, Recall, F-measure, and others. Their work focused on the integration of knowledge acquisition with machine learning techniques for the ontology creation.

We purpose it is evaluate semantic relationships with evidence in the domain corpus through of latent semantic analysis method. For the evaluation, we use the mesure of accuracy.

3   LSA

Latent Semantic Analysis (LSA) is a computational model used in natural language processing, considered in its beginnings as a method for representing knowledge [6]. LSA is considered an unsupervised dimensionality reduction tool, such as principle component analysis (PCA) [7]. The rationale behind this model indicates that words in the same semantic field tend to appear together or in similar contexts [8,9].

LSA has its origin in an information retrieval technique called Latent Semantic Indexing (LSI) whose purpose is to reduce the size of an array of document terms using a linear algebra technique called Singular Value Decomposition (SVD). The difference with LSA is that it uses a word-context matrix. The context can be a word, a sentence, a paragraph, a document, a test, etc.

Venegas [6] considers that LSA is characterized for being a mathematical-statistical technique that allows the creation of multidimensional vectors for the semantic analysis of the relationships that exist among the different contexts.

The purpose of dimensionality reduction in LSA is to eliminate noise present in the relationships between terms and contexts, since it is usually possible to express the same concept with different terms.

LSA does not consider the linguistic structure of contexts, but the frequency and co-occurrence of terms. However, it has been possible in some cases to identify semantic relationships such as synonymy using LSA [8].

This technique is based on the principle that the words in a same context tend to have semantic relationships, and consequently, indexing of documents with similar contexts should be included by the words that appear in similar contexts even if the document does not contain that words.

4   Clustering By Committee

The Clustering By Committee algorithm (CBC) allows automatic discovery of concepts from text [10,11]. Initially it discovers a set of strict groups called committees that are scattered in the space of similarity. The feature vector that represents a group is the centroid of the committee members, and the clustering method proceed to assign elements to their most similar groups.

The CBC algorithm consists of three phases:

1. To find the most similar elements. In order to calculate the most similar words of a word w, first the characteristics of the word w are ranked according to their mutual information with w.
2. To discover the committees. Each committee that is discovered in this phase defines one of the final groups for the output of the algorithm.
3. To assign elements to the groups. Each element is assigned to the group containing the most similar committee.

CBC has also been used to find the meanings of a word w [12] (algorithm in its flexible version), and for clustering texts (algorithm in its strong version) [13].

Other authors, such as Chatterjee and Mohan [14], have successfully used this algorithm in its flexible version for the discovery of word meanings, including *Random Indexing* to reduce the dimensionality of the context matrix.

5   The Proposed Approach

The proposed approach uses the method of latent semantic analysis, with the purpose of identifying the semantic relationships between the concepts existing in the ontology and looking for evidence in the domain corpus for further evaluation.

LSA points out that words in the same semantic field tend to appear together or in similar contexts, therefore, we considered that the concepts that are semantically related can be in the same sentence or in different sentences sharing information in common.

Based on this assumption, we present the following algorithm that takes into account two variants: (a) Cosine similarity and (b) Grouping by committees (CBC) that assign a weight w to each evaluated relation of the domain ontology. The algorithm performs the following steps:

1. *Pre-processing the domain corpus and domain ontologies.* The domain corpus is divided into sentences and the empty words (such as prepositions, articles, etc.) are removed. The Porter stemming algorithm is applied to the words contained in these sentences [15]. The concepts are also extracted from the ontology[1]. The same process is applied to each one of the concepts of the ontology in order to maintain consistency in the terminology representation (empty words elimination and the Porter stemming algorithm).
2. Application of the LSA algorithm to reduce the dimensionality of the context matrix. In this case, we use the S-Space[2] package and the LSA algorithm[3]. The algorithm receives as parameters the sentences of the corpus of Domain and K dimensions (we use 300 dimensions). The output of the LSA algorithm are semantic vectors of dimension K for each word identified by LSA in the corpus.
3. Construction of concepts. The words obtained by the LSA method are clustered by using the cosine similarity to form the concepts of the ontology.
4. Dimension reduction of vocabulary (vectors) in the LSA matrix. Only the concepts obtained in the previous step are kept in the next step, the rest of the words of the original matrix are removed.
5. Application of variants. At this point two variants are used: cosine similarity for each relation and CBC algorithm to cluster concepts.
   - Similarity cosine

---

[1] We used Jena for extracting concepts and semantic relations (http://jena.apache.org/).

[2] https://github.com/fozziethebeat/S-Space.

[3] http://code.google.com/p/airhead-research/wiki/LatentSemanticAnalysis.

   (a) Calculation of cosine similarity. The concepts obtained in the previous step are used to determine the degree of similarity between each pair of concepts that form the class-inclusion and non-taxonomic relations.

   (b) Calculation of threshold u and weight w assigned to the relation. The threshold u is calculated as the sum of the similarities between the total of relationships divided by 2. If the value of the degree of similarity of the relation is greater than the threshold u, the relation takes the weight of $w = 1$, otherwise $w = 0$.

  – CBC Algorithm

   (a) Application of the CBC algorithm in its flexible version. The concepts formed by similarity, in the previous step, are the input to the CBC algorithm. The output of the algorithm are the clustered concepts.

   (b) Identification of the concepts that form the relationship in the clusters generated by CBC. If the pair of concepts that form the relation (class-inclusion and non-taxonomic) are in the cluster, the relation takes the weight $w = 1$ otherwise it receives the weight $w = 0$.

6. Ontology evaluation. We used the metric of accuracy for evaluating the concepts and semantic relations obtained with our approach for each input domain ontology.

The next section, we present the obtained results with this approach.

6   Experimental Results

Below, we present the dataset and the results obtained with the aforementioned approach.

6.1   Dataset

The domains used in the experiments are Artificial Intelligence (AI)[4], standard e-learning SCORM [16] and OIL taxonomy.

   In Table 1 we present the number of concepts (C), class-inclusion relations (S) and non-taxonomic relations (R) of the ontology evaluated. The characteristics of its reference corpus are also given in the same Table: number of documents (D), number of tokens (T), vocabulary dimensionality (V), and the number of sentences (O)

6.2   Results

The number of vectors or words retrieved by the LSA algorithm from the domain corpus are shown in Table 2. After concepts discovering by employing the cosine similarity, the approach reduces the matrix to the total of concepts of the domain ontology. Por example, from 1,659 words obtained by LSA for the ontology IA, the matrix is reduced to 276 concepts included in the ontology (see Table 1).

---

[4] The ontology together with its reference corpus can be downloaded from http://azouaq.athabascau.ca/goldstandards.htm.

Table 1. Datasets

Domain	Ontology			Corpora			
	C	S	R	D	O	T	V
AI	276	205	61	8	475	11,370	1,510
SCORM	1,461	1,038	759	36	1,621	34,497	1,325
OIL	48	37	–	577	546,118	10,290,107	168,554

Table 2. Vocabulary obtained by the LSA algorithm for each domain

Domain	LSA_V
AI	1,659
SCORM	1,473
OIL	168,762

The LSA method with cosine similarity obtained favorable results for the three domains ontologies evaluated, finding more than 70% of the class-inclusion relations (see Table 3). The CBC method obtained the best results in the OIL ontology with 54% of accuracy.

Table 3. Experimental results of the LSA approach to class-inclusion relation in each domain ontology

Ontology	Total	Variant	Enc	Accuracy
AI	205	LSA-cosine	179	87.32%
		LSA-CBC	32	15.61%
SCORM	1038	LSA-cosine	908	87.48%
		LSA-CBC	194	18.69%
OIL	37	LSA-cosine	26	70.27%
		LSA-CBC	20	54.05%

In Table 4 we show the total of concepts that integrate a class-inclusion relation (CO) in the domain ontology and the total of these obtained by the LSA approach (Enc) for this type of relation.

The accuracy of the concepts found by the LSA method is greater than 79% with the cosine similarity variant (see Table 4). However, the CBC variant does not report satisfactory results for the first two ontologies. In the case of the OIL ontology it obtained a better behavior by achieving 62% accuracy, but without exceeding the result of the cosine variant (79%). The CBC method does not cluster all the concepts, so it was expected that most of the relations would not be found.

Table 4. Experimental results of concepts that maintain only a class-inclusion relation using the LSA approach for each domain ontology

Ontology	Variant	CO	Enc	Accuracy
AI	LSA-cosine	233	219	93.99%
	LSA-CBC	233	47	20.17%
SCORM	LSA-cosine	1154	1069	92.63%
	LSA-CBC	1154	285	24.70%
OIL	LSA-cosine	43	34	79.07%
	LSA-CBC	43	27	62.79%

In the case of non-taxonomic relations, the results obtained by the approach are presented in Table 5. Again, the cosine variant obtains better results (78% accuracy) than the CBC variant for this type of relation. As the CBC variant failed to cluster all concepts (see Table 6), the approach does not achieve a satisfactory accuracy in such relations. A first approximation of this approach is presented in [17] reporting only the concepts found by LSA.

Table 5. Experimental results of the LSA approach to non-taxonomic relations in each domain ontology.

Ontology	Total	Variant	Enc	Accuracy
AI	61	LSA-cosine	51	83.61%
		LSA-CBC	16	26.23%
SCORM	759	LSA-cosine	594	78.26%
		LSA-CBC	113	14.89%

In the case of concepts, the cosine variant obtains 85% accuracy in comparison with that obtained with the CBC variant (see Table 6).

Table 6. Experimental results of concepts that keep a non-taxonomic relation using the LSA approach for each domain ontology

Ontology	Variant	CO	Enc	Accuracy
AI	LSA-cosine	69	61	88.41%
	LSA-CBC	69	21	30.43%
SCORM	LSA-cosine	570	485	85.09%
	LSA-CBC	570	123	21.58%

7 Conclusions

The LSA method has been widely used in the state of the art to represent semantic at the context level, and with the proposed approach it was possible to obtain more than 70% of the semantic relations of each domain ontologies.

The results of the LSA approach, considering only the cosine similarity variant, obtained satisfactory results. But when the CBC variant was employed, it was not possible to find in the clustered concepts all the ontology relations (approximately only 10% of the total concepts).

The CBC method is very costly at runtime and did not produce satisfactory results. We consider that this is because we do not have enough information from each domain ontology, that this variant can process.

The LSA based approach requires a robust corpus (in terms of domain and size), including a large vocabulary that this allows more terms to be clustered. However, the accuracy offered is acceptable for one of the variants presented.

As future work we consider to increase the number of documents processed by the approach, as well as, the reviewing of other alternatives of concept clustering for the evaluation of domain ontologies.

References

1. Gruber, T.R.: Toward principles for the design of ontologies used for knowledge sharing. Technical report KSL-93-04, Knowledge Systems Laboratory, USA (1993)
2. Novelli, A.D.P., de Oliveira, J.M.P., Maria, J.: Simple method for ontology automatic extraction from documents. Int. J. Adv. Comput. Sci. Appl. (IJACSA) **3**(12), 44–51 (2012). http://ijacsa.thesai.org/
3. Cederberg, S., Widdows, D.: Using LSA and noun coordination information to improve the precision and recall of automatic hyponymy extraction. In: Proceedings of the Seventh Conference on Natural Language Learning at HLT-NAACL (CONLL 2003), Stroudsburg, vol. 4, pp. 111–118. Association for Computational Linguistics (2003)
4. Sheena, N., Jasmine, S.M., Joseph, S.: Automatic extraction of hypernym and meronym relations in English sentences using dependency parser. Procedia Comput. Sci. **93**, 539–546 (2016). Proceedings of the 6th International Conference on Advances in Computing and Communications
5. Sankat, M., Thakur, R., Jaloree, S.: Design of ontology learning model based on text classification for domain concept taxonomy. IJSRSET **2**, 138–142 (2016). Themed Section: Engineering and Technology
6. Venegas, V.R.: Análisis Semántico Latente: una panorámica de su desarrollo. Revista signos **36**, 121–138 (2003)
7. Sidorov, G.: Non-linear Construction of n-grams in Computational Linguistics: Syntactic, Filtered, and Generalized n-grams. Instituto Politécnico Nacional, Mexico (2013)
8. Landauer, T.K., Dutnais, S.T.: A solution to platoś problem: the latent semantic analysis theory of acquisition, induction, and representation of knowledge. Psychol. Rev. **104**, 211–240 (1997)

9. Vázquez Pérez, S.: Resolucin de la ambigedad semntica mediante mtodos basados en conocimiento y su aportacin a tareas de PLN. Ph.D. thesis, Universidad de Alicante (2009)
10. Lin, D., Pantel, P.: Concept discovery from text. In: Proceedings of the 19th International Conference on Computational Linguistics (COLING 2002), Stroudsburg, vol. 1, pp. 1–7. Association for Computational Linguistics (2002)
11. Pantel, P.A.: Clustering by committee. Ph.D. thesis, University of Alberta (2003)
12. Pantel, P., Lin, D.: Discovering word senses from text. In: Proceedings of the Eighth ACM SIGKDD International Conference on Knowledge Discovery and Data Mining (KDD 2002), pp. 613–619. ACM, New York (2002)
13. Pantel, P., Lin, D.: Document clustering with committees. In: Proceedings of the 25th Annual International ACM SIGIR Conference on Research and Development in Information Retrieval (SIGIR 2002), pp. 199–206. ACM, New York (2002)
14. Chatterjee, N., Mohan, S.: Discovering word senses from text using random indexing. In: Gelbukh, A. (ed.) CICLing 2008. LNCS, vol. 4919, pp. 299–310. Springer, Heidelberg (2008). doi:10.1007/978-3-540-78135-6_25
15. Porter, M.F.: Readings in Information Retrieval. Morgan Kaufmann Publishers Inc., San Francisco (1997)
16. Zouaq, A., Gasevic, D., Hatala, M.: Linguistic patterns for information extraction in ontocmaps. In: Blomqvist, E., Gangemi, A., Hammar, K., del Carmen Suárez-Figueroa, M. (eds.) WOP. CEUR Workshop Proceedings, vol. 929. CEUR-WS.org (2012)
17. Tovar, M., Pinto, D., Montes, A., González, G., Ayala, D.V., Beltrán, B.: Validación de conceptos ontoógicos usando métodos de agrupamiento. Res. Comput. Sci. **73**, 9–16 (2014)

Semantic Similarity Analysis of Urdu Documents

Rida Hijab Basit[1], Muhammad Aslam[1(✉)], A.M. Martinez-Enriquez[2],
and Afraz Z. Syed[3]

[1] Department of Computer Science and Engineering,
University of Engineering and Technology, Lahore, Pakistan
`ridahijab@gmail.com`, `maslam@uet.edu.pk`
[2] Department of Computer Science, CINVESTAV-IPN, Mexico, D.F., Mexico
`ammartin@cinvestav.mx`
[3] Information Technology Program (ITP),
Lambton College of Applied Science and Technology,
Sarnia, Canada
`Afraz.Syed@lambtoncollege.ca`

Abstract. Semantic similarity analysis is an emerging research area and plays an important role in document classification, text summarization, and plagiarism identification. Moreover, digital data are increasing tremendously over the Internet. Such unstructured data need efficient tools to find any relevant topic or related content optimally. Thus, many systems have been developed for various languages (English, Arabic, Hindi, Turkish, etc.) to retrieve documents based on semantic similarity but no such work has been done on Urdu language. For optimal search of Urdu digital documents, there is a need of such a system that finds semantically similar documents. This paper focuses on studying the existing systems and proposing an approach for Urdu documents providing a better semantic similarity score. Our proposed system - Semantic Similarity System for Urdu (**TripleS4Urdu**) provides good results that have been compiled after evaluation.

Keywords: Semantic similarity analysis · Latent semantic analysis · LSA

1 Introduction

Semantic similarity (SS) analysis has become an important research area within natural language processing (NLP). Due to tremendous increase in online data, it is quite difficult to retrieve relevant content optimally. Therefore, SS measures are needed which analyze the data based on their meaning and provide the required data, saving time and energy. There exist various SS measures like Euclidean distance, cosine semantic similarity or Latent Semantic Analysis (LSA) [1]. SS analysis is not just confined to finding similarity among documents; it also plays an important role in text summarization, plagiarism detection, and document classification [2,3]. Research shows that existing systems of different languages

© Springer International Publishing AG 2017
J.A. Carrasco-Ochoa et al. (Eds.): MCPR 2017, LNCS 10267, pp. 234–243, 2017.
DOI: 10.1007/978-3-319-59226-8_23

use various techniques to determine SS among documents. These include modified cosine similarity measure, latent semantic indexing, WordNet relations, and pattern mining [4–6].

Most of the SS measuring systems use lexical databases (WordNet) to extract synonymous relations among words for effective similarity analysis. These databases are custom-designed and each language has its own lexical database. Urdu language, on the other hand, does not have a mature lexical database due to which WordNet based SS measures cannot be used to determine SS among Urdu documents.

Urdu is a morphologically rich language [7] that entails certain challenges - cursive script, complex word structure, use of diacritics, limited resources, etc. These issues must be addressed while dealing with Urdu language. Moreover, literature survey shows that most of the systems use statistical measures for SS analysis. On the other hand, systems for translation, summarization, interpretation, etc. [8,9] are being offered for Urdu language but no system has been proposed for SS analysis. Keeping all this in mind, we have proposed a system which processes Urdu documents and determines SS among them. Our approach proposes an extended and improved formula for generating an initial matrix of LSA representing the documents. This results in better and more accurate SS scores. Our system, **TripleS4Urdu** has been evaluated and results have been compiled.

The rest of the paper is structured as follows: Sect. 2 reviews the existing semantic similarity measuring systems, Sect. 3 discusses the details of **TripleS4Urdu**, Sect. 4 highlights the experimentation setup and system results, whereas Sect. 5 concludes the paper and draws future work.

2   Related Work

Latent semantic analysis (LSA) and singular value decomposition (SVD) have been employed to determine semantically similar Arabic documents [1] using an extended Term Frequency - Inverse Document Frequency (TF-IDF) formula to construct the initial matrix for LSA. The results given by this matrix are quite overwhelming. Wu and Palmer method has been used by Grammar-based system to find similar English sentences making use of grammar links and its subtypes [10]. It is a topological based approach, using depths from the lexical database for SS analysis. Another system calculates SS of Hindi words [11] based on three different methods: LCH, WUP, and Resnik. Arabic Word Semantic Similarity (AWSS) measuring algorithm uses topological approach to calculate SS of Arabic words [12]. Arabic WordNet has been used to extract depths and lengths of the words for SS analysis.

Turkish documents have been classified using modified Wu-Palmer method which is a topological based approach [4]. Latent Semantic Indexing (LSI) has been used to cluster English documents [5]. LSI is termed as a statistical based approach. SS among Web documents have been found using Vector Space Model (VSM) [6]. WordNet based semantic model has been proposed to cluster English documents [13] whereas another system finds semantically similar word pairs using statistical based approach: information content [14]. Three different methods have been used to determine SS between words and sentences [15]: association rule mining, support vector machine, and sequential clustering which can be termed as statistical and data mining based approaches. Taxonomy based system also uses a statistical based approach - cosine similarity to identify similar Turkish documents [16]. Turkish WordNet is used to extract lexical information. Machine learning based method has been employed to classify Arabic sentences as similar or not similar [17]. It uses five different classifiers to perform classification based on lexical, semantic, and syntactic-semantic features. SS analysis also reduces the dimensionality of text by merging semantically similar Arabic sentences [18]. It uses cosine distance as a SS measure.

Majority of the systems use statistical based approaches namely, cosine similarity, latent semantic analysis, information content, and vector space model. The main reason to use these measures is because of accurate SS scores given by them. Research also shows that SS measuring systems performing analysis at sentence or document level have a preprocessor to remove irrelevant text that does not influence SS analysis. Every language has its own preprocessor module depending upon the language properties.

Systems that have used LSA show that accurate SS score is directly related to initial matrix of LSA representing the documents. If this matrix is constructed carefully then it results in better and improved SS analysis. Keeping all this in mind, we have proposed a system which measures semantic similarity among Urdu documents using statistical based measures - LSA and Cosine Similarity. **TripleS4Urdu** proposes an improved formula to construct initial matrix of LSA as it is the key to better results.

3   Semantic Similarity System for Urdu (TripleS4Urdu)

TripleS4Urdu, after reviewing the existing systems, has been developed with a modular architecture using statistical measures to calculate SS scores among Urdu documents. Here, architectural and system overviews of **TripleS4Urdu** have been discussed.

3.1   Architectural Overview

TripleS4Urdu has a modular architecture, as discussed earlier. Each module has been designed for a specific purpose and is divided into sub-modules (see Fig. 1). The principle advantages of our architectural design approach are: reusability, flexibility, reliability, standardization, and less complexity.

1. **Preprocessor module** is an important part of Natural Language Processing (NLP) applications for treating symbols, punctuations, diacritics, un-wanted characters, and unimportant text. This module includes four sub-modules: text cleaner, stop word remover, stem word extractor [19], and synonymous relation extractor.

(a) **Text Cleaner** sub-module removes all unimportant text from incoming Urdu documents. Unimportant text refers to text which does not play any role in SS analysis. It includes diacritics, punctuations, symbols, and digits (both roman and Urdu digits). Urdu uses the same set of diacritics as that of Arabic - Zabar (Arabic Fathah), Pesh (Arabic Dammah), Zer (Arabic Kasrah), Jazam (Arabic Sakun), and Shad (Arabic Tashdid). Symbols of Urdu language have also been borrowed from Arabic language like Arabic sign Sallallahou Alayhe Wassallam, Arabic sign Alayhe Assallam, Arabic sign Takhallus, etc. All these symbols do not play any role in SS analysis, therefore, they are removed before further processing.

(b) **Stop Word Remover** deletes stop words from incoming Urdu documents. Stop words are the most common words in a language which have insignificant importance during NLP. There is no definite list of stop words and every system has its own list depending upon the requirements of analysis. For SS analysis of Urdu documents, we have selected a list of stop words which include conjunctions, case markers, tense auxiliary, cardinals, ordinals, fractions, multiplicatives, names of week days, months, seasons, day times, times, greetings, titles, directions, pronouns, and some other words. This sub-module checks for stop words in each sentence and removes them accordingly.

(c) **Stem Word Extractor** replaces each word with its stem/root word. This process is termed as stemming - a process of generating root words from derived and inflected words. Urdu words undergo complex morphological processes resulting in a number of forms for each root word. Due to this, it needs complex lexicons and sets of rules in order to get a stem/root word. **TripleS4Urdu** uses a simple lexicon which consists of a stem word along-with various forms of that word. For each incoming word, it checks through the lexicon and replaces that word with its stem/root word.

(d) **Synonymous Word Extractor** sub-module extracts synonyms for each word in the document. Urdu WordNet has not been used here because it is not mature enough and does not contain all sorts of words. Therefore, **TripleS4Urdu** maintains a list containing lexical information. For each word, its synonyms have been written which are then added in a sentence for each word. This makes processing easy and performance of the system high.

2. **Feature Extractor module** extracts useful terms from the documents and constructs vectors out of them to be used by the next module for SS score calculation. It further consists of two sub-modules: LSA Initial Matrix Generator and LSA Applier.

(a) **LSA Initial Matrix Generator** sub-module uses our proposed formula to generate term-document matrix of Urdu documents. The formula given in [1] has been changed from simple occurrence frequency to weighted occurrence frequency (in our case). Our proposed formula is given in Eq. (1):

$$a_{i,j} = \frac{1}{2} + \frac{WOF_{i,j} * log(\frac{|N|}{DF_i})}{2 * max(WOF_{i,j}) * log(M)} \tag{1}$$

Here, $WOFi,j$ represents weighted occurrence frequency of word i in document j (Word's frequency/Sum of sentences' length in which that word occurs). DFi indicates the number of documents in which the word i occurs, **M** is the number of all documents whereas **N** represents the number of unique terms across all documents. In [1], $WOFi,j$ represents simple occurrence frequency of word i in document j.

Our proposed formula constructs a term-document matrix with rows representing unique terms in the documents and columns representing the documents. These documents are the candidate documents with which the SS of test document has to be calculated. The main purpose of using weighted occurrence frequency instead of simple occurrence frequency is to give some weight to the terms in the documents. Upon evaluation, this formula shows improvements in SS scores.

(b) **LSA Applier** Latent Semantic Analysis (LSA) is then applied on the matrix constructed in the previous sub-module. It is an NLP technique which analyzes useful relationships among terms in the documents. Singular Value Decomposition (SVD) is applied on LSA initial matrix as part of LSA technique. It reduces the dimension of the matrix keeping semantic relations the same. This technique generates two matrices - one containing information about the unique terms in the documents and other having document level information. These two matrices are then used by the last module of **TripleS4Urdu** to calculate semantic similarity score.

3. **Similarity Calculator module** reads a test document and constructs a vector out of it using the matrices given by the previous module. It then calculates the similarity score among the test document vector and vectors containing document level information using cosine similarity measure. It measures the angle between two non-zero vectors determining their similarity. Values closer to 0 indicate no or less similarity whereas values closer to 1 indicate high similarity.

3.2   System Overview

TripleS4Urdu first reads candidate documents and preprocesses them one by one using the preprocessor module. The treated documents are then processed by the Feature Extractor module to extract useful patterns and construct vectors for these documents. These vectors are then used by the third module of our system to calculate their similarity with the test document. It provides a numeric value indicating the similarity between the candidate document and test document. System overview has been shown in Fig. 2.

4   Experiment and Results

First of all, we discuss some examples of SS analysis with Urdu documents. Here, we present three documents - two candidate documents discussing two

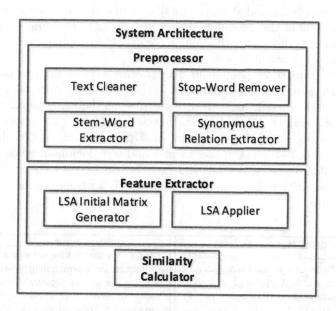

Fig. 1. System architecture of TripleS4Urdu

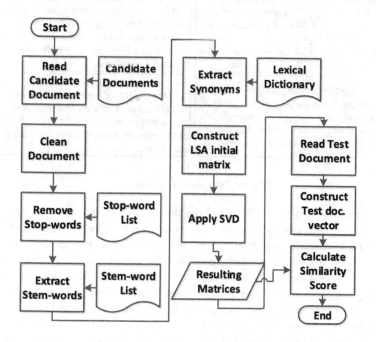

Fig. 2. System overview of TripleS4Urdu

different topics (firing and dengue virus) and one test document having the same topic as one of the candidate documents (firing). The similarity between the two documents discussing the same topic should be high as compared to the ones that discuss different topics.

Snapshots from three documents have been given in Fig. 3 showing some portion of the documents. We can observe that candidate document 1 is quite similar to test document, as both of these discuss the same topic. Candidate document 2 discusses a different topic. However, they share some words but the main theme of document 2 is very different. Upon SS analysis, **TripleS4Urdu** gives a score of **0.984** between document 1 and test document whereas it gives a score of **0.345** between document 2 and test document. The score of 0.984 shows high similarity whereas a low score of 0.345 shows less similarity which is exactly the case.

Candidate Document 1 – Firing	Candidate Document 1 – Firing
۔۔۔ افغانستان میں مقامی حکام کا کہنا ہے کہ قندھار کے ائیر پورٹ پر نامعلوم مسلح افراد نے فائرنگ کر کے پانچ خواتین سکیورٹی اہلکاروں کو ہلاک کر دیا ہے۔ قندھار کے گورنر کی ترجمان نے بتایا کہ سنیچر کی صبح خواتین سکیورٹی اہلکار اپنی ڈیوٹی پر ائیر پورٹ آ رہی تھیں کہ موٹر سائیکل پر سوار دو نامعلوم مسلح افراد نے اس وین پر حملہ کیا ۔۔	...Officials say that at Kandhar airport Afghanistan, some armed men have opened fire killing five women security officials. Kandhar Governer further told that these women security officials were on their way to the airport when unknown armed men on a motorcycle attacked their van...
Candidate Document 2 – Dengue Virus	Candidate Document 2 – Dengue Virus
۔۔ پاکستان میں ڈینگی وبا کے پھیلاؤ کے خلاف جنگ ہسپتالوں کی بجائے اب موبائل فونز کے ذریعے لڑی جا رہی ہے۔ پچھلے چار سالوں کے اعداد و شمار بتاتے ہیں کہ ٹیکنالوجی کے استعمال کے ذریعے یہ جنگ زیادہ موثر انداز میں کامیابی کے ساتھ جاری ہے ۔۔	...In Pakistan, the war of dengue virus has been fought with mobile phones instead of hospitals. According to the survey of past four years, this war can be fought effectively through technology...
Test Document – Firing	Test Document – Firing
۔۔ نامعلوم مسلح افراد نے پیٹرولنگ پولیس اہلکاروں پر فائرنگ کی اور اس کے بعد ایک قدیم قلعے میں ایک پولیس سٹیشن کو نشانہ بنایا۔ حکام کے مطابق قلعے کے اند کئی سیاح پھنسے ہوئے تھے جنہیں باہر نکال لیا گیا ۔۔	...Unknown armed men opened fire at patrolling police officials and then attacked a police station near some old castle. According to the officials, many tourists are stuck in the old castle...

Fig. 3. Documents' snapshots

Two columns each consisting of three boxes have been shown in Fig. 3. First column shows Urdu documents and the boxes next to first column boxes lying in the second column give translation of the Urdu documents for non-native readers. The SS analysis between candidate document 1 and test document has been given in Fig. 4 below. It shows the preprocessed documents where removed stop words are represented as strike-through words, each word is replaced with its stem word, and added synonyms are indicated with a slash among words. Same shaped or same style underlined words represent same meaning words between the two documents which play an important role in SS analysis. Upon feature extraction and similarity calculation, these same meaning words result in higher SS score between these two documents i.e. 0.984.

Texts in Fig. 4 have also been tested with a baseline method - Cosine similarity measure with vectors constructed using Bag of Words (BoW) technique. SS score given by it is much less than the one given by **TripleS4Urdu** i.e. 0.702 indicating poor performance as compared to our system.

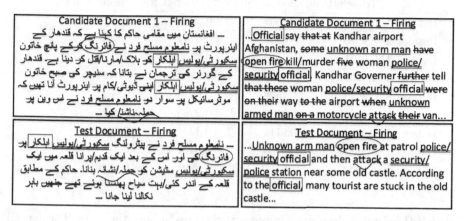

Fig. 4. SS analysis b/w candidate document 1 & test document

4.1 Experimentation Data and Setup

For evaluating **TripleS4Urdu**, data documents have been constructed by gathering data from various resources. The resources which have been used are BBC Urdu website (http://www.bbc.com/urdu) and Urdu digest website (http://urdudigest.pk/). Total 27 documents have been constructed, each consisting of different number of sentences indicating different sizes.

To ensure diversity, different domains have been included so that generality of the system can also be tested. Domains that have been incorporated are culture, sports, science, health, news, technology, entertainment, and religion.

27 constructed documents have been divided into 9 sets, each set consisting of 3 documents. Two of these three documents are candidate documents from two different domains whereas the remaining one document is considered as the test document having the same domain as one of the candidate documents of that set. It means that in one set, SS score is calculated between test document and candidate document from same domain, and between test document and candidate document from other domain. This process has been repeated for all 9 sets.

The formula given in [1] is the modification of simple TF-IDF and it gives best results as claimed by the author. Therefore, we have compared our technique with it using all 9 sets. Results have been given in Table 1.

Second and third columns show SS scores for same domain documents whereas fourth and fifth give SS scores for different domain documents using TripleS4Urdu and the discussed system. Same domain documents should have high similarity giving values closer to 1 whereas different domain documents should have less similarity giving values closer to 0. Table 1 shows that

Table 1. Comparative results

Datasets	Same domain documents & test documents		Different domain documents & test documents	
	TripleS4Urdu	Formula in [1]	TripleS4Urdu	Formula in [1]
Dataset 1	0.8915	0.8509	0.5636	0.6672
Dataset 2	0.9379	0.8848	0.4251	0.5389
Dataset 3	0.9841	0.9973	0.286	0.2722
Dataset 4	0.9759	0.9601	0.3362	0.4227
Dataset 5	0.8614	0.7529	0.6039	0.7621
Dataset 6	0.9836	0.9635	0.2963	0.4026
Dataset 7	0.9725	0.9714	0.3251	0.3413
Dataset 8	0.9838	0.9755	0.3451	0.418
Dataset 9	0.9314	0.9196	0.4924	0.5738

TripleS4Urdu has given more improved SS scores in both the cases. The overall accuracy of our system is around 88.89% whereas precision is recorded as 90%.

5   Conclusion

Many systems exist that find semantically similar documents of different languages but none has been proposed for Urdu documents. This paper has proposed a system that determines semantic similarity (SS) of Urdu documents using statistical measures - LSA and cosine similarity. We have proposed a formula to generate LSA initial matrix giving better semantic similarity scores. Our system (**TripleS4Urdu**) has been evaluated with data from different domains and results show an overall accuracy of 88.89%.

Further work that can be carried out in this area would be the improvement and construction of mature Urdu language resources especially Urdu WordNet by adding more words and relationships like IS-A, hypernyms etc. so that WordNet-based measures can also be used for SS analysis of Urdu documents. Moreover, the weighting scheme presented in this paper can be enhanced further by adding more weight to useful terms in the documents resulting in improved SS scores.

References

1. Hussein, A.S.: Arabic document similarity analysis using n-grams and singular value decomposition. In: IEEE 9th International Conference on Research Challenges in Information Science (RCIS), pp. 445–455 (2015)
2. Al-Saleh, A.B., Menai, M.E.B.: Automatic Arabic text summarization: a survey. Artif. Intell. Rev. **45**(2), 203–234 (2016)
3. Saloot, A.M., Idris, N., Mahmud, R., Ja'afar, S., Thorleuchter, D., Gani, A.: Hadith data mining and classification: a comparative analysis. Artif. Intell. Rev. **46**(1), 113–128 (2016)

4. Yucesoy, B., Oguducu, S.G.: Comparison of semantic and single term similarity measures for clustering Turkish documents. In: Proceedings of IEEE 6th International Conference on Machine Learning and Applications (ICMLA), pp. 393–398 (2007)

5. Han, C., Choi, J.: Effect of latent semantic indexing for clustering clinical documents. In: Proceedings of IEEE/ACIS 9th International Conference on Computer and Information Science (ICIS), pp. 561–566 (2010)

6. Ensan, A., Biletskiy, Y.: Matchmaking through semantic annotation and similarity measurement. In: IEEE 25th Canadian Conference on Electrical & Computer Engineering (CCECE), pp. 1–5 (2012)

7. Humayoun, M., Hammarstrm, H., Ranta, A.: Urdu morphology, orthography and lexicon extraction. In: Proceedings of 2nd Workshop on Computational Approaches to Arabic Script-Based Languages (2007)

8. Syed, A.Z., Aslam, M., Martinez-Enriquez, A.M.: Associating targets with SentiUnits: a step forward in sentiment analysis of Urdu text. Artif. Intell. Rev. **41**(4), 535–561 (2014)

9. Syed, A.Z., Aslam, M., Martinez-Enriquez, A.M.: Sentiment analysis of Urdu language: handling phrase-level negation. In: Batyrshin, I., Sidorov, G. (eds.) MICAI 2011. LNCS, vol. 7094, pp. 382–393. Springer, Heidelberg (2011). doi:10.1007/978-3-642-25324-9_33

10. Lee, M.C., Chang, J.W., Hsieh, T.C.: A grammar-based semantic similarity algorithm for natural language sentences. Sci. World J. (2014)

11. Singh, J., Sharan, A.: Lexical ontology-based computational model to find semantic similarity. In: Mohapatra, D.P., Patnaik, S. (eds.) Intelligent Computing, Networking, and Informatics. AISC, vol. 243, pp. 119–128. Springer, New Delhi (2014). doi:10.1007/978-81-322-1665-0_12

12. Almarsoomi, A.F., Oshea, D.J., Bandar, Z., Crockett, K.: AWSS: an algorithm for Arabic word semantic similarity. In: Proceedings of IEEE International Conference on Systems, Man, and Cybernetics, pp. 504–509 (2013)

13. Shehata, S.: A wordnet-based semantic model for enhancing text clustering. In: IEEE International Conference on Data Mining Workshops, pp. 477–482 (2009)

14. Wagh, K., Kolhe, S.: Information retrieval based on semantic similarity using information content. Int. J. Comput. Sci. Issues **8**(4), 364–370 (2011)

15. Adhikesavan, K.: An integrated approach for measuring semantic similarity between words and sentences using web search engine. Int. Arab J. Inf. Technol. **12**(6), 589–596 (2015)

16. Madylova, A., Oguducu, S.G.: A taxonomy based semantic similarity of documents using the cosine measure. In: Proceedings of IEEE 24th International Symposium on Computer and Information Sciences (ISCIS), pp. 129–134 (2009)

17. Wali, W., Gargouri, B., hamadou, A.B.: Supervised learning to measure the semantic similarity between Arabic sentences. In: Núñez, M., Nguyen, N.T., Camacho, D., Trawiński, B. (eds.) ICCCI 2015. LNCS, vol. 9329, pp. 158–167. Springer, Cham (2015). doi:10.1007/978-3-319-24069-5_15

18. Awajan, A.: Semantic similarity based approach for reducing Arabic texts dimensionality. Int. J. Speech Technol. **19**(2), 191–201 (2016)

19. Daud, A., Khan, W., Che, D.: Urdu language processing: a survey. Artif. Intell. Rev. 1–33 (2016)

Mining the Urdu Language-Based
Web Content for Opinion Extraction

Afraz Z. Syed[1], A.M. Martinez-Enriquez[2], Akhzar Nazir[3],
Muhammad Aslam[3(✉)], and Rida Hijab Basit[3]

[1] Information Technology Program (ITP),
Lambton College of Applied Science and Technology, Sarnia, Canada
Afraz.Syed@lambtoncollege.ca
[2] Department of CS, CINVESTAV-IPN, D.F. Mexico, Mexico
ammartin@cinvestav.mx
[3] Department of CS and E, University of Engineering and Technology,
Lahore, Pakistan
akhzarn@yahoo.com, maslam@uet.edu.pk,
ridahijab@gmail.com

Abstract. People prefer to share and express opinions in their own language. Internet is a biggest repository for sharing opinions. Opinion mining uses Natural Language Processing (NLP), text analysis and computational linguistics to identify and extract subjective information in data. Opinion mining for Urdu language is not a well explored area. Therefore, an approach has been proposed which identifies and extracts adji-units and decisions from the given text using lexicon-based approach focusing on Urdu language. Adji-units are the expressions which contain subjective text in a sentence. Our proposed approach uses two-step lexicon to extract opinions from text chunks. Moreover, for Urdu language no such lexicons exist. The main aim is to develop a diverse two-step lexicon and highlight the linguistic as well as technical aspects of this multi-dimensional research problem. The performance of the proposed system is evaluated on multiple texts and the achieved results are quite satisfactory.

Keywords: NLP · Opinion mining · Sentiment analysis · Urdu lexicon · Adji-units

1 Introduction

World Wide Web has emerged as the largest repository of the user generated texts consisting of opinions. Suggestions from different people exist on the Internet and Internet users, nowadays, use forums, news blogs, discussion groups or review sites for opinions and suggestions even while taking a smallest decision e.g. buying a routine device [1]. The opinions can be defined as the subjective expressions that describe people's feelings [2], sentiments, or appraisals towards objects, procedures, events and their characteristics.

Sites, forums or discussion groups gathering opinions consist of bulks of data making difficult for a person to search for relevant opinions manually. Moreover,

J.A. Carrasco-Ochoa et al. (Eds.): MCPR 2017, LNCS 10267, pp. 244–253, 2017.
DOI: 10.1007/978-3-319-59226-8_24

survey companies may also need such data to carry out a research about any product, person or political party. Hiring individuals for this job would be costly and time consuming. Therefore, a system is needed which automatically mines through such data to get relevant opinions or suggestions about any specific thing.

Different approaches exist for opinion mining like supervised, unsupervised, lexicon based approaches. These techniques have been used for mining opinions of different languages like English, Persian, Hindi, Turkish but none has been used for Urdu language. Our proposed system focuses on Urdu language as it is a major language spoken and understood around the globe with 80 million speakers in the subcontinent [3]. It poses certain challenges due to its complex morphology and orthography. These challenges have to be overcome during Urdu language processing.

Different forums and social media sites are localizing their content by allowing users to comment and chat in their native language. Such forums are tremendously increasing for Urdu language as well where people can add their suggestions in Urdu. Due to this, we have proposed an approach for Urdu language which analyzes the data and highlights positive and negative opinions. In this work, Lexicon-based approach has been implemented.

Section 2 reviews related work in opinion mining, Sect. 3 briefly describes the implementation of two-step lexicon based opinion mining model for Urdu (LOMMU), Sect. 4 discusses the evaluation results of LOMMU, whereas, Sect. 5 concludes the paper along-with some future directions.

2 Literature Survey

Many different approaches for opinion mining have been proposed by different researchers. Learning methods that are supervised, unsupervised, and semi-supervised in nature have been used by some of them. Unsupervised learning methods have been increasingly successful in recent NLP research mainly because it takes unlabeled data as input. Moreover, unsupervised learning results in better understanding of modeling methods, optimization of algorithms and conversion of domain knowledge into structured models. Sentiment analysis also uses lexicon based approach with un-supervised learning method. Three different approaches are used to construct sentimental lexicon - manual, dictionary-based or corpus-based approach.

Naïve Bayes algorithm is the most widely used supervised classification model [4]. It estimates the probabilities of opinions (as positive or negative) using the joint probabilities of a set of words in a given category. Support Vector Machine (SVM) is a non-probabilistic binary classification method proposed by Vladimir Vapnik. It looks for a hyper plane with the maximum margin between positive and negative examples of the training opinions. In addition to the above, K-Nearest Neighbor (KNN) classification (KNN) is based on the assumption that the classification of an instance is most similar to classification of other instances that are nearby in the vector space. In comparison to the other classification methods such as Naïve Bayes, KNN does not rely on prior probabilities and is computationally efficient [5]. Naïve Bayes, SVM, and KNN classifiers discussed above have been used for English language opinion mining. All these are termed as supervised learning methods. Another technique has been proposed which performs classification based on some fixed syntactic patterns that are likely to be used to express opinions [6]. Lexicon-based approaches have also been used by English language. Comprehensive lexicons have been constructed for English language like Senti-WordNet 3.0 which is publically available and is used by different researchers for opinion extraction [7].

Cross-domain sentiment analysis has been done by many researchers [8] experimented with German emails. German emails are converted to English for calculating sentiment orientations, after which they are again converted to German. Precision of this system is satisfactory but recall is recorded to be quite poor. In [9], a slightly different problem has been attempted by using a maximum entropy-based EM algorithm. It jointly learns two monolingual sentiment classifiers by treating the sentiment labels in the unlabeled parallel text as unobserved latent variables.

Urdu is a morphologically complex language having a different writing style due to which using cross-domain sentiment analysis technique for Urdu opinion mining would be quite difficult. Moreover, Urdu data available online is unlabeled and less data is available for analysis. Therefore, less work related to lexicon implementation has been done using corpus. One of the most comprehensive Urdu language lexica is available at http://www.cle.org.pk [3]. This data is XML based, as per the annotation schema, containing about 20 etymological, phonetic, morphological, syntactic, semantic and other parameters of information about a word. Another lexicon proposed by [10] has been constructed from news Urdu corpus having 1.5 million words. It has been tokenized on space and punctuation marks, keeping the diacritics. Extracted lexicon contains 9,126 total words and 4,816 unique words. These lexicons do not contain enough data for decision making. Moreover, accuracy of these lexicons is not as good as described by the researchers.

Urdu lexicon development involves decisions regarding parts-of-speech (POS) tags and their respective features, lemmas, transcription, and lexicon format. POS tagger used for Urdu lexicon development tags sentence on the basis of noun, verb, adjective, adverb, numeral, postpositions, conjunctions, pronouns, auxiliaries, case markers, harf, etc. Most of the on-going works have used XML based lexicon formats [11, 12]. Construction of such lexicons is time consuming as each scenario has a detailed information attached to it.

An alternate solution to XML based would be a Java Script Object Notation (JSON) based two-step lexicon approach as it is easy to implement and is less time consuming. It consists of different keys and each key has corresponding values associated with it. Proposed lexicon-based opinion mining model for Urdu (LOMMU) using JSON format is described in the next section.

3   Lexicon-Based Opinion Mining Model for Urdu Language (LOMMU)

Developing a lexicon for opinion mining is quite critical. LOMMU can be implemented for any operating system (OS) but our work has been tested with Macintosh OS. It uses an algorithmic approach to develop a two step lexicon. JSON format based lexicon structure is shown in Fig. 1.

```
{
  "Orthography": "مردوں",
  "ENTRY": [
    {
      "NOM": {
        "Case": "oblique",
        "Number": "plural",
        "Gender": "masculine"
      },
      "LEMMA": "مردر",
      "PHONETIC": "m @ r - d_d o~"
    },
    {
      "NOM": {
        "Case": "oblique",
        "Number": "plural",
        "Gender": "invariant"
      },
      "LEMMA": "مردہ",
      "PHONETIC": "m U r - d_d o~"
    }
  ]
}
```

Fig. 1. JSON Format-based Lexicon Structure

JSON format given above gives detailed information about the word "مردوں" {Mardon, Men}. It contains gender information, number, phonetics, case, and lemma of the candidate word.

Raw corpus has been annotated using a POS tagger and then adji-units are extracted from the given text. All the decisions which are made during opinion mining use adji-units. Negations have also been handled in our system by using them as polarity shifters. LOMMU uses a two-step lexicon consisting of positive and negative lexemes. Extracted adji-units from the text under consideration are compared with the lexemes and in case of negations attached with adji-units, the polarity of the sentence shifts. System overview has been given in Fig. 2.

Here, we define our problem of Urdu opinion mining. Let "O" be the Urdu text consisting of sentences which can be factual or opinionated. So we can say that "O" is a union of factual and opinionated sentences:

O = {set of factual sentences} U {set of opinionated sentences}

LOMMU differentiates opinionated sentences from factual ones because of the significance of opinionated sentences in opinion mining. The main tasks of LOMMU can be described as follows:

- **Convert Gathered Data into UTF-16 Format for Processing:** Gathered data is in different forms and hence, cannot be processed as it is. Therefore, it is converted to UTF-16 format for further processing.

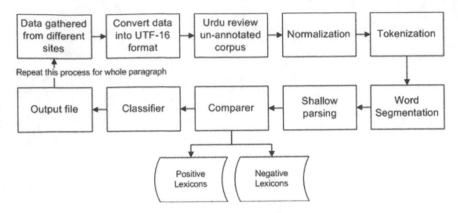

Fig. 2. Two Step Lexicon based Opinion Mining Model for Urdu Language

- **Normalization of Data:** The converted data is then normalized by removing dots, punctuation marks, commas, dashes or any irrelevant symbols. This step can be referred to as a preprocessing step.
- **Tokenization:** Extracting each word from a sentence is known as the process of tokenization. Tokens are just separated by spaces and may not be complete meaningful words. Some examples of tokens for the following sentence are given below:

{Apple computers are very beautiful} ایپل کے کمپیوٹرز بہت خوبصورت ہوتے ہیں

<Token>	ایپل {Apple}	<Token>
<Token>	کمپیوٹرز {computers}	<Token>
<Token>	بہت {very}	<Token>
<Token>	خوبصورت {beautiful}	<Token>
<Token>	ہوتے {are}	<Token>

- **Segmentation:** It is a process of extracting meaningful words. Some tokens are not meaningful words as said in the previous step; therefore, segmentation is needed to get a complete meaningful segment of a word. Examples of segments for the same sentences given above are shown below:

<Word>	ایپل {Apple}	<Word>
<Word>	کمپیوٹرز {computers}	<Word>
<Word>	بہت {very}	<Word>
<Word>	خوبصورت {beautiful}	<Word>

- **Shallow parsing:** Adji-units are extracted for opinion mining using shallow parsing after annotating the corpus. Any POS tagger e.g. CRULP POS tagger can be used for annotating the corpus. Phrase level negations are also handled as part of shallow parsing. Examples of shallow parsing are given below with reference to the sentence given above.

<NP>	ایپل کے {Apple's}	<NP>
<Noun>	کمپیوٹرز {computers}	<Noun>
<Verb>	ہوتے {are}	<Verb>
<Adji-	بہت خوبصورت {very beauti-	<Adji-
unit>	ful}	unit>

- **Adji-units Analysis:** Adji-units are then compared with the positive and negative lexemes in the lexicon. Due to this, it is known as two-step lexicon based opinion mining model for Urdu language. The presence of the word (بہت {very}) in a sentence enhances the intensity of that sentence (either positively or negatively). Overall polarity of the sentence is then calculated. Adji-units which do not match with either positive or negative entries in the lexicon have been entered manually for efficient processing. lexicons.

4   Evaluation of LOMMU

LOMMU has been evaluated by using sample text files consisting of sentences and 10,000 tagged words downloaded from http://www.cle.org.pk. First of all, tag-set has been selected to extract adji-units. Secondly, extracted adjectives have been compared with positive and negative lexemes in the lexicon. Finally, results have been discussed along-with the system accuracy. Figures 3 and 4 show complete working of LOMMU.

Developed LOMMU reads a text file for which polarity has to be calculated. It extracts the list of adji-units from the tagged words by retaining the words with a tag <JJ> . Negations and other factors that may increase or decrease polarities are stored at backend to be used while calculating final results. Extracted adji units are then compared with the entries in the lexicon and sentence-by-sentence analysis is conducted for making the overall decision.

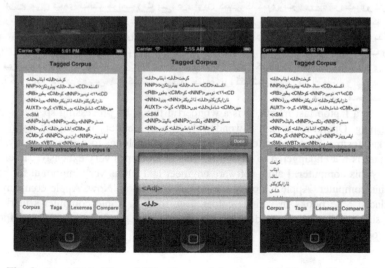

Fig. 3. Read Urdu Tagged File, Tag-set Selection and Adji-units Extraction

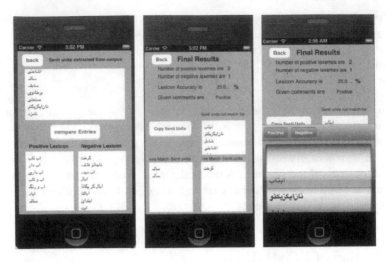

Fig. 4. Lexicon having Negative and Positive Lexemes; and Extracted Adji-units

Figure 3 shows adji-units extracted from the given file and results given by LOMMU.

LOMMU has been tested with test data and results obtained are satisfactory. Some sample texts have been discussed here along-with the results given by our system when these texts are passed through it.

Sample 1: ایپل کمپیوٹرز {Apple Computers}

پچھلے مہینے میں نے ایک لیپ ٹاپ خریدا ہے. یہ ایک شاندار رفتار بہت <u>حیرت</u> انگیز ہے. آپریٹنگ چیز ہے. اس کی پروسیسنگ سسٹم <u>بہترین</u> ہے. اگرچہ بیٹری <u>دیر</u>یا نہیں ہے، جو میرے لئے قابل قبول ہے. یہ لیپ ٹاپ دیکھنے میں بھی بہت <u>خوبصورت</u> ہے. یہ ایپل کمپنی کا کمپیوٹر ہے. اس کمپیوٹر میں وائ☐رس آنے کا <u>خطرہ</u> نہیں ہے. یہ کمپیوٹر بہت <u>مہنگا</u> ہے. ہر شخص اس کمپیوٹر کو نہیں خرید سکتا. میں ایک سوفٹ ویئر انجینئر ہوں اور میرے لیئے اس کمپوٹر کو خریدنا بہت ضروری ہے. ایپل کے لیپ ٹاپ کو میک بک بولا جاتا ہے. اب ایپل کمپنی نے مختلف رنگوں میں میک بک متعارف کروائی ہیں. اس کی وجہ سے ایپل کمپیوٹر کی مارکیٹ بڑی <u>تیزی</u> سے آگے بڑھ رہی ہیں. اس کمپنی کی جتنی بھی پروٹکٹس ہیں وہ <u>اچھی</u> اور <u>پائیدار</u> ہیں. جن آئی فون سر فہرست ہیں.

{I bought a laptop last month. It is a <u>wonderful</u> thing. Its processing speech is quite <u>astonishing</u>. Operating system is <u>wonderful</u>. Although battery is not <u>long-lasting</u>, but it is okay for me. This laptop looks very <u>beautiful</u>. It belongs to Apple company. This computer is not <u>endangered</u> to any virus. This computer is very <u>expensive</u>. Everyone can buy this computer. I am a software engineer and it was very important for me to buy this computer. Apple's laptop is called as Mac book. Now, Apple company has introduced Mac books in various colors. Due to this, Apple computer's market in increasing <u>rapidly</u>. All the products of this company are <u>good</u> and <u>long-lasting</u>. Iphone is one of them.}

Sample 1 contains reviews about laptop taken from a discussion forum. More than 400 words have been minimized to around 160 words making a complete paragraph. This data has then been tagged using an existing POS tagger. LOMMU has read data word by word and extracted adji-units as shown by underlined words in the text. Negations attached with any of the extracted adji-unit have been stored at backend for final decision-making.

Table 1. Sample 1 results

Sample 1 - Final results
System extracts 4 out of 8 positive lexemes and 2 out of 2 negative lexemes. So, total matched lexemes are 6 out of 10. Therefore, lexicon accuracy is 60%. After using negations as polarity shifters, the overall opinion about given data is positive.

When we pass this text from our system it matches 4 positive lexemes and 2 negative lexemes but skips others. Positive lexemes have been shown by simple underlined words whereas negative lexemes have been shown with dotted underlined words. Skipped lexemes can be manually added into the existing lexicon list for future use. Overall results for sample 1 have been discussed in Table 1 below. For the given text in sample, the accuracy of LOMMU lexicon is 60%.

Sample 2: میرا دوست {Mera Dost}

> میرے بہت سے دوست ہیں۔ لیکن محسن میرا سب سے اچھا دوست ہے۔ محسن ایک ڈیزائنر ہے اور ایک بہت بڑی کمپنی میں بہت اعلی درجے پر فائز ہے۔ محسن ایک اعلی تعلیم یافتہ نوجوان ہے۔ محسن ہمیشہ سچ بولتا ہے اور اپنا کام محنت اور لگن سے کرتا ہے محسن کے دفتر میں ہر شخص اس کی شخصیت سے متاثر ہے۔ محسن اپنے زمانہ طالب علمی سے ہی بہت محنتی ہے۔ محسن ہمیشہ میرے مشکل وقت کا ساتھی ثابت ہوا ہے۔ وہ کھانے میں اکثر بے احتیاطی کرتا ہے جسکی وجہ سے وہ اکثر بیمار رہتا ہے اس کی اس عادت کی وجہ سے اس کے دوست اور گھر والے بہت تنگ ہیں۔ محسن نے کبھی اپنی صحت کا خیال نہیں رکھا۔ وہ اگر محنت کے ساتھ ساتھ اپنی صحت کا بھی خیال رکھے تو وہ مزید خوشگوار زندگی گزار سکتاہے۔ کیونکہ صحت مند جسم ہی صحت مند دماغ ہوتا ہے۔
>
> {I have many friends. But Mohsin is my best friend. Mohsin is a designer and works in a very big company at a very good post. Mohsin is a well-educated young man. Mohsin always speaks truth and does his work with dedication and passion. Everyone in Mohsin's office is impressed by his personality. Mohsin is very hard-working from his student life. Mohsin has always proven to be my partner in my difficult times. He often shows carelessness in eating due to which he always remains ill. His friends and family members are fed up of this habit of his. Mohsin has never taken care of his health. If his takes care of his health along-with the hardwork he does, he can live a more happy life. Because, healthy body is a healthy mind.}

Sample 2 discusses reviews about an employee of a software firm. Here, 350 words have been reduced to 250 words making a complete paragraph. This sample data has also been converted to tagged data using POS tagger. Extracted adji-units have been shown by the underlined words in the text above. Simple underlined words are positive lexemes whereas dotted underlined words are negative lexemes.

Sample 2, when passed through our system, matches 8 positive lexemes and 2 negative lexemes. In this case, the overall accuracy of LOMMU lexicon is recorded as 55.55% which has been given in Table 2 below.

Table 2. Sample 2 results

Sample 2 - Final results
LOMMU extracts 8 out of 14 positive lexemes and 2 out of 4 negative lexemes in case of sample 2. Therefore, lexicon accuracy is 55.55% as 10 out of 18 lexemes have been matched. After using negations as polarity shifters, overall opinion about given data is positive.

Entire experimentation has been conducted using different corpuses having around 100,000 words. This experiment has given a decreased accuracy of 50–52%. The main reason for this accuracy decline is that our lexicon is not mature enough and contains only 15,000 words (adji-units). Adji-units can be added manually for efficient processing. Increasing the number of adji-units in the lexicon would definitely increase the LOMMU accuracy.

LOMMU presents a sentiment-annotated lexicon for mining opinionated positive and negative expressions of any given Urdu text. It is an integral basis of Urdu text based sentiment analysis. LOMMU gives an accuracy of about 50–52% with just 15,000 adji-units in the lexicon. Increasing this further would definitely increase the LOMMU lexicon accuracy.

Moreover, all the existing sentiment analysis systems are for Windows platform. Our system, on the other hand, provides a platform for Macintosh users.

5   Conclusion and Future Work

Two-step lexicon based opinion mining model has been proposed for Urdu language which uses a JSON based approach for constructing the lexicon. For each word in the lexicon, detailed information has been given. It has been tested with different corpuses having about 100,000 words. The system gives an accuracy of about 50-52% as our lexicon consists of only 15000 words.

Future work associated with it would be the enhancement of developed lexicon by adding more words so that high system accuracy can be achieved.

References

1. Syed, A.Z., Aslam, M., Martinez-Enriquez, A.M.: Associating targets with SentiUnits: a step forward in sentiment analysis of Urdu Text. Artif. Intell. Rev. **41**(4), 535–561 (2014)
2. Dave, K., Lawrence, S., Pennock, D.M.: Mining the peanut gallery: opinion extraction and semantic classification of product reviews. In: Proceedings of 12th International Conference on World Wide Web, pp. 519–528 (2003)
3. Hussain, S.: Resources for Urdu language processing. In: Proceedings of 6th Workshop on Asian Language Resources IJCNLP, pp. 1–10 (2008)
4. Xia, R., Zong, C., Li, S.: Ensemble of feature sets and classification algorithms for sentiment classification. Inf. Sci. J. 1138–1152 (2011)
5. Han, E.H.S, Karypis, G., Kumar, V.: Text categorization using weight adjusted k-nearest neighbor classification. In: Pacific-Asia Conference on Knowledge Discovery and Data Mining, pp. 53–65 (2001)
6. Turney, P.D.: Thumbs up or Thumbs down? Semantic orientation applied to unsupervised classification of reviews. In: Proceedings of the 40th Annual Meeting on Association for Computational Linguistics, pp. 417–424 (2002)
7. Pang, B., Lee, L.: Opinion mining and sentiment analysis. Foundations Trends Inf. Retrieval **2**(1–2), 1–135 (2008)
8. Kim, S.M., Hovy, E.: Identifying and Analyzing Judgment Opinions, In: Proceedings of the Main Conference on Human Language Technology Conference of the North American Chapter of the Association of Computational Linguistics, pp. 200–207 (2006)
9. Lu, B., Tan, C., Cardie, C., Tsou, B.K.: Joint bilingual sentiment classification with unlabeled parallel corpora. In: Proceedings of the 49th Annual Meeting of the Association for Computational Linguistics – Human Language Technologies, vol, 1, pp. 320–330 (2011)
10. Humayoun, M., Hammarström, H., Ranta, A.: Urdu morphology, orthography and lexicon extraction. In: Proceedings of 2nd Workshop on Computational Approaches to Arabic Script-based Languages (2007)
11. Ijaz, M., Hussain, S.: Corpus based Urdu lexicon development. In: Proceedings of Conference on Language and Technology (CLT), pp. 1–10 (2007)
12. Syed, A.Z., Muhammad, A.: Lexicon based sentiment analysis of Urdu text using senti-units, In: Proceedings of 10th Mexican International Conference on Advances in Artificial Intelligence, pp. 32–43 (2010)

Applications of Pattern Recognition

Morphological Analysis Combined with a Machine Learning Approach to Detect Utrasound Median Sagittal Sections for the Nuchal Translucency Measurement

Giuseppa Sciortino[1], Domenico Tegolo[1,2(✉)], and Cesare Valenti[1]

[1] Dipartimento di Matematica e Informatica,
Università degli Studi di Palermo, Palermo, Italy
giuseppa.sciortino@gmail.com, {domenico.tegolo,cesare.valenti}@unipa.it
[2] CHAB-Mediterranean Center for Human Health Advanced Biotechnologies,
Palermo, Italy

Abstract. The screening of chromosomal defects, as trisomy 13, 18 and 21, can be obtained by the measurement of the nuchal translucency thickness scanning during the end of the first trimester of pregnancy. This contribution proposes an automatic methodology to detect mid-sagittal sections to identify the correct measurement of nuchal translucency. Wavelet analysis and neural network classifiers are the main strategies of the proposed methodology to detect the frontal components of the skull and the choroid plexus with the support of radial symmetry analysis. Real clinical ultrasound images were adopted to measure the performance and the robustness of the methodology, thus it can be highlighted an error of at most 0.3 mm in 97.4% of the cases.

Keywords: Mid-sagittal section · Neural network · Nuchal translucency · Symmetry transform · Wavelet analysis

1 Introduction

The nuchal translucency term was coined by Professor K. Nicolaides who is considered as a pioneer in the study of prenatal trisomy 21 at the Fetal Medicine Foundation [7]. It is a fluid collection located in the nuchal region of the fetus that may also extend along the spine (see Fig. 1) which appears as an anechoic region surrounded by two thin hyperechoic regions. The terms hyperechogenic, hypoechogenic, isoechogenic and anechogenic express the echogenicity properties of tissues not only based on their physical properties but also in relation to the surrounding tissues. An anechogenic tissue tends to be black in terms of gray values while echogenic tissues have a greater reflective power and consequently appear brighter. The small differences between the skin and the amniotic fluid are not easy to distinguish, since during the prenatal phase both these structures appear as thin membranes. It is visible from the first weeks of pregnancy and

© Springer International Publishing AG 2017
J.A. Carrasco-Ochoa et al. (Eds.): MCPR 2017, LNCS 10267, pp. 257–267, 2017.
DOI: 10.1007/978-3-319-59226-8_25

increases in thickness together with the growth of the fetus, reaching its maximum thickness between the eleventh and the thirteenth weeks; after this period it tends to dwindle. The nuchal translucency thickness is defined as the maximum thickness of the translucent space between the two echogenic lines and put in evidence as a dark area. Since the '90s, the nuchal translucency is the subject of a thorough analysis during the first trimester of pregnancy, after having noticed a correlation between its thickness and the incidence of chromosomal abnormalities: the greater the thickness of the translucency, the greater likelihood that the fetus present anomalies. Chromosomopathies related to a translucency thickness higher than the reference values are relate to the Down syndrome (trisomy 21), Edwards syndrome (trisomy 18), Patau syndrome (trisomy 13) and Turner syndrome (gonadal dysgenesis) but also other abnormalities affecting the heart such as the omphalocele or diaphragmatic hernia [18].

Ultrasound imagery is strongly operator-dependent and a number of qualities have to be endowed such as special manual skills, spirit of observation, image interpretation and clinical experiences to identify the artifacts. The non-invasive nature of ultrasound allowed to elect this approach as the best and high sensitivity diagnostic method and it becomes one of the main investigation techniques during the entire gestation period. Nonetheless the absence of accurate and automated tools deprived of non-operator's objectivity is tied both to the identification of the median sagittal section and to the problem of measuring the nuchal translucency. The identification of median sagittal sections in the literature was mainly addressed to volumetric acquisitions and a few studies were carried out in the two-dimensional case. The proposed methods have shown their power to support the diagnosis by the physician, thus reducing human intervention.

2   Materials and Methods

Though the literature on medical image analysis in general is huge and covers many topics, not so much work was dedicated to automatic fetal measurements in ultrasound images. This is due to the fact that these images are quite difficult to deal with. The system described in [3] helps the user determining the borders of the NT with some basic image processing. The borders are identified manually selecting two points and entirely determined via a flood-fill operation. The NT thickness is then measured using the same algorithm adopted in [4]. Less efforts by the operator are required for the method described in [11]: the image is preprocessed to reduce speckle noise, typical of ultrasound images. The method described in [4] present a different cost function from [11] to overcome drawbacks due to the previous method, making the algorithm general and not dependent on weights that need to be tuned. The authors of [5] propose a hierarchical structural model for the automatic detection of the NT area. In that paper single SVM classifier for the NT produces a hit rate of about 55%. The performances are improved to a hit rate of 59–60% by introducing another model that denotes the spatial relationships among NT, body and head of the foetus, identified using three ad hoc classifiers.

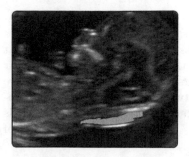

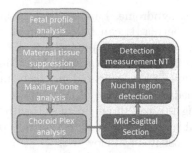

Fig. 1. Example of nuchal translucency detected by our methodology.

Fig. 2. Diagram of the functional modules of the proposed methodology.

This contribution summarizes the results described already by some of the authors in [2,15] and the proposed method is sketched in Fig. 2.

2.1 Fetal Ultrasound

Prenatal ultrasound is among the most complex and comprehensive diagnostic techniques and it has become one of the main investigation approaches during the entire gestation period. In order to search for specific markers to identify eventual genetic diseases, ultrasound examinations in the first trimester provide essential information for the entire progression of pregnancy. The thickness of the nuchal translucency can be measured among 11–13 weeks. During the second trimester (20–30 weeks) the focus is on the observation of some morphological aspects and during the third quarter (30–33 weeks) the focus is on all those aspects related on childbirth. Currently, ultrasound results highly dependent on the operator and for this reason the ability of the worker affects the quality of graphics and acquisition information.

2.2 Nuchal Translucency to Detect Abnormal Chromosomes

The nuchal translucency (NT) term was coined by Nicoladeis, who studied prenatal trisomy 21 for the Fetal Medicine Foundation [7]. NT is a liquid collection located in the nuchal region of the fetus which may also extend along the spine. The NT ultrasound region appears as a dark flap or as an anechogenic area, bordered by two thin hyperechogenic regions. It is visible from the first weeks of pregnancy and increases in thickness together with the growth of the fetus, reaching its maximum thickness during the eleventh and thirteenth weeks. Since the '90s, the NT became the subject of a thorough analysis during the first trimester of pregnancy, after having found a correlation between the thickness and the incidence of chromosomal and other abnormalities: the greater is the thickness of the translucency, the greater is the probability that the fetus will present some anomalies. Chromosomopathies related to the thickness translucency are primarily Down syndrome, Edwards syndrome, Patau syndrome and

Turner syndrome. Further abnormalities such as the omphalocele or diaphragmatic hernia [1] can be highlighted too. Studies [13] show that 70% of fetuses in the absence or hypoplasia of the nasal bone with a frontal bone fairly flat are suffering by Down syndrome. Moreover in the last two decades a number of studies were dedicated on biochemical markers in maternal serum; they have reached a percentage of correctness ranging between 50% and 70% [20]. Combining the measurement of NT with biochemical tests and maternal age, it is possible to achieve a detection rate of 85%–90% and to reduce the number of false positives to below 5%.

2.3   Criteria Proposed by the Fetal Medicine Foundation

The Fetal Medicine Foundation [7] is a non-profit organization that aims to improve the health of pregnant women and their babies through research and training in fetal medicine. The protocol drawn up by the FMF includes the following recommends also:

- the ultrasound machine should be of high resolution;
- the fetal crown-rump length should be 45–84 mm;
- a good mid-sagittal section must be acquired;
- the vertical branch of the maxilla, which branches off from the upper jaw to the nasal bone, must not be visible;
- the plexus should not be visible, that is the region is uniformly echogenic.

We need to address all these criteria in the measurement of nuchal translucency because leaving out some items leads to misdiagnosis. For example, the measurement can be increased by 0.6 mm in the case of fetal neck hyperextension (Fig. 3a) while it can be underestimated by 0.4 mm when the neck is flexed (Fig. 3b). Therefore, non-median sagittal sections cause false measurements (Fig. 3c) where you can see the thickness of the translucency visibly increased in comparison to correct median sagittal sections (Fig. 3d).

2.4   Median Sagittal Sections

As suggested by the FMF, median sagittal sections represent a critical point for a good measurement of NT. Therefore an automatic tool to locate the exact position of the ultrasound probe can help the physician to acquire a correct NT measurement. The sickle brain divides the two cerebral hemispheres, thus a good mid-sagittal section will be obtained by positioning the probe on this evidence. Two further elements have to be considered: the choroid plexus and the maxillary bone (Fig. 4). The evidence of an uniformly echogenic zone in the choroid plexus allows a good positioning of the ultrasound probe, it is a liquid substance thus its presence will appear as an anechoic area (Fig. 5). In addition, the absence of the jawbone vertical branch (Fig. 5) allows to identify a good median sagittal section.

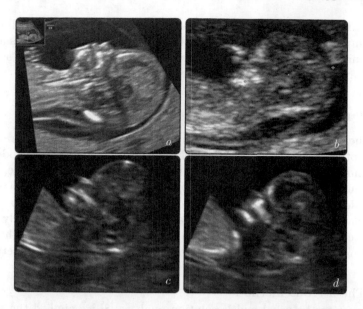

Fig. 3. Examples of hyperextended (a) and inflected (b) heads. The NT has a the correct size in a mid-sagittal section (c) and a greater thickness in a non-median sagittal section (d).

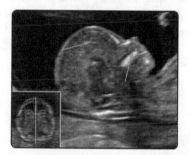

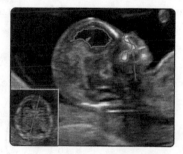

Fig. 4. Mid-sagittal section: the arrows indicate the echogenic plexus and the absence of the jawbone vertical branch.

Fig. 5. Non-median sagital section: the arrows indicate the anechogenic plexus and the jawbone vertical branch.

In brief, some elements of fetus skull must be put in evidence and others have to be not considered. Our method adopts anisotropic filters and the á trous transform to identify the versus of the profile, neural network methods to detect jaw bone area, and a morphological methodology to identify the choroid plexus.

2.5   Dataset of Ultrasound Images

An expert physician organized our image dataset between the 11st and the 13rd weeks of pregnancy. It hosts 10 video sequences representing 10 different subjects with both the left and the right profiles. All digital files were stored with the lowest compression ratio of the H.264 codec to avoid as many artifacts as possible. We uniformly extracted 3000 frames from the video sequences in a random way and saved them in a lossless raw format with 640×480 pixels.

2.6   Pre-processing and Wavelets Analysis

Due to the speckle noise that usually affects ultrasound images, our methodology adopts a variation [9] of the widely used anisotropic filter introduced by Perona and Malik [14] to pre-process our data. As highlighted before, sagittal sections must be median; this means that the ultrasound probe must be placed in correspondence at the falx cerebri which divides the choroid plexus into two symmetrical halves. To locate the jaw bone and other components (already used to identify the profile versus) we apply the à trous wavelet transform that highlights the main parts of the face of the fetus to be analyzed by the neural networks.

2.6.1   The à trous method

With respect to the usual multiresolution analysis [8] we applied the so-called à trous algorithm [16,17] because the former method returns wavelet planes of decreasing sizes (therefore useful for image compression) while the latter produces wavelet planes of the same size as the original image (which is better for image segmentation). We perform the following sequence of low-pass and high-pass filters:

$$I_0(\mathbf{p}) = I(\mathbf{p}), \quad I_i(\mathbf{p}) = I_{i-1}(\mathbf{p}) \circledast \ell_i$$

where the non-zero elements of ℓ_i are given by the isotropic kernel [10]:

$$\ell = \frac{1}{16} \begin{pmatrix} 1\,2\,1 \\ 2\,4\,2 \\ 1\,2\,1 \end{pmatrix}, \quad \ell_i(2^{i-1}\mathbf{q}) = \ell(\mathbf{q})$$

The pixel $\mathbf{q}$ spans the 3×3 neighborhood of each pixel $\mathbf{p}$. The high-pass filter is defined as the difference between two consecutive spatial scales:

$$W_i(\mathbf{p}) = I_{i-1}(\mathbf{p}) - I_i(\mathbf{p})$$

Small objects are enhanced in the first planes while bigger components are present in the last ones. We experimentally verified that a threshold based on the average μ and standard deviation σ of the luminosity of $W_{4,5,6}$ puts in evidence the main components of the face of the fetus (Fig. 6):

$$C_i(\mathbf{p}) = \{\mathbf{p} : W_i(\mathbf{p}) \geq \mu(W_i) + 2\sigma(W_i)\}, \quad C = \bigvee_{i=4,5,6} C_i$$

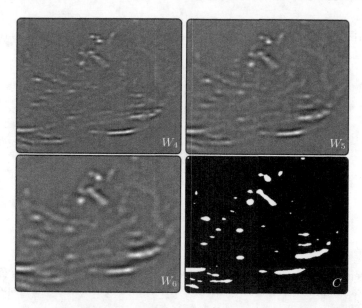

Fig. 6. The wavelet planes W_4, W_5 and W_6 show the main structures of various sizes. The main components are detected through the binary mask C.

2.6.2 Neural Networks

The output of the previous section constitutes the input of this step in which two neural networks are involved. The first network was trained to recognize the presence of vertical branch (Fig. 5) which is a perpendicular bone that extends from jaw bone (dental arcade) to the nasal bone: when the vertical branch is visible in the image, then the section is not median. The network takes in input an image and outputs a probability in the range 0.1–0.9, indicating the membership degree to the classes. The input layer has 640×480 elements (size of the image), 10 hidden layers and an output layer with 2 elements that represent the two classes. The training set is composed of 1500 images (1000 for training and 500 for validation) and the two classes are equally represented. The second network was trained to recognize the jaw bone (Fig. 4). This network takes in input an image and outputs a probability in the range 0.1–0.9, indicating the membership degree to 4 classes (chin, jaw bone, nose, etc.). The input layer has 640×480 elements and an output layer with 4 elements, that represent the classes. The training set is composed of 2000 images and the four classes are equally represented.

The networks are defined with a feed forward model and a back-propagation process minimizes the overall error in accordance with the standard general equations [6]. The physician labeled manually in all the frames the regions representative of the classes nasal bone, the mandible, the chin and 'other'. Moreover representative areas of 3×3 pixels were used to evaluate the mean μ and standard deviation σ of plexus area and to calculate its probability distributions of

Table 1. Confusion matrices of the neural networks to classify the vertical branch (left) and to distinguish among the nasal bone, mandible, chin and 'other' classes (right).

		Physician			
		mandible	chin	nose	other
Methodology	mandible	98.56%	0.20%	0.10%	0.63%
	chin	0.85%	99.80%	0.10%	3.54%
	nose	0.00%	0.00%	99.70%	20.60%
	other	0.59%	0.00%	0.10%	75.23%

	Physician	
	yes	no
Methodology yes	94.96%	11.46%
Methodology no	5.04%	88.54%

Vertical branch test   Face components test

the echogenicity. The performances are reported as confusion matrices (Table 1) in percentages and therefore they include the values of sensitivity and specificity.

2.6.3   Analysis of the Frontal Region

The choroid plexus is located in the cranial region, whose morphology can be approximated to a circumference which can be located by the fast detector of circular objects defined in [12]. The underlying idea consists in the observation that the contours delimiting the objects in an image are obtainable by higher values in the gradient magnitude image. Therefore, amplifying the contribution of gradient vectors which lie along a circular shape highlights the center of the circle. We modified this symmetry detector to consider only bright sectors with pre-determined radii (85–90 pixels) and angles (60°–120°) and we experimentally fine-tuned these parameters taking into account small variations in size of the head in the images acquired by our ultrasound equipment.

Once the skull is identified, the image is considered a valid mid-sagittal section if the number of anechogenic and echogenic pixels inside the circular sector fulfills the following predetermined test, where #*overall* refers to doubtful zones:

$$\frac{\#anechogenic}{\#overall} \leq 0.43 \quad \bigwedge \quad \frac{\#echogenic}{\#overall} \geq 0.11$$

2.7   Analysis of the Nuchal Translucency

Simple anatomical observations on both the skull and the frontal region let us locate, with a correctness rate equal to 99.95%, a bounding box (Fig. 7) which contains the nuchal translucency region. The mathematical morphology edge enhancer ρ [19] on the superimposed binary image C lets obtain a good contour of the nuchal translucency. According to the FMF protocol, the thickness is

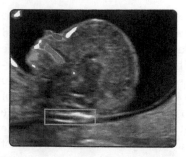

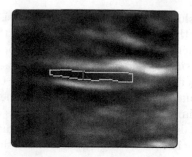

Fig. 7. The positions of the frontal region components let locate the bounding box.

Fig. 8. The nuchal translucency thickness is defined as the maximum diameter.

defined as the maximum distance between the pixels belonging to the lower and upper edges (Fig. 8). In order to verify the correctness of the methodology we compared the automatic measurement against the ground truth provided by an expert physician, thus verifying that 23.3% of the solutions present no error while 97.4% of the solutions show an error up to 0.3 mm (Fig. 9).

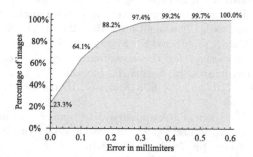

Fig. 9. Nuchal translucency thickness: error (with respect to the manual measurement) versus percentage of the whole dataset.

3   Conclusions

The study of fetal images is a difficult task in general and just a few works concern even the semi-automatic analysis of ultrasound fetal images. Although this is still a key area of research, new efforts are now fostering to provide a complete diagnosis with further non-invasive and complementary techniques. Some examples in the gene field are given by the Polymerase Chain Reaction and the Next Generation Sequencing, but with greater costs, requirements and turnaround times with respect to the proposed approach.

In this paper we presented an unsupervised methodology able to choose proper mid-sagittal sections from ultrasound video streams, to locate the nuchal translucency through wavelet analysis and neural networks and finally to measure the translucency thickness on the enhanced edges obtained by standard mathematical morphology. At the best of our knowledge, this is the first time that such a complete system is proposed in the literature. We validated the overall performance against the ground truth created by an expert physician who manually classified different components of the skull of the fetus in dataset of real images. Anatomical observations allow us to detect the bounding box of the nuchal translucency and to make a measurement almost in accordance with the physician (on average we introduce the error of at most just 0.3 mm in 97.4% of the measurements).

References

1. Alfirevic, Z., Sundberg, K., Mujezinovic, F.: Amniocentesis and chorionic villus sampling for prenatal diagnosis. Cochrane Database Syst. Rev. **3** (2003). doi:10.1002/14651858.CD003252
2. Anzalone, A., Fusco, G., Isgrò, F., Orlandi, E., Prevete, R., Sciortino, G., Tegolo, D., Valenti, C.: A system for the automatic measurement of the nuchal translucency thickness from ultrasound video stream of the foetus. In: International Symposium on Computer-Based Medical Systems, pp. 239–244. IEEE (2013)
3. Bernardino, F., Cardoso, R., Montenegro, N., Bernardes, J., Marques De Sà, J.: Semiautomated ultrasonographic measurement of fetal nuchal translucency using a computer software tool. Ultrasound Med. Biol. **24**(1), 51–54 (1998)
4. Catanzariti, E., Fusco, G., Isgrò, F., Masecchia, S., Prevete, R., Santoro, M.: A semi-automated method for the measurement of the fetal nuchal translucency in ultrasound images. In: Foggia, P., Sansone, C., Vento, M. (eds.) ICIAP 2009. LNCS, vol. 5716, pp. 613–622. Springer, Heidelberg (2009). doi:10.1007/978-3-642-04146-4_66
5. Deng, Y., Wang, Y., Chen, P.: Automated detection of fetal nuchal translucency based on hierarchical structural model. In: International Symposium on Computer-Based Medical Systems, pp. 78–84. IEEE (2010)
6. Egmont-Petersen, M., Ridder, D., Handels, H.: Image processing with neural networks - a review. Pattern Recogn. **35**(10), 2279–2301 (2002)
7. FMF. Fetal Medicine Foundation nuchal translucency. www.fetalmedicine.org
8. González-Audícana, M., Otazu, X., Fors, O., Seco, A.: Comparison between Mallat's and the 'à trous' discrete wavelet transform based algorithms for the fusion of multispectral and panchromatic images. Int. J. Remote Sens. **26**(3), 595–614 (2005)
9. Guastella, D., Valenti, C.: Cartoon filter via adaptive abstraction. J. Visual Commun. Image Represent. **36**, 149–158 (2016)
10. Jain, R., Kasturi, R., Schunck, B.: Machine Vision. McGraw-Hill, New York (1995)
11. Lee, Y., Kim, M., Kim, M.: Robust border enhancement and detection for measurement of fetal nuchal translucency in ultrasound images. Med. Biol. Eng. Comput. **45**(11), 1143–1152 (2007)
12. Loy, G., Zelinsky, A.: Fast radial symmetry for detecting points of interest. IEEE Trans. Pattern Anal. Mach. Intell. **25**(8), 959–973 (2003)

13. Orlandi, F., Rossi, C., Orlandi, E., Jakil, M., Hallahan, T., Macri, V., Krantz, D.: First-trimester screening for trisomy-21 using a simplified method to assess the presence or absence of the fetal nasal bone. Am. J. Obstet. Gynecol. **194**(4), 1107–1111 (2005)
14. Perona, P., Malik, J.: Scale-space and edge detection using anisotropic diffusion. IEEE Trans. Pattern Anal. Mach. Intell. **12**(7), 629–639 (1990)
15. Sciortino, G., Orlandi, E., Valenti, C., Tegolo, D.: Wavelet analysis and neural network classifiers to detect mid-sagittal sections for nuchal translucency measurement. Image Anal. Stereology **35**(2), 105–115 (2016)
16. Sciortino, G., Tegolo, D., Valenti, C.: Automatic detection and measurement of nuchal translucency. Comput. Biol. Med. **82**, 12–20 (2017)
17. Shensa, M.: The discrete wavelet transform: wedding the à trous and Mallat algorithms. IEEE Trans. Sig. Process **40**(10), 2464–2482 (1992)
18. Snijders, R., Noble, P., Sebire, N., Souka, A., Nicolaides, K.: UK multicentre project on assessment of risk of trisomy 21 by maternal age and fetal nuchal translucency thickness at 1014 weeks of gestation. Lancet **6**(9125), 343–351 (1998)
19. Soille, P.: Morphological Image Analysis: Principles and Applications. Springer, Heidelberg (2010)
20. Wald, N., George, L., Smith, D., Densem, J., Pettersonm, K.: Serum screening for Down's syndrome between 8 and 14 weeks of pregnancy. Br. J. Obstet. Gynecol. **103**(5), 407–412 (1996)

BUSAT: A MATLAB Toolbox for Breast Ultrasound Image Analysis

Arturo Rodríguez-Cristerna[1]([⊠]), Wilfrido Gómez-Flores[1],
and Wagner Coelho de Albuquerque-Pereira[2]

[1] Center for Research and Advanced Studies of the National Polytechnic Institute,
Ciudad Victoria, Tamaulipas, Mexico
{arodriguez,wgomez}@tamps.cinvestav.mx
[2] Biomedical Engineering Program, COPPE, Federal University of Rio de Janeiro,
Rio de Janeiro, Brazil
wagner.coelho@ufrj.br

Abstract. This paper presents the Breast Ultrasound Analysis Toolbox (BUSAT) for MATLAB, which contains 62 functions to perform image preprocessing, lesion segmentation, feature extraction, and lesion classification. BUSAT is useful to codify programs for computer-aided diagnosis (CAD) purposes in reduced time; hence, to replicate several approaches proposed in literature is feasible. We provide the implementation of a CAD system to classify breast lesions into benign and malignant classes and an example to evaluate the classification performance. BUSAT could be downloaded from the following permanent link: http://www.tamps.cinvestav.mx/~wgomez/downloads.html.

Keywords: Breast ultrasound · Image analysis · Computer-aided diagnosis · MATLAB

1 Introduction

Breast cancer is the most frequently diagnosed cancer and the leading cause of cancer death among women worldwide [1]. Hence, early diagnosis is a crucial factor in breast cancer treatment, where medical images are important sources of diagnostic information. Currently, breast ultrasound (BUS) is an important coadjuvant technique to mammography (x-ray) in patients with palpable masses and normal or inconclusive mammogram findings [2]. Also, BUS images are particularly effective in distinguishing cystic from solid lesions and are useful for differentiating between benign and malignant tumors [3].

In order to assist radiologists in the BUS image interpretation, computer-aided diagnosis (CAD) systems have emerged as a 'second reader' for analyzing the images by using computational approaches. Generally, the pipeline of a CAD system involves four basic stages: image preprocessing, lesion segmentation, feature extraction, and lesion classification [4]. Then, radiologists can take the CAD outcome as a second opinion and make a more conclusive diagnosis for reducing unnecessary biopsies in benign cases [5].

© Springer International Publishing AG 2017
J.A. Carrasco-Ochoa et al. (Eds.): MCPR 2017, LNCS 10267, pp. 268–277, 2017.
DOI: 10.1007/978-3-319-59226-8_26

Image preprocessing commonly increases the contrast between the lesion region and its background, and also considers low-pass filtering to reduce the speckle artifact. Next, BUS segmentation procedure separates the lesion region from its background and other tissue structures. Thereafter, from segmented lesions, morphological and texture features are usually computed and to improve the between-class discrimination, relevant features are selected. These features represent the classifier inputs for distinguishing the lesions into benign and malignant classes [4].

In literature, a plethora of approaches have been proposed to address each stage of CAD systems for BUS images. In this sense, Cheng et al. [4] and Huang et al. [6] presented comprehensive surveys related to BUS image analysis. Despite the large quantity of proposed approaches, to get useful computational implementations for research purposes is usually difficult, because the source codes or programs are not commonly shared by the authors.

Hence, we introduce a MATLAB (The MathWorks, Natick, Massachusetts, USA) [7] toolbox for BUS image analysis, aiming to share with the research community the efforts that we made to implement several methods to develop CAD systems for breast ultrasound. The toolbox is composed of 62 functions divided into four sections: image preprocessing, lesion segmentation, feature extraction, and classification. This toolbox could be downloaded from our permanent link http://www.tamps.cinvestav.mx/~wgomez/downloads.html.

2 Toolbox Organization

The Breast Ultrasound Analysis Toolbox (BUSAT) has 62 functions oriented to image preprocessing (contrast enhancement, despeckling, and domain transformation), lesion segmentation (semi-automatic and fully-automatic methods), feature extraction (morphological, texture, and BI-RADS lexicon), and classification (linear and non-linear classifiers). Figure 1 illustrates the general organization of BUSAT and the list of available functions.

It is worth mentioning that all the functions were codified by our research group based on several articles from literature; hence, all the implemented methods have theoretical basis. In addition, several functions take advantage of some methods developed by other research groups to guarantee the quality of the results, for instance, LIBSVM to train Support Vector Machines [8], minimum redundancy maximum relevance (mRMR) for feature selection [9], etc.

On the other hand, the main BUSAT directory contains the following six subfolders:

- **Data:** contains data files and test images to run the examples of the toolbox.
- **Preprocessing:** 13 functions for contrast enhancement, speckle filtering, and domain transformation.
- **Segmentation:** four functions for lesion segmentation.
- **Features:** 29 functions for computing morphological, texture, and BI-RADS features.

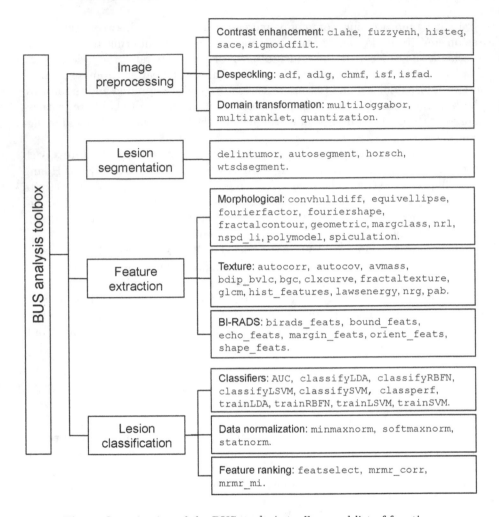

Fig. 1. Organization of the BUS analysis toolbox and list of functions.

- **Classification:** 16 functions for lesion classification in benign and malignant classes.
- **C functions:** 21 compiled C code functions that are used by several functions of the toolbox.

3 Toolbox Usage

3.1 Installation

To start using BUSAT, the script RUN_ME_FIRST should be firstly run to add all the toolbox directories to the MATLAB search path.

3.2 Help Topics

To display the organization BUSAT, type in the MATLAB Command Window the statement help Contents. Note that every listed function has a hyperlink to its own help topics. Also, the user can consult the help topics of a specific function by typing the statement help followed by the name of the function as illustrated in Fig. 2. Observe that help topics are displayed in three parts: the syntax explanation of the function, an illustrative example, and the reference or bibliography for theoretical details. Also, hyperlinks to similar functions are showed.

```
>> help isfad
  isfad Interference-based speckle filter followed by anisotropic diffusion.
    J = isfad(I,W,MAXITER) performs the interference-based speckle filter
    followed by anisotropic diffusion for despeckling a breast ultrasound
    image I, where W is the window size of the median filter, and
    MAXITER is the maximum number of iterations for diffusion.

    Example:
    --------

    I = imread('BUSreal.tif');
    J = isfad(I,10,500);
    figure;
    subplot 121; imshow(I); title('Original Image');
    subplot 122; imshow(mat2gray(J)); title('Filtered Image');

    See also adf adlg chmf isf


    Reference:
    ----------

    F. M. Cardoso, M. M. S. Matsumoto, S. S. Furuie "Edge-preserving speckle
    texture removal by interference-based speckle filtering followed by
    anisotropic diffusion," Ultrasound in Medicine and Biology, vol. 38,
    no. 8, pp. 1414-1428, 2012.
```

Fig. 2. Example of help topics for a specific function.

3.3 Running Examples

Every function in BUSAT could be tested by running the example provided in the help topics. This could be performed by copying and pasting the example text on the Command Window. In the case of *image preprocessing* and *lesion segmentation* functions, both the original and the processed images are showed. For instance, images showed in Fig. 3 are displayed after running the example code in Fig. 2.

3.4 Special Considerations

Two special considerations should be taken into account:

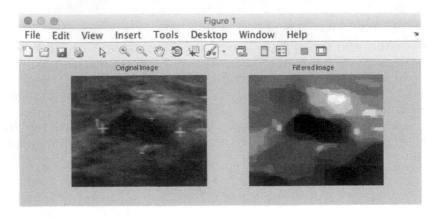

Fig. 3. Example of a BUS image despeckled by `isfad` function.

1. **C code functions:** despite BUSAT provides compiled C code functions (called *mex functions*) for Linux, Mac OS and Windows using 64-bits processors, in some operative systems they should be recompiled from the source codes by using the MATLAB `mex` function. These source codes are provided within the directory `Source_C_codes`.
2. **Parallel Computing Toolbox:** to speed-up the execution of the functions `autosegment`, `trainLSVM`, `trainSVM`, `trainRBF`, and `featselect`, the *parallel pool* is automatically open if the MATLAB *Parallel Computing Toolbox* is available, otherwise, the functions are sequentially executed.

4 Practical Examples

4.1 Building a CAD System

BUSAT is useful to quickly build a CAD system by following the pipeline in Fig. 4. Note that distinct functions of contrast enhancement, speckle filtering, lesion segmentation, feature extraction, and lesion classification could be combined to create a specific CAD system.

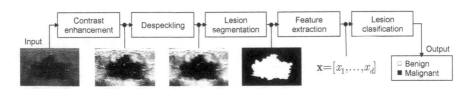

Fig. 4. Conventional pipeline of a CAD system for BUS images.

Herein, BUSAT is used to exemplify the implementation of a CAD system that uses five morphological features and linear classification [10]. The implemented CAD system uses the `wtsdsegment` function to segment the breast lesion. This function already considers the image preprocessing, where contrast enhancement is performed by sace function, whereas speckle filtering is performed by chmf function. Thereafter, the segmentation algorithm based on watershed transformation is applied to get the lesion contour [11]. Next, five morphological features are computed: elliptic-normalized skeleton, lesion orientation, number of substantial protuberances and depressions, depth-to-width ratio, and overlap ratio. Finally, classifyLDA function classifies the lesion in benign and malignant classes by using linear discriminant analysis (LDA). Obviously, the LDA classifier should be previously trained with the trainLDA function to create the prediction model. Then, the MATLAB program that implements the CAD system is written as follows:

```
% Read breast ultrasound image
I = imread('bus_image.tif');
% Load training data: normalization data and LDA model
load('cad.mat');
% Image preprocessing and lesion segmentation
BW = wtsdsegment(I);
% Compute five morphological features:
% 1. Elliptic normalized-skeleton
x1 = equivellipse(BW,'ens');
% 2. Lesion orientation
x2 = equivellipse(BW,'angle');
% 3. Number of substantial protuberances-depressions
x3 = nspd_li(BW,'nspd');
% 4. Depth-to-width ratio
x4 = geometric(BW,'dwr');
% 5. Overlap ratio
x5 = convhulldiff(BW,'or');
% Complete feature vector
x = [x1 x2 x3 x4 x5];
% Normalization: mean (mn) and standard deviation (sd)
xn = softmaxnorm(x,[mn;sd]);
% Lesion classification with LDA model
C = classifyLDA(xn,lda_model);
```

4.2   Evaluating a CAD System

When a CAD system is developed, it is necessary to evaluate its classification performance in terms of some indices such as accuracy, sensitivity, specificity, area under the ROC curve, etc.

Let $\mathcal{X} = \{\mathbf{x}_1, \ldots, \mathbf{x}_n\}$ be a feature space with n observations, where the ith observation is a d-dimensional feature vector denoted by $\mathbf{x}_i = [x_{i,1}, \ldots, x_{i,d}]$. Also, the observation $\mathbf{x}_i$ is associated to a class label $y_i \in \{1, 2\}$, where 1 and 2 denote benign and malignant lesions, respectively. Note that this kind of labeling is required by the training functions, although depending on the classifier the labels are adjusted. For instance, for the SVM classifier, the label $y = 1$ becomes $y = -1$ and the label $y = 2$ becomes $y = +1$.

Then, to perform CAD assessment, from the $\mathcal{X}$ set, training and test sets should be created, where the former is used to generate the prediction model and the latter is used to evaluate the classifier generalization. In addition, if the classifier requires hyperparameters, a grid-search scheme and k-fold cross validation method are automatically performed by the training functions to tune such parameters. For instance, the function `trainSVM` adjusts both the soft margin parameter C and the Gaussian kernel parameter γ, if they are not introduced in the input arguments of the function.

BUSAT contains the `classperf` function to evaluate the classification performance of a CAD system. Suppose that a user generates a feature matrix **X** of size $n \times d$ and a target vector **Y** of size $n \times 1$. Also, suppose that the CAD's classifier is based on SVM with Gaussian kernel. Then, the following MATLAB program implements the evaluation of a CAD system:

```
% Split feature space into training and test sets
ho = crossvalind('HoldOut',Y,0.2);  % 20% for test set
% Normalize training set
[Xtr,mn,sd] = softmaxnorm(X(ho,:));
% Normalize test set
Xtt = softmaxnorm(X(not(ho),:),[mn;sd]);
% Get training targets
Ytr = Y(ho,:);
% Get test targets
Ytt = Y(not(ho),:);
% Train SVM prediction model
svm_model = trainSVM(Xtr,Ytr);
% Classify test set with the SVM prediction model
C = classifySVM(Xtt,svm_model);
% Evaluate classification performance
p = classperf(C.Labels,Ytt);
```

5   Experimental Results

BUSAT contains three classifiers for distinguishing between benign and malignant lesions: linear discriminant analysis (LDA), support vector machine (SVM) with Gaussian kernel, and radial basis function network (RBFN). These classifiers are evaluated within a CAD system to determine which method performs better in terms of the indices Matthews correlation coefficient (MCC), area under the ROC curve (AUC), accuracy (ACC), sensitivity (SEN), and specificity (SPE) [12].

The BUS dataset considered 1,128 cases from 659 female patients acquired during routine breast diagnostic procedures at the National Cancer Institute

(INCa) of Rio de Janeiro, Brazil. All the cases were histopathologically proven by biopsy, where 781 images presented benign lesions and 347 images had malignant tumors. The images were collected from three ultrasound scanners with linear transducer arrays with frequencies between 7.5 and 12 MHz: Logiq 7 (GE Medical System Inc.), Logiq 5 (GE Medical System Inc.), and Sonoline Sienna (Siemens).

The entire dataset was segmented by the wtsdsegment function. Next, 25 morphological and texture features were computed, which are summarized in Table 1. The feature space was randomly split in training (90%) and test (10%) sets, which were normalized by the softmaxnorm function. Thereafter, LDA, SVM, and RBFN classifiers were trained by the functions trainLDA, trainSVM, and trainRBFN, respectively. It is worth mentioning that trainSVM and trainRBFN functions perform grid-search and k-fold cross validation method (with $k = 10$) to tune their parameters. In the case of the SVM, the C and γ parameters are adjusted, whereas for the RBFN, the number of hidden units is determined. Finally, the test set was classified by the functions classifyLDA, classifySVM, and classifyRBFN, and the classification performance of each classifier was evaluated by the classperf function. For statistical analysis, 50 independent runs of training-testing procedure was performed.

Table 1. Computed features for lesion classification. $\mathcal{M}$ and $\mathcal{T}$ denote morphological and texture features, respectively. Symbol # denotes number of features.

Type	Technique	Function	#
$\mathcal{M}$	Equivalent ellipse	equivellipse	5
$\mathcal{M}$	Fractal analysis	fractalcontour	2
$\mathcal{M}$	Geometry	geometric	1
$\mathcal{M}$	Anfractuosity	margclass	1
$\mathcal{M}$	Signature	nrl	1
$\mathcal{T}$	Autocorrelation	autocorr	1
$\mathcal{T}$	Gray-level average	avmass	2
$\mathcal{T}$	Gray-level co-occurrence matrix	glcm	9
$\mathcal{T}$	Laws' energy measures	lawsenergy	2
$\mathcal{T}$	Posterior acoustic behavior	pab	1

Table 2 summarizes the classification performance results obtained by the three evaluated classifiers. Besides, Table 3 shows the one-way analysis of variance (ANOVA) results to test whether the mean values between compared classifiers are different at $\alpha = 0.05$. Also, the Scheffe's method determines if there is statistical significance between two classifiers.

It is notable that the three classifiers did not present statistical differences in terms of MCC and AUC indices, that is, they are capable of distinguishing adequately between benign and malignant cases. However, the SVM classifier

Table 2. Classification performance results (mean ± standard deviation).

Classifier	MCC	AUC	ACC	SEN	SPE
LDA	0.75 ± 0.05	0.89 ± 0.02	0.88 ± 0.02	0.85 ± 0.05	0.90 ± 0.03
SVM	0.76 ± 0.05	0.89 ± 0.02	0.89 ± 0.02	0.90 ± 0.04	0.89 ± 0.03
RBFN	0.76 ± 0.06	0.90 ± 0.02	0.86 ± 0.03	0.79 ± 0.06	0.94 ± 0.02

Table 3. p-values of the statistical comparison between classifiers. Symbol $(-)$ denotes that groups are not statistically significant different (i.e., $p > 0.05$), contrarily symbol $(+)$ indicates that groups are statistically significant different (i.e., $p < 0.05$).

Comparison	MCC	AUC	ACC	SEN	SPE
LDA vs. SVM	$0.28^{(-)}$	$0.67^{(-)}$	$0.03^{(+)}$	$0.00^{(+)}$	$0.09^{(-)}$
LDA vs. RBFN	$0.70^{(-)}$	$0.20^{(-)}$	$0.09^{(-)}$	$0.00^{(+)}$	$0.00^{(+)}$
SVM vs. RBFN	$0.75^{(-)}$	$0.67^{(-)}$	$0.00^{(+)}$	$0.00^{(+)}$	$0.00^{(+)}$

outperformed its counterparts in terms of sensitivity (SEN = 0.90) and accuracy (ACC = 0.89), whereas the RBFN classifier obtained the best results in terms of specificity (SPE = 0.94). These results pointed out that the SVM classifier is adequate to be implemented within a CAD system for BUS images.

6   Conclusions

This paper presented the Breast Ultrasound Analysis Toolbox (BUSAT) for MATLAB, which contains several approaches proposed in literature to perform image preprocessing (contrast enhancement and speckle filtering), lesion segmentation (semi-automatic and fully-automatic methods), feature extraction (morphological, texture, and BI-RADS lexicon), and classification (linear and non-linear classifiers).

We presented the experimental results of the evaluation of three classifiers (LDA, SVM, and RBFN) to distinguish between benign and malignant cases, where SVM presented an adequate classification performance. Obviously, the configuration of the CAD system could lead to different classification results, that is, the image preprocessing techniques, the segmentation method, and the computed features impact on the lesion classification. Thus, the potential of BUSAT is the versatility to build and evaluate different configurations of CAD systems in reduced time.

To the best of our knowledge, BUSAT is the first toolbox intended to provide to the research community an easy and quick way to codify programs for computer-aided diagnosis for breast ultrasound. In addition, because the source codes are available to the users, it is possible to modify the functions in order to enhance the implemented methods or reuse code in new functions. Feature work considers to increase the number of implemented methods, for instance, new multiclass classifiers for BI-RADS categorization.

References

1. Ferlay, J., Soerjomataram, I., Dikshit, R., Eser, S., Mathers, C., Rebelo, M., Parkin, D., Forman, D., Bray, F.: Cancer incidence and mortality worldwide: sources, methods and major patterns in globocan 2012. Int. J. Cancer **136**(5), E359–E386 (2015)
2. Kelly, K.M., Dean, J., Comulada, W.S., Lee, S.J.: Breast cancer detection using automated whole breast ultrasound and mammography in radiographically dense breasts. Eur. Radiol. **20**(3), 734–742 (2010)
3. Stavros, A.T., Thickman, D., Rapp, C.L., Dennis, M.A., Parker, S.H., Sisney, G.A.: Solid breast nodules: use of sonography to distinguish between benign and malignant lesions. Radiology **196**(1), 123–134 (1995)
4. Cheng, H.D., Shan, J., Ju, W., Guo, Y., Zhang, L.: Automated breast cancer detection and classification using ultrasound images: a survey. Pattern Recogn. **43**, 299–317 (2010)
5. Drukker, K., Gruszauskas, N.P., Sennett, C.A., Giger, M.L.: Breast us computer-aided diagnosis workstation: performance with a large clinical diagnostic population. Radiology **248**(2), 392–397 (2008)
6. Huang, Q., Luo, Y., Zhang, Q.: Breast ultrasound image segmentation: a survey. Int. J. Comput. Assist. Radiol. Surg. **12**(3), 493–507 (2017)
7. MathWorks: Matlab. the language of technical computing
8. Chang, C.C., Lin, C.J.: LIBSVM: a library for support vector machines. ACM Trans. Intell. Syst. Technol. (TIST) **2**(3), 27 (2011)
9. Peng, H., Long, F., Ding, C.: Feature selection based on mutual information: criteria of max-dependency, max-relevance, and min-redundancy. IEEE Trans. Pattern Anal. Mach. Intell. **27**(8), 1226–1238 (2005)
10. Gómez, W., Pereira, W.C.A., Infantosi, A.: Improving classification performance of breast lesions on ultrasonography. Pattern Recogn. **48**(4), 1125–1136 (2015)
11. Gómez, W., Leija, L., Alvarenga, A.V., Infantosi, A.F.C., Pereira, W.C.A.: Computerized lesion segmentation of breast ultrasound based on marker-controlled watershed transformation. Med. Phys. **37**(1), 82–95 (2010)
12. Sokolova, M., Lapalme, G.: A systematic analysis of performance measures for classification tasks. Inf. Process. Manage. **45**(4), 427–437 (2009)

Fiber Defect Detection of Inhomogeneous Voluminous Textiles

Dirk Siegmund$^{(\boxtimes)}$, Timotheos Samartzidis, Biying Fu, Andreas Braun,
and Arjan Kuijper

Fraunhofer Institute for Computer Graphics Research (IGD),
Fraunhoferstrasse 5, 64283 Darmstadt, Germany
{dirk.siegmund,timotheos.samartzidis,biying.fu,
arjan.kuijper,andreas.braun}@igd.fraunhofer.de

Abstract. Quality assurance of dry cleaned industrial textiles is still a mostly manually operated task. In this paper, we present how computer vision and machine learning can be used for the purpose of automating defect detection in this application. Most existing systems require textiles to be spread flat, in order to detect defects. In contrast, we present a novel classification method that can be used when textiles are in inhomogeneous, voluminous shape. Normalization and classification methods are combined in a decision-tree model, in order to detect different kinds of textile defects. We evaluate the performance of our system in real-world settings with images of piles of textiles, taken using stereo vision. Our results show, that our novel classification method using key point pre-selection and convolutional neural networks outperform competitive methods in classification accuracy.

1 Introduction

In recent years environmental awareness and need for cost reduction has increasingly influenced the use of reusable industrial-textiles. Nowadays more than one billion cleaning textiles in Europe are being leased and reused per year. Besides the big volume of processed pieces, quality assurance of used industrial textiles has remained a mostly manually operated task. Compared to humans automated systems can have several advantages, such as lower costs, higher reliability and accuracy. Quality assurance after dry cleaning is one of the most cost intensive operating processes. Lowering its costs will lead to an overall cost reduction and therefore may encourage more customers to start using reusable industrial textiles. With increasing performance of artificial intelligence, automatic fabric defect detection has become one of the most relevant areas in this domain. So far, recent work in the field of textile inspection deals mostly with continuous 2D textures. This is because fabric inspection algorithms are mostly used during furling in the production process. Compared to that we present a solution intended to be used by the cleaning industry, that handles textiles individually in an assembly-line work flow. Due to high flow rates of textiles, an automatized mechanism to spread out textiles mechanically while in movement has not yet

© Springer International Publishing AG 2017
J.A. Carrasco-Ochoa et al. (Eds.): MCPR 2017, LNCS 10267, pp. 278–287, 2017.
DOI: 10.1007/978-3-319-59226-8_27

been invented. In this paper we focus on an inspection of defects in a pile-like arrangement, where every item is still dealt with separately, on an assembly-line. The uneven surface, varying colors of sewing pattern and weaving of different textile fibers are some of the challenges in this task. Textiles furthermore differ in the composition of fibers which include cotton, linen, polyester or compositions. Previous research on outspread fabrics achieve high recognition rates but are not resistant against some effects caused by voluminous shapes. Shape, folds, edges, borders, overlapping edges and ambient occlusion are some of these effects. They have a negative impact on correct detection of fiber-defects using known methods. The voluminous shape of textiles also results in loss of focus. The here presented method can therefore be seen as a baseline for fiber defects recognition on uniform textured textiles in voluminous shape. There are furthermore differences in fiber-defects as shown in Fig. 5a. Our database contains most of the defects as they are defined by the textile industry [1]. This includes: stains, bonding, silicon relics, holes, enclosures, dropped stitches, press-offs or others. After washing a textile multiple times, fibers may have changed color and appearance (see Fig. 5b). Several relevant steps like preprocessing and classification are shown in our inspection pipeline in Sect. 3. In view of addressing the mentioned problems of voluminous shape, we present a novel approach for fiber defect recognition in Sect. 4.1. In our experiments we implemented and evaluated this method in comparison with other known methods as described in Sect. 4.3. Our approach is invariant to different fiber-weavings and fulfills the requirements of compatibility with different uniform colored textiles. The effectiveness of our method is shown in experiments at Sect. 5.

2   Related Work

Web inspection is a common application of automatized textile defects inspection. It is mostly performed on spread fabrics, which are carried out during their manufacturing process. Most recent work focused on defect detection and classification. Mishra [2] distinguishes woven, knitted and dyeing/finishing defects which occur during spooning or weaving. Textiles can be categorized generally into uniform and different kinds of textured materials (uniform, random or patterned) [3]. For detection of defects on uniform textured fabrics, three defect-detection techniques exist: statistical, spectral and model-based [4]. Defect detection on (un-spread) textiles in voluminous shape is a relatively new field. Our work focuses on the inspection of textured material which has an almost homogeneous color and uses a combination of a statistical and model based approach.

Neural Networks (NN), AdaBoost [5] and Support Vector Machines (SVM) [6] are notable machine learning techniques that were used in a number of articles in this field. Some approaches on flat, and spread-out 2D surface achieve success rates in fabric defects detection higher than 90% [4,7]. Compared to that, humans achieve detection rates of only 60–75% [8]. Supervised learning strategies achieved good performance using a counter-propagation NN, trained by a resilient back-propagation algorithm [9]. As several NN suffer from a high sensibility regarding changes in orientation and lightning, we decided to use other machine learning techniques. In case of un-spread (inhomogeneous) textile classification, we proposed a system for classification of textile fibers using LBP-features and local-interest points in our recent work [10]. The process was evaluated by the use of preselected image patches in order to reduce computational costs in the textile classification using SVM and AdaBoost. The most time consuming processes when using these supervised machine learning methods are the acquisition and labeling processes. Thousands of patches have to be acquired and labeled manually. In contrast our novel method reduces the required effort to be spent in labeling of the data and combines it with convolutional neural network classification. Two of the most challenging problems in fabric classification are ambient occlusion and folds. These effects are caused by the shape of the textile and the influence of illumination. In our recent work [11] a normalization method was introduced that reduces these effects, paid by a loss of information. This method, based on stereo-imaging is used in our work for preprocessing of the acquired images.

3 Methodology

We present an inspection system pipeline (see Fig. 1) that classifies dry-washed textiles in pile-like arrangement into the classes 'fiber defect' and 'no defect'. The system is intended to be used in an assembly line like environment, where every item is served individually. It is built on a hierarchical decision tree model in which a first classifier determines stain defects and excludes them from further classification (see Fig. 1c. A second, in Sect. 4.1 presented classifier (see Fig. 1e) recognizes fiber defects and makes the final decision on whether an image shows a defect textile. The parts: Fig. (1a) Image Acquisition and Fig. (1b) Preprocessing follow the stereo-normalization approach presented in [11], in order to reduce the effect of shading in the captured images. A disparity map is used to exclude areas showing folds or shadows. Other areas classified as 'stain defect' are also excluded from the image as shown in step (see Fig. 1d).

3.1 Capturing Environment and Database

Our approach is evaluated on an image database of dry cleaned woven cotton cleaning textiles as they are used in many different industrial applications. We use a soft-box with homogeneous illumination in the image acquisition, to guarantee a controlled capturing process with even lighting. Two synchronized CMOS color cameras with a CMOS 1/1.8" sensor and a resolution of 1280×1003 pixels are used for image acquisition. The database contains 910 images of 258 different textiles with and without fiber defects (see Table 1). Fiber-defects are defects that can originate through the manufacturing/furling process (e.g. dropped stitches, press-offs or broken ends) or intensive stress. Most defects of that defect category were caused by intensive use of these textiles in industrial environments and show mostly holes and cut like defects (see Fig. 5).

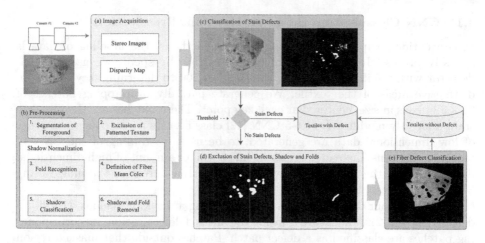

Fig. 1. Pipeline showing a combination of the proposed preprocessing steps (b) and (d) and the classification decision tree steps (c) and (e).

Because of diverse uses, the fibers also often show different levels of shading. Images are captured from a top-down perspective, therefore, some defects might be hidden (e.g. if they are in a fold or on the bottom side). Iterations in which the textile is physically reoriented and then classified could solve this problem.

Table 1. Quantities of textiles images.

Defect	Images	Depth Maps	Textiles
Fiber	310	155	98
Stain and fiber	300	150	88
None	300	150	72

4   Fiber Defect Recognition

In this Chapter we introduce a novel method for classification of fiber-defects in images of textiles in voluminous shape. As discussed in the state of the art, other methods perform well on fiber-defect recognition of outspread fabric. On textiles in inhomogeneous shape, most of these methods fail, as shadow and stain elements influence the classification negatively [11]. We propose a novel method using a combination of SURF key points and convolution neural network classification (see Sect. 4.1). This method requires images whereas areas of shadow and stain defects are normalized, as described in [11]. In Sect. 4.3, adoptions of competitive methods are presented, that have shown their effectiveness in similar classification tasks [10]. Furthermore an experiment, using an illumination normalization technique is presented.

4.1   CNN Classification Based on Keypoint Preselection

In conventional supervised machine learning methods, the labeling of data is a costly process. In order to reduce the effort to be spent, we use a SURF detector with a minimum Hessian threshold of 500 to determine key points on distinctive areas of the textile. We generate partially overlapping patches of 32×32 pixels in size, centered at each key point. These patches are then used to train a slightly modified LeNet-5 CNN [12] classifier. As a consequence patches of low dimensional data are generated, to be used in training of the CNN to recognize patterns. Instead of requiring an enormous amount of high dimensional data, we direct the network to key points of distinct areas of the image using SURF key points.

We labeled our database manually by defining a mask on regions with fiber-defects (see Fig. 2). If a feature is inside the masked defect-region, its corresponding patches are classified as a defect patch. Patches outside that masked region are classified as a non defect patch.

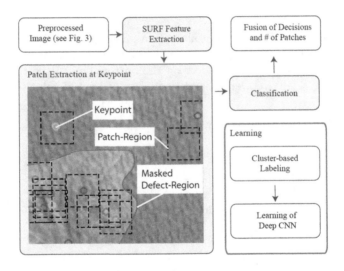

Fig. 2. Pipeline of micro-patch-based classification using CNN.

We use two convolution layers which are combined with a max-pooling layer (see Fig. 3 for a visualization of the customized network). After the inner-product layer we receive a fully connected layer to score our data. The loss layer is built using the soft-max function. To implement the network we use the caffe libraries from BVLC [13].

The so created database consists of approx. 58000 image patches and is divided in a training (80%) and testing set (20%). The data is separated and shuffled taking into account the individual images in such a way that no image is simultaneously present in the test and training set. It is trained with 1500 iterations and a batch size of 1000 features, which results in approx. 30 epochs. The

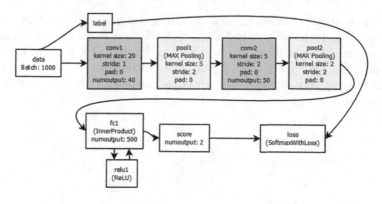

Fig. 3. Our adjusted LeNet-5 CNN.

classifier predicts whether a patch belongs to a region showing some fiber-defect or not. For each textile we receive as many predictions as there are key points.

4.2   Fusion

To make a final decision for each textile, we use the weighted sum combination rule to calculate a fused unified decision. It is based on two features, the first one is the number of key points detected. The second feature the difference between positive and negative decisions. We represent both values as scores and normalized them to a comparable range using min-max normalization, which can be formulated as:

$$S' = \frac{S - min\{S_k\}}{max\{S_k\} - min\{S_k\}} \qquad (1)$$

where $min\{S_k\}$ and $max\{S_k\}$ are the minimum and maximum values of existing scores in the data of the corresponding sources and S' is the normalized score. We used the weighted sum score fusion, where for each score source a weight is defined that indicates its relevance on the fused decision. The weight is calculated by 1-EER and fused by the weighted sum rule F for N score sources.

$$F = \sum_{k=1}^{N} w_k S_k, k = \{1, \dots, N\} \qquad (2)$$

4.3   Competitive Methods

Non-overlapping patches of 128×128 Pixel are used to reduce the complexity of the analyzed pattern. Two different sets of features were selected from related work and examined as fiber defect classification step (see Fig. 4). As in the normalization step only ambient occlusion and stains were removed from the input image, another illumination normalization method is applied to the image to enhance the contrast between fibers and exclude differences in shading. We used a technique introduced by Tan and Triggs [14] which is based on a series of steps to counter the effects of illumination variation, local shadowing and highlights. The preprocessing chain consists of: gamma correction, which uses differences of Gaussian filtering, masking and contrast equalization (see Fig. 4a for an example of a normalized image).

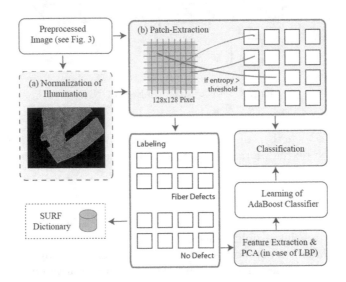

Fig. 4. Pipeline of patch-based classification methods.

The extracted patches contain all kind of different texture properties such as: seams, diverse edges and defects like cuts, open ends, holes, stains and others. During extraction, we used the Shannon-Entropy-Value to determine the amount of information in a patch. Our evaluation showed that patches with an entropy-value below 2000 do not contain enough information to be classified and are therefore rejected. Every patch is labeled manually by assigning them the class fiber-defect (see Fig. 5a) or none-defect (see Fig. 5b).

The local interest point descriptor SURF [15] has shown its effectiveness in many applications as local feature detectors and descriptors for non-rigid 3D objects [16]. They are scale-invariant and robust against rotation, translation and changing light conditions. A set of interest points is extracted into a 64-dimensional feature-vector, following a Bag of Words (BOW) approach [10].

Local binary patterns (LBP) features are a known technique, when dealing with textile fiber classification tasks [11]. The used LBP type [17] is invariant against rotation and pixel intensity variations and shows a relationship between a pixel and its neighborhood. It fulfills the requirements in regard to computational cost compared to other scale-invariant LBP approaches [18]. An evaluation on a subset of the database showed that a radius of 3 and a block size

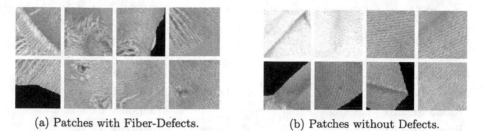

(a) Patches with Fiber-Defects.   (b) Patches without Defects.

Fig. 5. Examples of the different extracted features.

of 32 pixels is optimal for the given pictures. Histograms of rotation-invariant
binary patterns are calculated and concatenated to a feature vector. To reduce
the dimensionality of the feature vector, Principal Component Analysis (PCA),
trained on a subset of the data set, is performed. Experiments showed that a
reduction to 300 components gives best results.

An AdaBoost based machine-learning classifier is used for classification of the
extracted feature vectors, using the REAL boosting method with confidence-
rated predictions. Our evaluations showed that this classifier performed best
among different evaluated classifiers (SVM, Random Forrest, AdaBoost and
JRIP). All features are evaluated with- and without using the illumination normalization method.

5 Results and Discussion

We evaluated our proposed method shown in Sect. 4 on our novel database (see
Sect. 3) with different preconditions. The given 'Accuracy after Stain Filtering'
column describes the detection of fiber-defects on textiles of the defect categories: Fiber, Fiber and Stain and None (see Table 1). Patches showing other
defects were excluded from the database using the first (stain-defect) classifier
in the inspection process. The achieved accuracy (TP+TN/$\sum$Total population)
was calculated on the full-image-level, which concludes to the final decision on
whether a textile contains a fiber defect or not. In the calculation of that accuracy value the results of the first (stain-) detection classifier is not considered
(see Sect. 3.4) (Table 2).

The results using the SURF BOW approach in combination with the
AdaBoost classifier showed an accuracy of 87.99%. Though using TanTriggs illumination normalization, appeared to us to be promising, the results showed to
be less accurate than without using it over all tested methods. Rotation invariant unified Local Binary Patterns showed overall less accurate results than the
bag of words approach. We suspect the reason for the bad results to be the
rotational invariant variant which is less information conserving then the SURF
BOW approach. Our approach using a combination of key point pre-selection
and convolutional neural networks (CNN) classification achieved best results
on the database. We found that the defined CNN is especially suitable to this

Table 2. Achieved Accuracy Rates of Fiber-Defect Detection Methods.

Methodology	Patchsize	Pre-processing	Classifier	Accuracy after stain filtering
SURF BoW	128 × 128	–	AdaBoost	87.99 ± 2.2%
SURF BoW	128 × 128	TanTriggs	AdaBoost	77.81 ± 2.7%
RI unified LBP	128 × 128	–	AdaBoost	84.94 ± 3.7%
RI unified LBP	128 × 128	TanTriggs	AdaBoost	72,31 ± 2.7%
SURF+CNN	Full Image + 32 × 32	–	CNN	90.15 ± 1.4%

recognition task. One reason might be the combination of dominant structural characteristics from a number of extracted patches that fit well to the NN characteristics, and the pre-selection of distinct regions in the image. Based on the huge amount of found SURF feature points, we were able to generate a huge data set used to train a deeper network. The best total rejection accuracy was achieved by that method gaining an accuracy of 90.15%. All approaches were evaluated using 4-fold cross validation with a regular distribution of defect-classes in each fold.

6   Conclusion

We presented a novel method for detection of fiber defects in textiles, that can be used when textiles are in an inhomogeneous, voluminous shape. Well performing aspects of the established methods: SURF key points, LBP, AdaBoost and CNNs were combined for an evaluation of this novel computer vision application. Our database showed textiles with different kind of fiber-defects such as holes and cuts in a pile-like arrangement, recorded with a stereo vision camera setup. Best results were achieved by the novel method using key points and CNN, which outperformed other recent methods used in the classification of voluminous textile fibers [10]. We described and proved in Sect. 4.1 how CNNs in combination with SURF-features were combined to effectively recognize and classify distinctive features by using low dimensional data as an input. A brief description of the performance of all proposed methods is given in Sect. 5. The future work may include a new normalization of image areas showing ambient occlusion, which will improve the performance of our system.

References

1. Hong Kong Productivity Council: Textile Handbook 2000. The Hong Kong Cotton Spinners Association (2000)
2. Mishra, D.: A survey-defect detection and classification for fabric texture defects in textile industry. Int. J. Comput. Sci. Inf. Secur. **13**, 48 (2015)
3. Kumar, A.: Computer-vision-based fabric defect detection: a survey. IEEE Trans. Ind. Electron. **55**, 348–363 (2008)

4. Ngan, H.Y., Pang, G.K., Yung, N.H.: Automated fabric defect detectiona review. Image Vis. Comput. **29**, 442–458 (2011)
5. Borghese, N.A., Fomasi, M.: Automatic defect classification on a production line. Intell. Ind. Syst. **1**, 373–393 (2015)
6. Murino, V., Bicego, M., Rossi, I.A.: Statistical classification of raw textile defects. In: ICPR, pp. 311–314 (2004)
7. Rebhi, A., Benmhammed, I., Abid, S., Fnaiech, F.: Fabric defect detection using local homogeneity analysis and neural network. J. Photonics **2015**, 9 (2015)
8. Schicktanz, K.: Automatic fault detection possibilities on nonwoven fabrics. Melliand Textilberichte **74**, 294–295 (1993)
9. Islam, M.A., Akhter, S., Mursalin, T.E., Amin, M.A.: A suitable neural network to detect textile defects. In: King, I., Wang, J., Chan, L.-W., Wang, D.L. (eds.) ICONIP 2006. LNCS, vol. 4233, pp. 430–438. Springer, Heidelberg (2006). doi:10.1007/11893257_48
10. Siegmund, D., Kaehm, O., Handtke, D.: Rapid classification of textile fabrics arranged in piles. In: Proceedings of the 13th International Joint Conference on e-Business and Telecommunications, pp. 99–105 (2016)
11. Siegmund, D., Kuijper, A., Braun, A.: Stereo-image normalization of voluminous objects improves textile defect recognition. In: Bebis, G., et al. (eds.) ISVC 2016. LNCS, vol. 10072, pp. 181–192. Springer, Cham (2016). doi:10.1007/978-3-319-50835-1_17
12. LeCun, Y., Bottou, L., Bengio, Y., Haffner, P.: Gradient-based learning applied to document recognition. Proc. IEEE **86**, 2278–2324 (1998)
13. Jia, Y., Shelhamer, E., Donahue, J., Karayev, S., Long, J., Girshick, R., Guadarrama, S., Darrell, T.: Caffe: convolutional architecture for fast feature embedding. In: Proceedings of the 22nd ACM International Conference on Multimedia, pp. 675–678. ACM (2014)
14. Tan, X., Triggs, B.: Enhanced local texture feature sets for face recognition under difficult lighting conditions. IEEE Trans. Image Process. **19**, 1635–1650 (2010)
15. Bay, H., Tuytelaars, T., Gool, L.: SURF: speeded up robust features. In: Leonardis, A., Bischof, H., Pinz, A. (eds.) ECCV 2006. LNCS, vol. 3951, pp. 404–417. Springer, Heidelberg (2006). doi:10.1007/11744023_32
16. Zeng, K., Wu, N., Wang, L., Yen, K.K.: Local visual feature detection and description for non-rigid 3d objects. Adv. Image Video Process. **4**, 01 (2016)
17. Zhao, G., Pietikäinen, M.: Improving rotation invariance of the volume local binary pattern operator. In: MVA, pp. 327–330 (2007)
18. Li, Z., Liu, G., Yang, Y., You, J.: Scale- and rotation-invariant local binary pattern using scale-adaptive texton and subuniform-based circular shift. IEEE Trans. Image Process. **21**, 2130–2140 (2012)

Language Proficiency Classification During Computer-Based Test with EEG Pattern Recognition Methods

Federico Cirett-Galán[1]([✉]) [iD], Raquel Torres-Peralta[1], and Carole R. Beal[2]

[1] Universidad de Sonora, Hermosillo, Mexico
{fcirett,rtorres}@industrial.uson.mx
[2] University of Florida, Gainesville, USA
crbeal@coe.ufl.edu

Abstract. The answering of any test represents a challenge for students; however, foreign students whose first language is not English have to deal with the difficulty of the understanding of a series of questions written on a different language in addition of the effort required to solve the problem. In this study, we recorded the behavior of the brain signals of 16 students, 10 whom first language was English and 6 who were English learners, and used two supervised classification algorithms in order to identify the students' language proficiency. The results shown that in both approaches, harder problems which required longer time to be responded had a higher accuracy rate; however, more tests are needed in order to understand the physical processing of written math text problem and the difference among both groups.

Keywords: Electroencephalography · Machine learning · Data mining · Pattern recognition · Intelligent Tutoring Systems · Physiology · Behavior

1 Introduction

Education systems require constant evaluation in the form of written texts. In any country, the solving of those represents a challenge. For foreign students, however, the difficulty of the understanding of a series of questions written on a different language requires intellectual work, in addition to the effort required to solve the problem itself, putting them in a position of disadvantage. The language proficiency of a student may be taken in to account when there are time restrictions to answer an exam, but until now there is no sufficient evidence to support this argument. Also, the use of online tutoring systems and tests make this practice a very common resource for professors and teachers all around the globe, creating the need of systems to measure the state of the student during online sessions in order to understand the limits and reaches of these tools, regardless of the language limitations and helping to design more efficient systems for a better education where all students have the same opportunities.

Considerable progress has been made over the last decade in integrating models generated from behaviors such as keyboard clicks and interaction latencies with real-time sensors indicating users' affective states. These models have produced significant

© Springer International Publishing AG 2017
J.A. Carrasco-Ochoa et al. (Eds.): MCPR 2017, LNCS 10267, pp. 288–296, 2017.
DOI: 10.1007/978-3-319-59226-8_28

advances in Intelligent Tutoring Systems (ITS) research, leading to more adaptive systems that should ultimately be able to intervene to optimize learning for individuals. However, when considering the typical classroom situation, feasibility of data collection becomes an important constraint to consider: Although more information about the user's state would clearly be better in terms of creating accurate student models, there is also a limit to the instrumentation that we can apply to the user, at least outside a laboratory situation.

In this study, we analyzed the behavior of the brain signals of 16 students, 10 whom first language was English and 6 who were English learners to find a difference in their cerebral activity in order to understand the physical processing of written text, previous to the solving of a math problem while using an online tutoring system.

2   Prior Work

Researchers in the ITS community have investigated several approaches to acquire information of the state of the learner. One of them is to use behaviors such as the time between clicks and rapid activation of instructional scaffolding (e.g., repeatedly clicking on the "help" button) to estimate whether the student is actually engaged with trying to solve a problem or is avoiding effort, perhaps by "gaming the system" by deliberately entering wrong answers in order to move on quickly [3]. Beal, Mitra and Cohen [4] used a Hidden Markov Model (HMM) to infer the level of engagement of the student to predict behavior in the next problem. The results showed that students had distinct trajectories of engagement, and that the HMM estimates were strongly related to independent estimates of individual students' mathematics motivation, based on students' self-report and reports provided by their teachers, and mathematics proficiency (grades, test scores). Johns and Woolf [12] also reported that an HMM provided good predictions about students' motivation while solving a series of problems in a math tutoring system, predicting a student correct response 72% of the time (versus 62% of the baseline) with a Dynamic Mixture Model based on Item Response Theory. Arroyo, Mehranian and Woolf [2] tested 600 students on an ITS for mathematics, estimating in real time the effort the student invested on solving each problem and using the results to choose the next problem with the same level or greater level of difficulty if the student is engaged, and an easier one otherwise. Students using the experimental effort-based selector scored significantly better than those that got problems served by a random problem selector (57% vs. 42% accuracy in post-tests).

Technology for capturing electroencephalography (EEG) signals has progressed considerably, to the point where the user can wear a lightweight recording unit that transmits data for analysis. The recording unit is sufficiently non-intrusive and it is being used in a variety of tasks that require sustained attention and cognitive effort, including long-haul truck driving, missile tracking and submarine systems control. Berka et al. [6] found that officers tracking missiles in a simulated environment Aegis Combat System had a high or extreme cognitive workload 25 to 30% of the time and achieved a detection efficiency of almost 100%. Education researchers have begun to use this type of device to track students' cognitive activity during problem solving.

In an effort to understand and improve the reading capabilities of English learners, Pegory and Boyle [15] suggest that those students must have different motivations for reading, since the effort they make is substantially higher than their classmates. Other studies focus on the differences between the experiences of English learners versus their schoolmates', describing the problem from a social perspective, mentioning the limitations in access to different resources, inequality issues [11], or emphasizing their needs [10].

In previous work [7], we found that English Language Learners using an Intelligent Tutoring System were less likely to answer correctly, had more incorrect answer attempts, took longer on each problem, and were more likely to use multimedia help features.

In this work, we try to explore and detect the patterns of the brain signals of both English learners and English primary students in order to recognize their differences. We compared the performance of two supervised machine learning algorithms (AdaBoost and Classification Tree C4.5) for classification when recognizing if a brain signal corresponded to an English Primary or an English Learner student.

3   Methodology

In this study, we invited a number of students (n = 16) to solve a series of easy and difficult math problems in an Intelligent Tutoring System (ITS). We asked them their primary language and we obtained a total of 10 English Primary (EP) speakers and 6 English Language Learners (ELL).

The verbal description of each problem selected for the experiment had different levels of complexity, since the composition of the text for each question varies in extension, and some of them may be more elaborated than the rest. It was necessary to differentiate these levels in order to test if the wording of the problem affected the ELL students' performance. We obtained the readability level of each question used in this study using the Collins-Thompson and Callan [9] method, using algorithms with automatic modeling and predicting of the reading difficulty of texts (See Table 1).

The EEG data was recorded as students solved a series of easy and difficult math problems. There were 16 participants in the study (8 males, 8 females). Participants were college students who were at least 18 years old and gave active written consent for participation. Each person participated in a 90-minute session, which included informed consent procedures, fitting the EEG headset, completing a 15-minute baseline calibration task, and solving eight multiple-choice math problems presented at the computer while wearing the EEG headset. Math problems were taken from a set of released SAT items; there were four easy problems and four hard problems, with difficulty level determined by information from the College Board. Each problem had four answer options. The items were presented to students within an online tutoring system that recorded the time on the problem (initial presentation on the screen to first answer selection) as well as the outcome (correct, incorrect answer chosen). Problems were presented in one of two sequences (easy, easy, hard, hard, easy, easy, hard, hard or hard, hard, easy, easy, hard, hard, easy, easy) across subjects.

Table 1. List of problems selected for the study, with their difficulty and reading difficulty level according to Collins et al.

Difficulty	Problem	Readability level	Problem description
Easy	Summation	2	Robert wants to know what is the sum of one plus two plus three plus four. What answer would you give him?
Easy	Bows	12	A piece of ribbon 4 yards long is used to make bows requiring 15 in. of ribbon for each. What is the maximum number of bows that can be made?
Hard	Triangle	6	A triangle has a perimeter 13. The two shorter sides have integer lengths equal to x and x + 1. Which of the following could be the length of the other side?
Hard	Fence	12	A straight fence is to be constructed from posts 6 in. wide and separated by lengths of chain 5 ft. long. If a certain fence begins and ends with a post, which of the following could not be the length of the fence in feet? (12 in. = 1 foot)
Easy	Village	4	In a certain village, m litres of water are required per household per month. At this rate, if there are n households in the village, how long (in months) will p litres of water last?
Easy	Classroom	12	In a class of 78 students 41 are taking French, 22 are taking German and 9 students are taking both French and German. How many students are not enrolled in either course?
Hard	Bus	8	Half the people on a bus get off at each stop after the first, and no one gets on after the first stop. If only one person gets off at stop number 7, how many people got on at the first stop?
Hard	Town	6	Amy has to visit towns B and C in any order. The roads connecting these towns with her home are shown on the diagram. How many different routes can she take starting from A and returning to A, going through both B and C (but not more than once through each) and not travelling any road twice on the same trip?

3.1 EEG Data Acquisition

The electroencephalogram (EEG) data were recorded from nine sensors integrated into a mesh cap covering the upper half of the head, along with two reference signals attached to the mastoid bones (behind the ears) and two sensors attached to the right clavicle and to the lowest left rib to record the heart rate (although the heart rate data were not used in the study). The location of each sensor was determined by the International 10–20 System [14] to ensure standardized reproduction of tests. This cap was equipped with a small wireless transmission unit. A small USB dongle received the wireless transmissions to a PC computer with Windows (XP/Vista/7) 32-bit operating system.

Each second, 256 EEG signals were transmitted and converted to Theta, Alpha, Beta and Sigma wave signals (ranging from 3 Hz to 40 Hz). These signals were processed by Advanced Brain Monitoring proprietary software from B-Alert to produce classifications of mental states, meaning the probability that the participant was in a particular state in epochs of one second. States included Engagement, Distraction, Drowsiness and Cognitive Workload [5]. Engagement includes estimates of cognitive activities such as information gathering, visual scanning and sustained attention, and Workload is a measure of effortful cognitive activity [16]. The Engagement and Workload data were selected for our analyses because levels of Drowsiness were almost non-existent in the present study, and Distraction is essentially the inverse of Engagement.

3.2   Data Processing and Classification

Of a total of 16 participants, 15 completed all eight problems and one completed seven of the eight items. The data set thus consisted of 127 completed math problems with its corresponding Engagement & Workload data. We refer to these as Engagement signal and Workload signal.

The signals were processed by converting each signal into one of three equal-sized bins, with limits set at 0.333 and 0.666 of the Cumulative Distribution Function. The count of signals below 0.333 were considered as the Low State, values between 0.333 and 0.666 as the Medium state, and those between 0.666 and 1.0 as the High state. By doing this, we assure a normalization between all the participants.

As a result, Table 2 shows that when measuring the Workload data, the participants solved problems in a High state around 65% of the time for both type of problems, easy and hard, in the Medium state around 24% of the time and in the Low state 11%, while processing the Engagement data shows that the predominant states where High and Low (both around 40% of the time) for the easy and hard problems.

Table 2. Percentage of time in three levels (low, medium and high) for Engagement and Workload states while solving math problems in the ITS. Student Workload inclines toward the High state in both Easy and Hard problems, while Engagement has Low and High states.

Problem type	Engagement			Workload		
	Low	Medium	High	Low	Medium	High
Easy	38%	16%	46%	11%	24%	65%
Hard	40%	18%	42%	10%	23%	67%

Using the language proficiency of the students, a table of average states was generated, comparing English Primary (EP) students versus English Language Learner (ELL) students. In this table it shows that ELL students spent more time in a high state of Engagement (visual acquisition) than the English primary students (Easy: 51% vs 44%, Hard: 46% vs 40%) and the EP students state averages for Workload show that they were in a High state (Easy: ELL 48% vs EP 74%, Hard: ELL 50% vs. EP 77%), as it can be seen in Table 3.

Table 3. Percentage of time in three levels (low, medium and high) for Engagement and Workload states while solving math problems in the ITS, for English Language Learners (ELL) and English Primary (EP) speakers. The state of the Workload signal is higher for English Primary (EP) speakers in both Easy and Hard problems.

Student	Easy problems						Hard problems					
	Engagement			Workload			Engagement			Workload		
	Low	Med	High	Low	Med	High	Low	Med	High	Low	Med	High
ELL	37%	12%	51%	15%	37%	48%	37%	17%	46%	15%	35%	50%
EP	38%	18%	44%	10%	16%	74%	42%	18%	40%	8%	15%	77%

In previous work [8] we used a transition probabilities table of the low, medium and high states of both signals for each problem solved, producing vectors of 9 features and a class tag (outcome of the problems). The results showed that it was possible to predict if the student was going to choose either a right or wrong answer with 83% of accuracy on easy problems. On this study, we use a similar approach, but instead of predicting the outcome, we will try to predict if the student was an English Learner or not. We used language proficiency (English Language Learner ELL or English Primary) to construct 8 files (one for each problem) to serve as training sets for to two different machine learning algorithms (AdaBoost and Classification Tree C4.5).

As the signals of Engagement and Workload are of different length for each user and problem, it is necessary to process them to represent them in a more manageable way.

We reduced the Workload signals into vectors of equal size with Piecewise Aggregate Approximation (PAA) [13], to use these vectors as features to a classifier. The signals are divided in N bins and the mean of each segment is calculated and stored in its bin, thus, describing the signal with a reduced representation. A file is generated for each problem, one user per register (16 users total) with language proficiency (ELL or

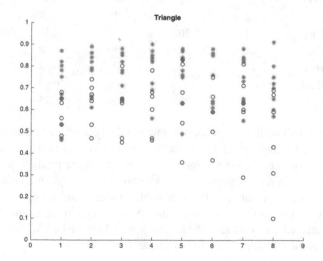

Fig. 1. Workload signal for problem 'Triangle' reduced to 8 bins using PAA. English Primary speakers (*) are mostly on the top of the chart.

EP) as the class, and the means as the features. A plot for the Workload signal for the problem 'Triangle' can be seen in Fig. 1.

The processed signals for each problem (8 for each 16 participants) are used as input to an AdaBoost classifier and a Classification Tree (C4.5) with random sampling, setting aside 90% of the records for training and 10% for testing. The average of 20 train/tests sessions is reported.

4   Results

The readability level of the word math problems appears to have no impact in the correct classification of the type of student, as well the average response time (See Table 4). The only factor that seems to affect the classification (using a Classifying Tree) is the difficulty type. Problems marked as 'Hard' had a better rating than the 'Easy' ones. 'Bus' had its students accurately classified 94% of the time, and 'Towns' and 'Triangle' show a 72% classification accuracy. Using AdaBoost with the SAMME algorithm and 50 estimators does not produce better results. Only 'Bus' and 'Triangle' show a classification accuracy better than 75%. 'Summation' and 'Village', being 'Easy' problems, classify no better than chance.

Table 4.   The classification accuracy for problems of type 'hard' is better than chance in 3 of the 4 cases with the Classification Tree algorithm.

Problem	Type	Readability level	Average response time (seconds)		Classification accuracy (random sampling)	
			English Primary	English Learner	AdaBoost	Classification Tree
Bows	Easy	12	105	147	0.67	0.84
Classroom	Easy	12	123	149	0.49	0.50
Summation	Easy	2	46	45	0.58	0.58
Village	Easy	4	202	126	0.46	0.42
Bus	Hard	8	115	130	0.78	0.94
Fence	Hard	12	270	206	0.54	0.45
Towns	Hard	6	236	189	0.69	0.72
Triangle	Hard	6	282	260	0.75	0.72

Results show that neither the response time nor the type of the problems are the key to distinguish the language proficiency for this set of users. Even when there are some problems where the accuracy is as high as 94%, in others, the prediction is as good as chance. However, when the complexity of the problem was high and the response time was larger, the accuracy was mostly acceptable. There could be different aspects to analyze in more depth, as the time the volunteers have been in the United States and at what age they arrived, as their use of the language on a daily basis. This is a first approach that has given some new directions to our research.

5   Conclusions

The average state of Engagement and Workload of English Primary speakers and English Language Learners shows that a difference between them exist and can be found under certain circumstances, where the complexity of the problem demands a higher workload for a longer period of time.

In this study, English learners were born in different countries, and all of them had the opportunity of study at a university in the United States. The time they spent in the country before the experiment was conducted is unknown, and probably that missing piece of information was relevant to the interpretation of the results. In a future experiment, we will test only foreign students who, even when they understand the language, are not used to speak English on a daily basis. Also, we will contemplate the incorporation of more participants as well including more challenging math problems.

Acknowledgments. The research was supported by National Science Foundation HRD 0903441. We would like to thank the staff at Advanced Brain Monitoring for their support, as well as the students who participated in the research.

References

1. Arroyo, I., Cooper, D.G., Burleson, W., Woolf, B.P., Muldner, K., Christopherson, R.: Emotion sensors go to school. In: Proceeding of the 2009 Conference on Artificial Intelligence in Education, pp. 17–24. IOS Press (2009)
2. Arroyo, I., Mehranian, H., Woolf, B.P.: Effort-based tutoring: an empirical approach to intelligent tutoring. In: Baker, R.S.J.d., Merceron, A., Pavlik, Jr., P.I. (eds.) Proceedings of the 3rd International Conference on Educational Data Mining (2010)
3. Baker, R.S.J.d., Walonoski, J., Heffernan, N., Roll, I., Corbett, A., Koedinger, K.: Why students engage in "gaming the system" behavior in interactive learning environments. J. Interact. Learn. Res. **19**, 185–224 (2008)
4. Beal, C., Mitra, S., Cohen, P.R.: Modeling learning patterns of students with a tutoring system using hidden markov models. In: Proceeding of the 2007 Conference on Artificial Intelligence in Education, pp. 238–245. IOS Press (2007)
5. Berka, C., Levendowski, D.J., Lumicao, M.N., Yau, A., Davis, G., Zivkovic, V.T., Olmstead, R.E., Tremoulet, P.D., Craven, P.L.: EEG correlates of task engagement and mental workload in vigilance, learning, and memory tasks. Aviat. Space Environ. Med. **78**(5 Suppl), B231–B244 (2007)
6. Berka, C., Levendowski, D.J., Ramsey, C.K., Davis, G., Lumicao, M.N., Stanney, K., Reeves, L., Regli, S.H., Tremoulet, P.D., Stibler, K.: Evaluation of an EEG workload model in an aegis simulation environment. In: Caldwell, J.A., Wesensten, N.J. (eds.) Society of Photo-Optical Instrumentation Engineers (SPIE) Conference Series, vol. 5797, pp. 90–99 (2005)
7. Cirett, F., Beal, C.R.: Problem solving by English learners and English primary students in an algebra readiness ITS. In: Guesgen, H., Murray, R.C. (eds.) Proceedings of the Twenty-Third International Florida Artificial Intelligence Research Society Conference, pp. 492–497. Flairs (Florida Artificial Intelligence Research Society), AIII Publications (2010). http://www.aaai.org/ocs/index.php/FLAIRS/2010/paper/view/1250

8. Cirett Galán, F., Beal, Carole R.: EEG estimates of engagement and cognitive workload predict math problem solving outcomes. In: Masthoff, J., Mobasher, B., Desmarais, Michel C., Nkambou, R. (eds.) UMAP 2012. LNCS, vol. 7379, pp. 51–62. Springer, Heidelberg (2012). doi: 10.1007/978-3-642-31454-4_5

9. Collins-Thompson, K., Callan, J.: Predicting reading difficulty with statistical language models. J. Am. Soc. Inf. Sci. Technol. **56**(13), 1448–1462 (2005). doi:10.1002/asi.20243, ISSN 1532-2882

10. Gándara, P., Maxwell-Jolly, J., Driscoll, A.: Listening to teachers of English language learners: a survey of california teachers' challenges, experiences, and professional development needs. Policy Analysis for California Education, PACE (NJ1) (2005)

11. Gandara, P., Rumberger, R., Maxwell-Jolly, J., Callahan, R.: English learners in California Schools: unequal resources, unequal outcomes. Educ. Policy Anal. Arch. **11**(36), 1–54 (2003)

12. Johns, J., Woolf, B.P.: A dynamic mixture model to predict student motivation and proficiency. In: Proceedings of the AAAI. IOS Press, Boston (2006)

13. Keogh, E., Chakrabarti, K., Pazzani, M., Mehrotra, S.: Dimensionality reduction for fast similarity search in large time series databases. Knowl. Inf. Syst. **3**(3), 263–286 (2001)

14. Niedermeyer, E.F.H., et al.: Electroencephalography Basic Principles, Clinical Applications, and Related Fields, 5th edn. Lippincott Williams and Wilkins, Philadelphia (2005)

15. Peregoy, S.F., Boyle, O.F.: English learners reading English: what we know, what we need to know. Theory Pract. **39**(4), 237–247 (2000)

16. Poythress, M., et al.: Correlation between expected workload and EEG indices of cognitive workload and task engagement. In: Augmented Cognition: Past, Present and Future, pp. 32–44 (2006)

Visual Remote Monitoring and Control System for Rod Braking on Hot Rolling Mills

Oleg Starostenko[1](✉), Irina G. Trygub[2], Claudia Cruz-Perez[1],
Vicente Alarcon-Aquino[1], and Oleg E. Potap[2]

[1] Department of Computing, Electronics and Mechatronics, Universidad de las Américas Puebla,
72820 Cholula, Mexico
{oleg.starostenko,claudia.cruzpz,vicente.alarcon}@udlap.mx

[2] Department Automation of Production Processes National Metallurgical Academy of Ukraine,
Dnipro, Ukraine
nmetau.trygub@gmail.com, potapoe@mail.ru

Abstract. In steel production the finishing process on hot rolling mill includes a set of essential operations managed by complex control mechanical, electrical and hydraulic equipment. However, accuracy of mill automating mechanisms and sensors is still low due to hot hostile environment with strong vibration and shock. The proposed solution is a computer vision application that exploits morphological filtering and discontinuity masks for detection and separation of rods on rolling mill and provides fast recognition and tracking rod front ends during their deceleration on cooler. The proposed algorithm has been implemented and evaluated in real time conditions achieving precision of rod front end recognition in range of 90–98% on artificial and daylight illumination, respectively.

Keywords: Computer vision · Morphological filtering · Pattern recognition and tracking · Steel manufacturing on rolling mill

1 Introduction

In area of steel manufacturing the control of processing hot metal sheets, strips and rods is not a trivial task due to nonlinearities and complexities of hot metal rolling within extremely hostile environment that frequently prevents the appropriate location of sensors to measure process features in real-time [1, 2]. Particularly, a forced braking process of rolled rods on the cooler is important and challenging task of automatic control in modern continuous small-section mills. The efficient solutions of management of steel production provide reducing the amount of scrap and increase a quality of rods with small cross-section profiles of diameter from 10 to 16 mm [3, 4].

Recently, the considerable efforts of researchers are addressed to develop high speed, accurate and low cost control equipment for metal processing on mills by using mechanical, electrical, hydraulic tool, smart optical, laser or X-ray sensors [5–8] However, the problems of accuracy of automating mechanisms used for stabilization of valve diverter of hot metal rods on cooler and control of their forced braking remain unsolved and require further research. The main technological factors that introduce disturbance to

© Springer International Publishing AG 2017
J.A. Carrasco-Ochoa et al. (Eds.): MCPR 2017, LNCS 10267, pp. 297–307, 2017.
DOI: 10.1007/978-3-319-59226-8_29

braking rods during their continuous rolling on mill are fluctuations of transfer velocity and variable feedback time delay of control tool for dynamic deceleration. Moreover, the velocity of metallic straight rods is affected by changes of their geometric dimensions, temperature, tension and strain [2, 6, 9]. So, the most difficult problem of automatic control is to provide such braking, when rod front ends after complete stoppage on cooler would be aligned in one plane (Fig. 1a).

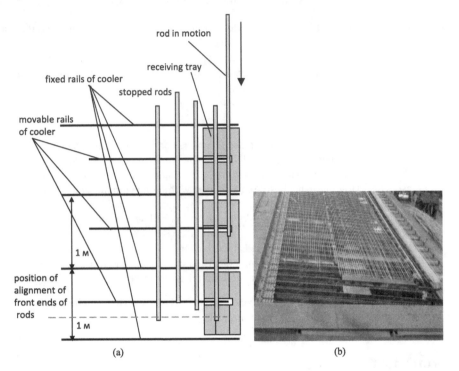

(a)   (b)

Fig. 1. (a) Arrangement of rods on the cooler (fragment, top view); (b) Rods on cooler of light-section mill 250-1 PJSC ArcelorMittal.

The principal goal of aligning the front ends of rods after their stoppage on cooler is significant reducing metal waste trimmings after cutting-to-length packages of rods on stationary shears. The aligning front ends of the rods in the package cannot be provided using fixed detent due to low bending stiffness of small rolled profiles.

In well-known mill machinery management systems the effective control action on the value of braking distance without interfering with the process of separation of adjacent rods on valve diverter is a forced deceleration generated by brake electromagnets placed in the receiving tray in the straightening trench of cooler [2, 7–9]. That also allows less severe thermal regime of the electric coils of brake electromagnets and does not require electric power supply of electromagnets to be mounted on the moving parts of mil [1, 3, 6]. However, disadvantages of recent implementations of mentioned control are the difficulty of sensor location near high temperature mill (up to 400°C), sensor unstable functionality and distortions of acquired from them information in environments with strong vibration and shock. The proposed solution of this problem exploits visual remote diagnostic, monitoring and control of rod braking process on hot mill based on specific methods of digital image processing and pattern recognition.

This paper is organized as it follows. In the Sect. 2 an overview of recent trends and new advances in computer vision is presented and some appropriate image processing approaches are selected. In the Sect. 3 the proposed algorithm for monitoring rod deceleration on hot rolling mill is introduced. Finally, the discussion of results obtained in tests and evaluation of the proposed approach are presented.

2   Image Processing Approaches for Industrial Applications

The main idea of alternative approach for controlling parameters of rod braking process on hot rolling mill consists in using methods of video processing acquired from remote vision systems. Recently, industrial use of optical measuring systems is common practice for monitoring production lines for steel manufacturing on rolling mills and cut-to-length lines [7, 8]. For solution of the particular task of managing technological operations on light-section mill 250-1 PJSC ArcelorMittal [10], the following technical requirements have to be taken into account. Hot metal rods arrive to cooler receiving tray in pairs with the speed about 0.1–0.2 m/s. The distance to provide forced braking rolled rods on cooler with deceleration 2.5–3.5 m/s^2 is less than 2 m. The accuracy of aligning front end of stopped rods must be less than 0.1 m. Stopped and aligned rods on the cooler are moved in the transverse direction for later formation of packs of 30–40 rods as it is shown in Fig. 1b).

The solution must solve: (1) real-time tracing thin rods in video by detection of their front ends; (2) generating continuous action signals for magnetic brakes on base of computing frame-by-frame deceleration; and (3) providing aligning rods by their standstill in one plane. In this paper the solution of 1st problem is presented.

There are some well-known techniques for estimation of motion characteristics such as displacement vector, velocity, acceleration, motion oscillation, errors of prediction, etc. This is a known practice to use optical flow or motion field for describing position of moving objects in images. The principal disadvantage of these approaches is the computational cost that makes difficult their utilization in real time applications [11, 12]. Another technique for quantitative estimation of motion characteristics is based on the block correspondence, when the same pattern is used within consecutive frames as reference. It allows overcoming a problem of progressive increment of compared patterns but aggregates accumulative error proportional to time function [13].

The most widely used approaches for pattern recognition may be subdivided in knowledge based, feature invariant, template matching and appearance-based methods. The methods of the 1st group may be used for detection and tracking of rods in motions because they consider a region of interest as whole entity and try to localize its core components. However, they provide high accuracy and speed only under controlled conditions in structured environment with well-established formalism for knowledge representation and reasoning [14].

The 2nd group of techniques uses local features of single pixel or small region (color, gradient, gray value, dimensionality, texture, etc.) invariant to variation of illumination, noise, scale, relative position, orientation and changes in viewing direction providing robust pattern recognition under different conditions. They are known as scale invariant feature transform (SIFT), difference of Gaussian points (DoG), entropy based salient region detector (EBSR), corner detectors, intensity and edge based regions, mathematical morphology and gradient based approaches. In is evident that for detection and tracking rods in motion this group of method is well fitted particularly, when recognition of patterns is not necessary as well [15, 16].

From other hand the template matching techniques involve evaluation of correlation of predefined template to test image. Unfortunately, for these approaches some factors like tolerance to deformation, robustness against noise and feasibility of template matching during image distortion must be taken into account, because they are sensitive to occlusions and variations within template [17, 18].

Finally, there are several prospect local or global appearance-based object recognition methods. The local approaches are used to process regions of interest characterized by corners, edges, shapes while the global ones use color, entropy, gradient moments and sometimes semantics. Particularly, the global approaches transform whole input image onto suitable lower dimensional data set that does not require complex time consuming processing. Unfortunately, the encoding size of dimensional data sets are enlarged considerably as well as used iterative training and evaluation processes limit their use in real-time applications [14, 19].

3   Proposed Algorithm for Detection and Tracking Rods on Mill

Due to specific technical requirements, the principal objective of this paper is to develop high-speed image processing algorithm for detection and feature extraction of rods in motion in images with variable illumination and presence of noise obtained from standard video camera with resolution no more than 1920×1080. The following proposed algorithm provides detection front-end of rods during their deceleration.

Input: RGB image I_{input} with resolution 1920×1080

Output: Image with separated rods and with their front-end position

1: $I_{600 \times 400} \leftarrow Resize_{600 \times 400}(I_{input})$

2: $I_{gray} \leftarrow Gray(I_{600 \times 400})$

3: $I_{sharpen} \leftarrow Sharpen(I_{gray})$

4: $I_{bin} \leftarrow Threshold(I_{sharpen})$

5: $I_{morpho} \leftarrow Morphology(I_{bin})$

6: $I_{discon} \leftarrow Discontinuity(I_{bin})$

7: $I_{separation\_mask} \leftarrow Discontinuity\_Separation\_Mask(I_{discon})$

8: $I_{difference} \leftarrow Differnce(I_{morpho}, I_{separation\_mask})$ for each blob in image

9: $I_{frontend\_mask} \leftarrow Discontinuity\_FrontEnd\_Mask(I_{difference})$

Return: $I_{frontend\_difference} \leftarrow Difference(I_{morpho}, I_{frontend\_mask})$

An original video must be adjusted to eliminate camera displacement if there is not fixed position of image capture. For this task a video stabilization algorithm was used, eliminating the black boxes generated by the stabilization algorithm reducing frame resolution to 600×400 pixels (Fig. 2a). The image frames $I_{600 \times 400}$ are converted to gray scale applying then sharpening filter producing image $I_{sharpen}$ shown in Fig. 2b).

(a)   (b)

Fig. 2. (a) Original image $I_{600 \times 400}$ with reduced resolution of 600×400 pixels; (b) Gray scale image $I_{sharpen}$ after applying sharpening filter.

After multiple experiments and due to very good image contrast of rods in motion over cooler tray (well separated objects in bimodal histogram of the scene), the binarization with empirically defined threshold is applied for generation a binary image I_{bin} shown in Fig. 3a). Some morphological filters are applied with the proposal to reduce the noise of image and separate connected rods. For that task the following structural elements SE (see Eqs. 1 and 2) were used for: (1) Opening with vertical structuring element (SE_ver); (2) TopHat with vertical structuring element (SE_hor); (3) Thickening with diagonal structuring element of type 1 (SE_diag1); (4) Thickening with diagonal structuring element of type 2 (SE_diag2).

$$SE\_hor = \left\{ \begin{matrix} 0 & 0 & 0 \\ 1 & 1 & 1 \\ 0 & 0 & 0 \end{matrix} \right\}; \quad SE\_ver = \left\{ \begin{matrix} 0 & 1 & 0 \\ 0 & 1 & 0 \\ 0 & 1 & 0 \end{matrix} \right\}; \tag{1}$$

$$SE\_diag1 = \left\{ \begin{matrix} 0 & 1 & -1 \\ 0 & 0 & 0 \\ -1 & 1 & 0 \end{matrix} \right\}; \quad SE\_diag2 = \left\{ \begin{matrix} -1 & 1 & 0 \\ 0 & 0 & 0 \\ 0 & 1 & -1 \end{matrix} \right\}; \tag{2}$$

The results of morphological operations are shown in image I_{morpho} in Fig. 3(b), where rods on cooler are not yet correctly separated. So, an algorithm for analysis of discontinuities between them is proposed. For better visual inspection the rods are considered as particular regions painted in different colors using fast blog generation method based applying L-shape connected components. In the Fig. 4(a) founded connected components are shown as sets of regions of different colors. For better rod separation each

Fig. 3. (a) Binary image I_{bin} obtained by applying empiricaly defined threshold; (b) I_{morpho} is a result of applying morphological operations to binary image.

detected vertical red line, which corresponds to particular rod, is evaluated looking for rod "adhesion" with adjacent ones. The pixels that represent adhesions (connection) between separated lines are detected and later are eliminated.

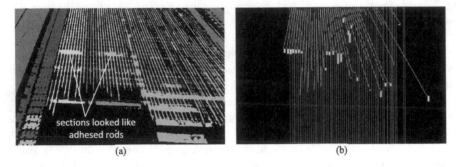

Fig. 4. (a) Some color regions with connected components; (b) I_{discon} with detected positions of horizontal sections with pixels indication connections between rods. (Color figure online)

During evaluation of pixels along each line that corresponds to particular rod, the sections with continuous horizontal sets of pixels of the same color are found indicating the presence of adhesion between adjacent rods as it is shown in Fig. 4(a). These horizontal sets of pixels may be considered as discontinuity of the red lines, which represent separated rods. In Fig. 4(b) the red lines indicate the start of search for each rod; the pink lines indicate the path while the absence of horizontal sections is detected and the yellow segments show the position, where there is adhesion (connection) between two rods. Yellow rectangles from Fig. 4(b) are used as masks representing the regions where rods have discontinuity. The positions found as discontinuities in Fig. 4(b) is shown as masks in Fig. 5(a). The difference I_{morpho} between the image obtained after application of morphological filters I_{morpho} and discontinuity mask $I_{separation\_mask}$ is computed and the regions forming individual rods are successfully separated as shown in Fig. 5(b). In Fig. 6 the image with color regions corresponding to detected and separated rods are shown.

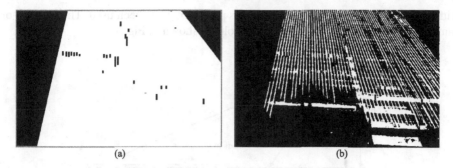

Fig. 5. (a) Mask $I_{separation\_mask}$ with detected discontinuities for connected rods; (b) Binary image $I_{difference}$ of successfully separated individual rods.

Fig. 6. Image with color regions corresponding to detected and separated rods. (Color figure online)

The same procedure for detecting rod front ends computing their discontinuity is used. In contrast to detection of horizontal sections of rod adhesion, for estimation of their front ends $I_{frontend\_difference}$ discontinuity is searched with respect to black background as shown in Fig. 6.

Obtained from images information is useful for generating control signals in rod deceleration braking system because at each moment the exact location of front ends of rods in motion is known. So, deceleration of rods now may be provided by iterative computing of their front ends until the final stop aligning them in one plane.

4   Discussion of Conducted Tests and Evaluation of Obtained Results

Several tests of the proposed algorithm and the designed visual system for monitoring of hot metal rods have been carried out on the continuous light-section mill 250-1 PJSC ArcelorMittal. For image sequence acquisition a conventional Cannon video camera with resolution 1920 × 1080 and 30 frames per second has been used. The location of camera with respect to receiving tray of cooler is shown in Fig. 7.

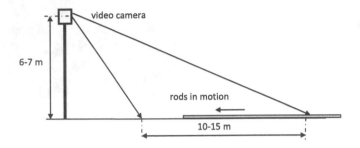

Fig. 7. Video camera location with respect to cooler in conducted tests.

Only two rods simultaneously arrive to cooler and must be stopped and removed from diverter valves of roller conveyor but in conducted tests all rods are processed.

A complete execution of the proposed algorithm is shown in Fig. 8 where the processing times for each step and finally for whole procedure are shown for different resolution of input images particularly, for 320 × 240 (with complete execution time equal to 0.012 s), 600 × 400 (0.055 s), 720 × 576 (0.61 s) and 1920 × 1080 (0.91 s).

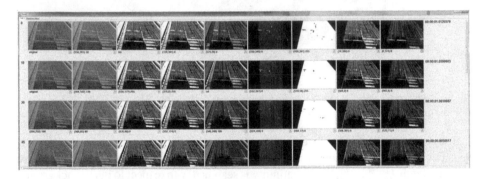

Fig. 8. Examples with algorithm execution and processing times for each step and for whole process for different resolution of input images (320 × 240, 600 × 400, 720 × 576 and 1920 × 1080).

The average time of detecting rod front end by the proposed algorithm with selected resolution of 600 × 400 is less than 0.055 s achieved on the computer with Core Duo processor of 3.3 GHz and 4 GB of RAM. Therefore, during one second the algorithm may process 18 frames. If the initial rod velocity on the cooler on the last meter of deceleration is about 0.1 m/s, the rod requires at least 5 s for complete stop. During these 5 s the algorithm processes 18 × 5 = 90 frames providing analysis of each 1.1 cm of rod displacement. The required maximum error of front end alignment after complete stop of rod is equal 10 cm; however, the capability of the proposed algorithm to operate with error about 1 cm is completely satisfactory. For processing images with resolution of

720×576 the algorithm provides correct analysis of rods in positions separated by 6.5 cm, even processing images with resolution of 1920×1080 gives acceptable alignment error of 10 cm.

For evaluation of precision of rod detection the matching strategy using Euclidean distance between coordinates of rod front end points found by algorithm and located in input images has been applied to more than 300 processed images for each of multiple recordings. The quantity of similarity of front end point coordinates between two images is defined by Eq. (3):

$$similarity(rod\_frontend\_position) = \sum match(I_{input}, I_{frontend\_difference}) \qquad (3)$$

The average precision of detection of rod front ends by algorithm for daylight and artificial illuminations and different resolution of input images is resumed in Table 1.

Table 1. Average precision of rod front end detection for different resolution of input images.

Resolution:	320×240	600×400	720×576	1920×1080
Illumination: Daylight	0.920	0.948	0.952	0.980
Artificial	0.898	0.921	0.930	0.963

It has been noted that the higher resolution of images does not provide significant improvement in the performance of algorithm but can interfere with timing parameters required for generation of the controlling signals for rod braking system. Despite of simplicity of the proposed algorithm, it provides quite acceptable low complexity processing however, the speed of rod detection and tracking during deceleration may be improved by eliminating double (horizontal and vertical) computing discontinuities and processing only small region of interest in image, where rods in motions are appeared. The improvement of the processing approach may also be done using second camera with viewing direction perpendicular to rod translation on a mill.

5 Conclusions

In this paper a specific computer vision application for inspection of steel manufacturing on hot rolling mills is presented. The proposed algorithm provides visual remote monitoring rod braking process on mill based on real-time detection and tracking thin rods in sequences of images by analysis of their front ends during deceleration. It exploits morphological filtering and discontinuity masks for finding and separation of rods on rolling mill as well as provides fast enough detection and tracking of rod front ends in cooler with precision in range of 90–96% on artificial and 92–98% on daylight illumination, respectively processing images with resolution from 320×240 to 1920×1080. The information obtained from analysis of rod deceleration during frame-by-frame image processing will be used to provide solutions for control of some processes particularly, for generation of continuous action signals for magnetic brakes as well as it will be used to align rods by their standstill in one plane reducing metal waste trimmings after cutting-to-length packages of rods.

Acknowledgments. This research is partially sponsored by European Grant: Customised Advisory Sustainable Manufacturing Services, EU FP7 PEOPLE IRSES.

References

1. Pittner, J., Simaan, M.A.: An initial model for control of a tandem hot metal strip rolling process. IEEE Trans. Industry Applications **46**(1), 46–53 (2010). doi:10.1109/TIA. 2009.2036539
2. Potapov O., Egorov A.P., et al: The patent 91194 Ukraine, IPC (2014.01) V21 V 37/00. The automatic leveling control rental on the small-mill refrigerator, SHEE "NSU". - № u 2014 00580; publ. 06.25.2014, Bull. Number 12 (2014)
3. Pittner J., Simaan M.A.: Advanced control using virtual processing for threading a hot metal strip mill. In: IEEE American Control Conference (ACC), MA, USA, pp. 1–8 (2016) doi: 10.1109/ACC.2016.7525421
4. Yildiz, S.K., Forbes, J.F., et al.: Dynamic modelling and simulation of a hot strip finishing mill. J. Appl. Mat. Modelling **3**, 3208–3225 (2009). doi:10.1016/j.apm.2008.10.035
5. Karpinski, Y.P.: WB system – system of creation and implementation. In: Karpinski Y.P., Kuva V.N., Kukushkin O.N., Chygrynskiy V.A. (eds.) Fundamental and Applied Problems of Ferrous Metallurgy, Proceeding of Science Reports, Dnipropetrovsk, National Academy of Sciences of Ukraine, vol. 17, pp. 301–310 (2008)
6. Jelali, M.: Performance monitoring of metal processing control systems. In: Jelali, M. (ed.) Control Performance Management in Industrial Automation, pp. 1–457. Springer, London (2013). doi:10.1007/978-1-4471-4546-2
7. Primetals Technologies: The new global competence in hot rolling, Japan, No.T04-0-N027-L4-P-V2-EN, Primetals Technologies Ltd. (2016). http://primetals.com/en/technologies/hot-rolling-flat/hot-strip-mill/
8. IMS-Group, Measuring systems in hot strip mills (2017). http://www.ims-gmbh.de/unternehmen/
9. Shore T.M.: US Patent US7207202 B1 Method of subdividing and decelerating hot rolled long products, (2007). http://www.google.com/patents/US7207202
10. ArcelorMittal: Maximising the processes that convert iron ore and coal into finished steel products (2017). http://corporate.arcelormittal.com/
11. Gibson, J., Marques, O.: Optical flow and trajectory methods in context. In: Gibson, J., Marques, O. (eds.) Optical Flow and Trajectory Estimation Methods. Springer Briefs in Computer Science, vol. 49, pp. 9–23. Springer, Heidelberg (2016)
12. Fitzner, D., Sester, M.: Field motion estimation with a geosensor network. Int. J. Geo-Inform. **5**(175), 1–22 (2016). doi:10.3390/ijgi5100175
13. Peng, Z., Yu, D., Huang, D., Heiser, J., Kalb, P.: A hybrid approach to estimate the complex motions of clouds in sky images. Sol. Energy **138**, 10–25 (2016)
14. Cruz-Perez, C., Starostenko, O., Alarcon-Aquino, V., Rodriguez-Asomoza, J.: Automatic image annotation for description of urban and outdoor scenes. In: Sobh, T., Elleithy, K. (eds.) Innovations and Advances in Computing, Informatics, Systems Sciences, Networking and Engineering. LNEE, vol. 313, pp. 139–147. Springer, Cham (2015). doi: 10.1007/978-3-319-06773-5_20

15. Deshmukh, A.C., Gaikwad, P.R., Patil, M.P.: A comparative study of feature detection methods. Int. J. Electr. Electron. Eng. Telecommun. **4**(2), 1–7 (2015). http://ijeetc.com/ijeetcadmin/upload/IJEETC_5524b6e0c6527.pdf

16. Starostenko, O., Cortes, X., Sanchez, J.A., Alarcon-Aquino, V.: Unobtrusive emotion sensing and interpretation in smart environment. J. Ambient Intell. Smart Environ. **7**(1), 59–83 (2015)

17. Fouda, Y.M.: A robust template matching algorithm based on reducing dimensions. J. Signal Inform. Process. **6**, 109–122 (2015)

18. Sharma, S.: Template matching approach for face recognition system. Int. J. Signal Process. Syst. **1**(2), 284–289 (2013)

19. Rivera-Rubio, J., Alexiou, I.I., Bharath, A.A.: Appearance-based indoor localization: a comparison of patch descriptor performance. Pattern Recogn. Letters **66**, 109–117 (2015)

Author Index

Printed in the United States
by Bookmasters